Immunology

NIPA® GENX ELECTRONIC RESOURCES & SOLUTIONS P. LTD.
New Delhi-110 034

Immunology

By:
Prof. Dushyant Kumar Sharma
Department of Zoology,
Govt. Model Science College
Gwalior, M.P.

and

Dr. Tripti Sharma
Assts Professor
Govt. Model Science College,
Gwalior, M.P.

NIPA® GENX ELECTRONIC RESOURCES & SOLUTIONS P. LTD.
New Delhi-110 034

NIPA® GENX ELECTRONIC
RESOURCES & SOLUTIONS P. LTD.

101,103, Vikas Surya Plaza, CU Block
L.S.C. Market, Pitam Pura, New Delhi-110 034
Ph : +91 11 27341616, 27341717, 27341718
E-mail: newindiapublishingagency@gmail.com
web: www.nipabooks.com

For customer assistance, please contact
Phone: + 91-11-27 34 17 17
Fax: + 91-11- 27 34 16 16
E-Mail: feedbacks@nipabooks.com

ISBN: 978-81-19215-41-6

Composed and Designed by NIPA®.

Preface

Immunology is a fast growing branch of biology which is directly concerned with the health and well-being of human beings. A basic understanding of the subject is very essential for every student of life science. It is a branch of biology which deals with the immune system of the body , the system which protects us from various types of infections.

Though a number of books are available on the subject but still there is a need for a book which is simple yet concise, covering all essential concepts. This book is a unique approach to understand the fundamentals and emerging trends of immunology. The aim is to provide up to date knowledge to the readers in a simple and lucid manner.

The book has been divided into twenty chapters covering all important areas of immunology which are included in the syllabi of various Universities and are essential for proper understanding of the subject. A large number of simple, hand drawn and self explanatory illustrations would help the students to grasp the concepts firmly and deeply. At the end of each chapter important facts have been summarized as *Points to Remember.*

We are thankful to our colleagues and friends for their valuable suggestions and cooperation. We express our sincere thanks to Mr. Sumit, New India Publishing Agency , New Delhi for his keen interest in the publication of this book.

We acknowledge our warm regards to our parents and family members for their everlasting blessings, love and moral support.

Though every effort has been made to make the book up-to-date and free from mistakes but still there may be some deficiencies in the book. We would appreciate valuable suggestions and constructive criticisms from the readers for further improvement in the book.

Finally, we hope that the book will be able to create an interest and zeal among the students for the subject.

Dr. Dushyant Kumar Sharma

Dr. Tripti Sharma

Contents

Chapter - 1

Introduction

Everyone on this earth wishes to have a healthy life free from all ailments, stress and strain. But it is a common observation that some of us are more susceptible to diseases and fall victim of many infections while others have the capacity to resist many such infections. They are not so much susceptible to diseases. Is there anything which distinguishes them from others? An organism is made up of organs and organs systems. Like many other organ systems, in the body there is a system which is concerned with the protection of the body from infections. This system which is also made up of cells, tissues and organs is called ***immune system*****(Fig.1.1).** The system functions in many ways to provide defense from various types of injurious foreign agents. The capacity of the body to fight against diseases is called ***immunity.*** The word immunity has been derived from Latin word *immunis* meaning 'exempt' or free from. Immunity refers to the resistance of the body to an infection. The study of the immune system in its various aspects is called ***immunology.*** Immunology is the branch of bioscience which deals with the defense mechanisms including all physical, chemical and biological properties of the organism that help it to combat its susceptibility to foreign agents and microorganisms. Immunology includes the development and functions of various components of immune system by which the body responds, destroys or neutralizes foreign agents or pathogens. It includes the study at cellular as well as molecular level. Immunology is an interdisciplinary subject which incorporates and requires the knowledge of many branches of biology like microbiology, biochemistry, molecular biology and clinical medicine.

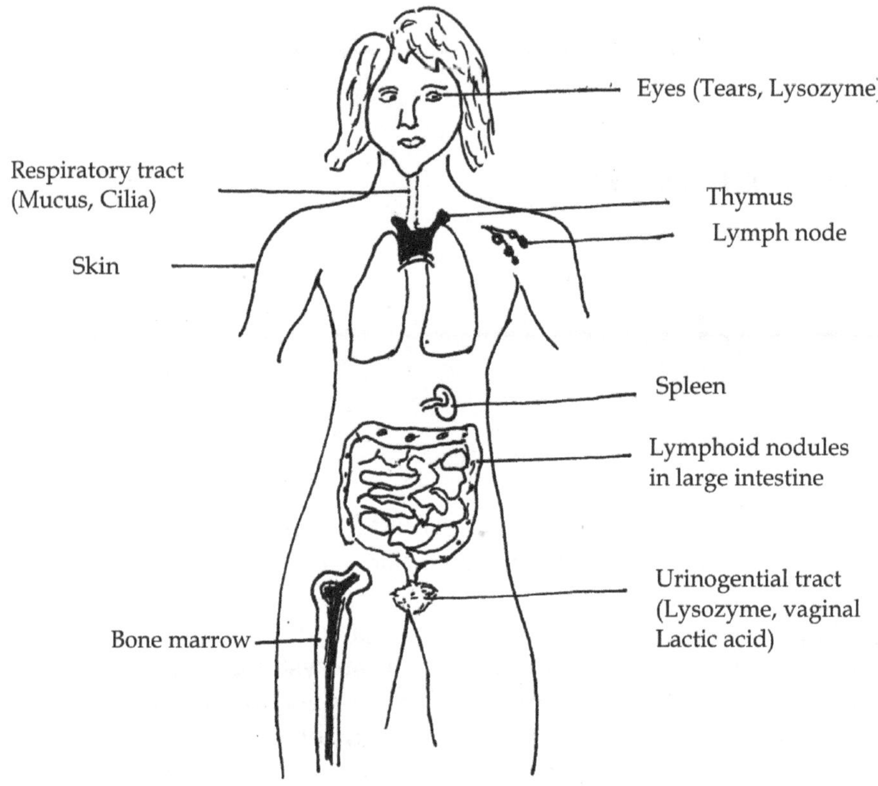

Fig. 1.1 : Some components of immune system.

The immune system has evolved to provide us protection from various kinds of pathogens. Thus, a strong immune system is the prerequisite to the survival of an individual. How does immune system accomplish this challenging task, is a question of great interest and research. Man has always tried to find out answers to this and many such questions.

HISTORICAL DEVELOPMENTS IN IMMUNOLOGY

Right from the beginning man has struggled to make his life safe and free from all diseases. As early as in 15th century Chienses and Turkes made attempt to induce immunity to prevent smallpox. They used to inhale the dried crust from smallpox pustules into the nostrils or inserted into small cuts in the skin to prevent diseases. The process was known as *variolation*.Lady Mary Montagu promoted the technique in Western countries. The practice of variolation spread rapidly throughout England in the 1740s and then to the American colonies.

In 1778,Edward Jenner deliberately inoculated a eight year boy,James Phipps with cowpox fluid to protect against smallpox. His method was a millstone in the field of vaccination.

In 1876, Robert Koch identified anthrax bacterium as the causative agent for anthrax and isolated *Bacillus anthracis* in a pure culture. During the same period, Louis Pasteur for the first time introduced attenuation of the microbes. He got attenuated form of chicken cholera bacillus. Pasteur called his treatment *vaccination*. Later on, he worked on anthrax and rabies and developed the first viable vaccine for them.This made a great impact in the field of medicine.The work of Louis Pasteur and the Robert Koch finally established germ theory of disease.

On the basis of his experiments on star fish, Metchnikoff, for the first time proposed the role of phagocytosis in immunity. His discovery provided support for cellular basis of immunity. Metchnikoff was a leading supporter of cellular immunity.

On the other hand,Paul Ehrlich proposed the idea of existence of antibodies. He suggested that it was the antigen that interacts with receptors borne by cells and results in the secretion of antibodies. This was the base for humoral immunity. He also gave the concept of immunological self/not-self discrimination. In 1890, Behring and Kitasto demonstrated that immunity could be transferred from one animal (who had been immunized to diphtheria) to another animal through serum. This was the technique of passive immunization. But what exactly was responsible for transferring the immunity through serum was still a mystery. Behring received Nobel Prize, in 1901, for serum anti-toxin. In 1901, Karl Landsteiner discovered ABO blood groups. This was an important discovery in immunology.

By the end of nineteenth century, it was well established by germ theory of disease that diseases are caused by bacteria and body can respond in two ways-by phagocytosis(cellular basis of immunity) or by generating antibodies (humoral immunity).

Though the existence of antibody was known but it was not known how antibody diversity was generated. It was assumed that antibodies could not be pre-formed; they were synthesized following exposure to an antigen. In the late 1950s, Jerne, Talmage and Burnet, working independently, developed the theory of *clonal selection.* According to this theory, each lymphocyte is responsive to a particular antigen by virtue of specific surface receptor molecules and upon contacting its appropriate antigen, the lymphocyte is stimulated to proliferate and differentiate. These differentiated cells, called 'effector' cells then secrete antibody. During 1959-62, Gerard Edelman and Rodney Porter elucidated the structure of antibody molecule. Edelmann demonstrated that antibody was made up of four polypeptide chains-two light and two heavy chains. They were awarded Nobel Prize in 1972. Dreyer and Bennet, in 1965, explained the genetic basis for the structure of immunoglobulins. According to them there are several genes which code for V region and a single gene codes for C region.

Although the clonal selection theory represented a theoretical breakthrough in the history of immunology, but it did not explain how lymphocytes actually recognize antigen. Snell, in collaboration with Peter Gorer, discovered histocompatibility genes in mice. Dausset worked on histocompatibility genes in humans. Snell, Peter Gorer and Dausset were together awarded Nobel Prize in 1980 for their work on major histocompatibility complex.

In 1974, Peter Doherty and Rolf Zinkernagel worked on the role of T lymphocytes in immune response to viral meningitis. The dendritic cells were identified by Ralph Steinman in 1978. Butcher, in1979, identified *adhesion molecules.* Production of monoclonal antibodies was an important advancement in immunology. Kohler and Milstein developed method for generation of monoclonal antibodies. They were awarded Nobel Prize for their work in 1984.

In 1986, Tim Mossmann and Bob Coffman discovered T helper subsets: T_{H1} and T_{H2}cells.

Research and discoveries have been a continuous process and work has been going on in various areas of immunology throughout the history. Some of the important contributions in the field of immunology have been enlisted in table 1.2.

Table 1.1 : Some important contributions in immunology

Year	Scientist	Discovery
1796	Edward Jenner	First demonstration of smallpox vaccination
1862	Ernst Haeckel	Phagocytosis
1877	Paul Ehrlich	Mast cells
1878	Louis Pasteur	Germ theory of disease
1881	Louis Pasteur	Anthrax vaccine
1883 - 1905	Elie Metchnikoff	Role of phagocytosis, Cellular theory of immunity
1885	Louis Pasteur	Attenuated" vaccine for rabies

Contd...

1890	Emil von Behring and Kitasato Shibasaburo	Beginning of humoral theory of immunity
1894	Richard Pfeiffer	Bacteriolysis
1896	Jules Bordet	Complement
1900	Paul Ehrlich	Theory of antibody formation
1901	Karl Landsteiner	Blood groups
1901	Jules Bordet and Octave Gengou	Complement fixation test
1902	Paul Portier and Charles Richet	Immediate hypersensitivity -Anaphylaxis
1903	Maurice Arthus	Intermediate hypersensitivity, the 'Arthus reaction'
1909	Paul Ehrlich	"Immune surveillance" hypothesis of tumor recognition and eradication
1917	Karl Landsteiner	Hapten
1938	John Marrack	Antigen-Antibody binding hypothesis
1940	Karl Landsteiner and Alexander Weiner	Rh antigens
1942	Karl Landsteiner and Merill Chase	Anaphylaxis
1942	Jules Freund and Katherine McDermott	Adjuvants
1946	George Snell and Peter A. Gorer	Mouse MHC
1957	Frank Macfarlane Burnet	Clonal selection theory
1957	Alick Isaacs and Jean Lindenmann	Interferon
1959-62	Gerald Edelman and Rodney Porter	Structure of antibody
1971	Peter Perlmann and Eva Engvall	ELISA
1973	Ralph M. Steinman	Dendritic Cells
1975	Georges Kohler and Cesar Milstein	Generation of the first monoclonal antibodies
1983	Luc Montagnier	Discovery of HIV
1990	E. Donnall Thomas and Joseph Murray	Nobel Prize for transplantation immunology
1996	Peter c. Doherty and Rolf M. Zinkernagel	Nobel Prize for Role of MHC in antigen recognition by T cells

Contd...

1996-97	Ruslan Medzhitov, Charlie Janeway and Jules Hoffmann	Role of Toll and Toll like receptors in immunity
2005	Ian Frazer	Development of human papilloma virus vaccine

IMMUNE RESPONSES

As mentioned in the beginning, the basic concern of the immune system is to provide defense from pathogens. Pathogens are the organisms which cause diseases. Pathogens can be divided into two categories- intracellular and extracellular. Intracellular pathogens are those which live inside the cells and infect individual cells e.g. virus. Many pathogens, like bacteria, do not penetrate inside the cells but they divide extracellularly within tissues or body cavities and damage the body. Such types of pathogens are called extracellular pathogens. The first crucial question before immune system is to recognize a pathogen as a pathogen and then to provide defense against it. The simplest way of protection is to stop the entry of a pathogen in the body. If we can prevent the entry of an organism then the problem is solved automatically but this is not possible always. In such circumstances when the body fails to avoid the entry of a pathogen there must be some mechanisms which act on the microbe which has got entered in the body. Our immune system acts in both ways. There are mechanisms which provide first line of defense and if it fails mechanisms of second line of defense come into operation.

FIRST LINE OF DEFENSE-INNATE IMMUNITY

The first line of defense is provided by the innate immunity. Innate immunity is the non-specific or less specific component of the immunity. It is a set of mechanisms which are not specific to a particular microorganism and such mechanisms exist prior to the exposure to a pathogen i.e. before the onset of the infection. These are the mechanisms that make one species innately susceptible and another resistant to certain infections. For instance, birds are naturally resistant to anthrax infection. In the same way, wild animals are rarely susceptible to rabies. These differences in the immunity among some species might be due to differences in their genetic makeup. Such differences are also found in different races and also at individual level. American Indians are more susceptible to tuberculosis than Europeans. Besides genetics, there are some other factors such as age, nutrition and environmental conditions which may affect the natural resistance of an individual to an infection.

Components of Innate Immunity

Innate immunity includes many components. These are: physical barriers, chemical barriers, phagocytosis, inflammation and fever.

Physical Barriers

The first step for a pathogen to cause a disease is to enter in the body of the host. Except a few pathogens which exist on the surface of skin (e.g. bacteria causing warts), nearly all pathogens must enter into the body for causing a disease. The first hurdle in their path is the skin. Intact skin and mucosa provide physical barrier to prevent the entry of the microorganisms. Intact skin checks the entry of most of the microorganisms. Infections may occur if there is any damage to skin. Damage may occur in many ways such as by burns, insect and animal bites and cuts and wounds. Besides preventing the entry of the pathogens, skin has several other functions. The sebaceous glands, present in the dermal layer of the skin secrete sebum which contains lactic acid and fatty acids. These acids maintain the pH of the skin between 3 and 5. Most of the microbes are unable to survive at this low pH. Respiratory, gastrointestinal and urinogenital tracts are lined by mucus membrane. Mucus is a sticky and slimy substance, secreted by this membrane. It entraps microorganisms which have entered these tracts. The respiratory tract is lined with small hair like structures called cilia. These cilia help in expelling the microbes out of the body. The resident flora (the bacteria that are normally present on or in the body but do not have pathogenic effects) of the epithelial cells of mucosal membrane themselves help in preventing the entry of pathogens. They compete with the pathogens for attachment site and food.

Chemical Barrier

Many secretions of the body act as antimicrobial substance and inhibit the growth and survival of microbes. Hydrolytic acid, secreted by the stomach, maintains the pH of the stomach between 1 to 3 .This pH is lethal to many bacteria. Saliva, sweat and tears contain *lysozymes* which break down pepidoglycans on the bacterial wall and act as bactericidal agent. Saliva and tears also contain *transferrins*, iron binding proteins, which inhibit bacterial growth. *Defensins* are antimicrobial peptides that are found in secretions of mucosa. *Collectins* are another group of antimicrobial substances which are found in serum. These proteins can bind on the sugars on microbial surfaces and promote their elimination. *Granulosyin* is found in the granules of Tc lymphocytes and many natural killer cells. These are known to enter the intracellular compartment of the microorganisms and kill them. *Interferons* are a group of proteins, secreted by viral infected cells. They act on neighbouring cells and stimulate antiviral mechanisms in them. Interferons may also act on the viruses. They may block protein synthesis and multiplication of virus. They are also known to activate macrophages and natural killer cells. Another group of proteins which form a very significant part of both innate and adaptive immunity is *complement*. It is a group of 20 heat labile plasma proteins which have several functions. Complement proteins are involved in cytolysis, opsonization and cell activation, inflammation and removal of immune complexes.

Phagocytosis

Phagocytosis is the ingestion of extracellular particulate material by phagocytes. Phagocytes are the cells that directly ingest and kill pathogens. There are many cells in the body which act as phagocytes such as neutrophils, monocytes, dendritic cells, monocytes and natural killer cells.

The process of phagocytosis has been depicted in **Fig.1.2.** The phagocytes adhere to the pathogen and spread membrane protrusions around the microbe. These membrane protrusions are called pseudopodia. Pseudopodia fuse with each other and the fusion of pseudopodia encloses the microbe within the membrane bound structure, called *phagosome*. The phagosome moves towards the cell interior and fuses with lysosome to form *phagolysosome*. Lysosome contains many hydrolytic enzymes which act on the microbe and digest it. The digested content is then eliminated.

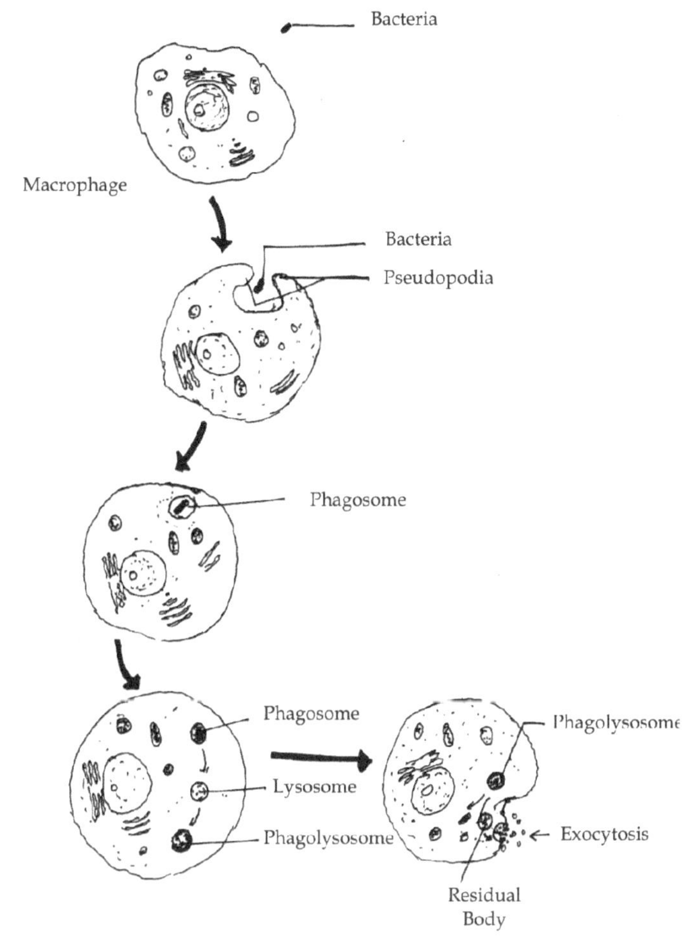

Fig. 1.2 : Mechanism of Phagocytosis

Fever

The normal body temperature of a normal healthy man is 98.6°F or 37°C. During infection, there is an abnormal rise in body temperature. This abnormal rise in the body temperature is called fever. The rise in temperature is due to the release of several chemicals in the body such as cytokines, interleukin-1 and αTNF . These substances act on the hypothalamus and stimulate the release of prostaglandins which elevate the body temperature. Increase in temperature helps the body in many ways. It stimulates the release of antimicrobial substances like transferrins and interferons. Transferrins decrease the iron availability to the microorganisms and help in inhibiting their growth.

Inflammation

Inflammation is a non-specific immune response to the tissue damage. Whenever there is tissue damage either by a physical agent (wound) or biological agent (microbial infection) it results in inflammation. Inflammation is characterized by local swelling (caused by increased capillary permeability which results in accumulation of fluid around the damaged tissue), tissue redness or erythema (due to increased blood flow) and increase in temperature and pain(also due to increased blood flow). During inflammation, dilation of blood vessels causes the increased blood flow to the affected area. Along with vasodilation, there is increase in capillary permeability. Increased permeability permits the large molecules to escape from the capillaries. It causes accumulation of fluid around the injured tissue which results in tissue swelling or edema. It also helps the migration of leukocytes out of the venules into the surrounding tissues.

The process begins with the adherence of phagocytes to the endothelial walls of the blood vessels. This is called *margination* and it is controlled by *chemoines,* a class of cytokines. Chemokines activate the circulating leukocytes to bind to endothelium and initiate their migration. Phagocytes migrate to the damaged tissue and accumulate at the site of infection. Phagocytes start damaging the invading microbes. As a result dead cells and debris accumulate around the damaged tissue. Finally, the debris is removed and tissue starts repairing or regenerating itself.

Inflammation is a defense mechanism to combat an invasion by pathogens. It allows the leukocytes , antibodies and complement to migrate to the tissue at the site of infection. Increase in acute phase proteins, activates the release of complement. **Figure 1.3** illustrates various steps, occurring during inflammation.

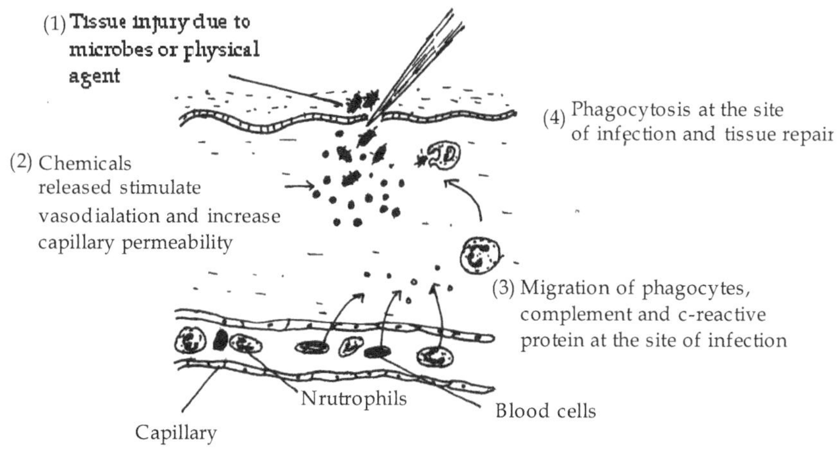

Fig. 1.3 : Steps involved in inflammation

Acute phase proteins

Acute phase proteins are some plasma proteins whose concentration dramatically rises during an infection or tissue damage. These are mainly synthesized in liver. During an infection macrophage release cytokines into the bloodstream, most notable of which are the interleukins IL-1, IL-6 and IL-8 and TNF-α which stimulate the synthesis of these proteins. Acute phase proteins include C-reactive protein (CRP), serum amyloid A protein, α_1 anti-trypsin, α_2 macroglobulin, fibrinogen, caeruloplasmin, C9 and factor B. Acute phase proteins activate complement and help in clearance of pathogens. The protein CRP binds to membranes of many bacteria and fungi.

Second Line of Defense-Adaptive Immunity

If a microbe escapes from the mechanisms of first line of defense, second line of defense mechanisms come into operation. These mechanisms, called the adaptive immunity, are highly specific which recognize a specific microorganism and act accordingly. This is also called acquired immunity as it develops after the birth of a person. There are some specific characteristic features which distinguish adaptive immunity from innate immunity (Table 1.2).

Table 1.2 : Comparison between innate and adaptive immunity

Characteristic	Innate immunity	Adaptive immunity
Specificity	Less	High
Response time	Fast	Slow
Memory cells	No memory cells	Memory cells formed

The two main characteristics of adaptive immunity are 'specificity' and 'memory'. Specificity is the ability to recognize and selectively eliminate a microbe or antigen. Antibodies are the molecules that are raised against specific antigens. They specifically recognize an antigen and are capable of distinguishing two different antigens even if they are closely related to each other. The body not only produces antibodies which are specific to a particular antigen but also generates tremendous diversity in antigen receptors in the antibodies. This diversity in antigen receptors allows recognition of billions of uniquely different structures to which our body can be exposed.

When our body encounters an antigen/ microbe for the first time, it not only responds to it by making antibodies(primary immune response) but also 'remembers' it. If the same antigen enters into the body again, in future, a faster and more heightened immune response is made (secondary immune response) **(Fig.1.4).** This is due to immunological memory. Another important feature of immune system is the recognition of self and non-self. The immune system has a unique feature of recognizing a substance as foreign and distinguishing it from self. It only responds against foreign molecules and ignores those that are self. Self recognition is very essential otherwise immune reaction would damage our own cells and tissues, though occasionally response is made against self antigens, as in autoimmune diseases.

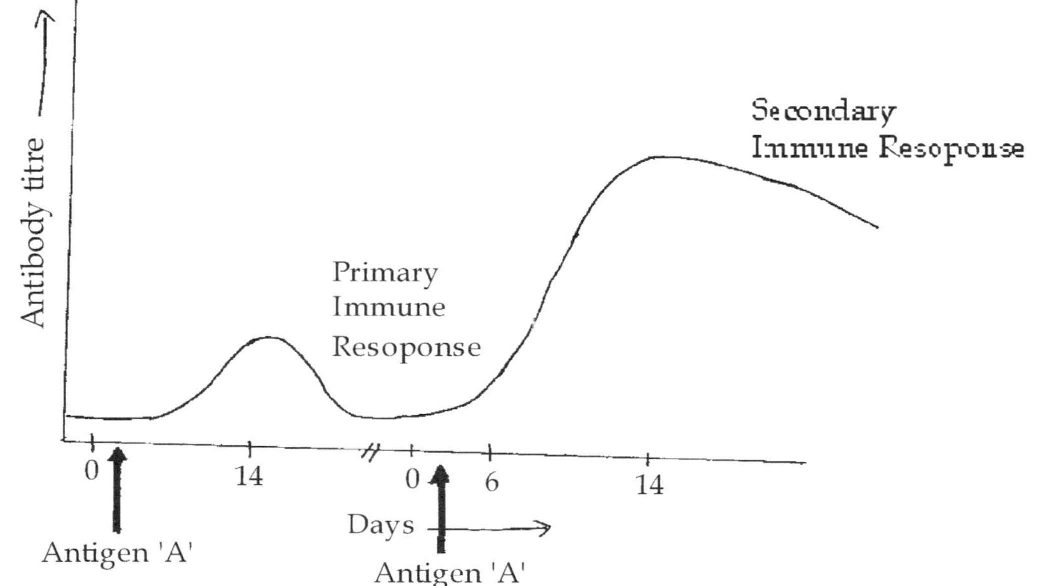

Fig. 1.4 : Primary and Secondary Immune Response (When an animal is exposed to an antigen for the first time body shows a primary immune response of low magnitude and of short duration. If the body is exposed to the same antigen, a more strong immune response develops with greater magnitude and of longer duration)

Humoral Immunity

Immunity, mediated by antibodies, is called humoral immunity. It is called humoral because antibodies are present in blood and immunity could be transferred with ' humour', that is liquid or blood. It is a major defense mechanism against extracellular microbes and toxins. Antibodies are made by B lymphocytes in the body. B cells recognize the antigens and antigen specific antibodies are formed. These antibodies act on the antigen and destroy and eliminate it by various ways such as agglutination, opsonization, activation of complement, neutralization and/or antibody dependent cell-mediated cytotoxicity.

Cell-mediated immunity

The immunity which is mediated by T lymphocytes is called cell mediated immunity. It protects against intracellular microbes that could survive inside a phagocytic cell or infect non-phagocytic cells. This type of immunity can be transferred through T cells. There are two main types of T cells- T_H cells and T_C cells. T_H cells secrete certain chemicals called cytokines which activate phagocytic cells to kill phagocytosed microorganisms. T_C cells directly kill the infected cells.

Clonal Selection Theory

The theory was proposed by Frank Macfarlane Burnet in 1957 to explain the formation of antibodies. According to this theory body contains a large number of clones of B and T lymphocyte with a specific antigenic specificity. A clone refers to identical cells which have originated from a single progenitor cell. Each clone has a specific antigen receptor. The antigenic specificity of each of these clones is determined by the specificity of antigen binding receptors present on the membrane of these clones of lymphocytes. This receptor specificity is produced during their maturation in the lymphoid organs (bone marrow or thymus) even before they encounter an antigen. When the body is exposed to an antigen, antigen interacts(or selects) with a particular B lymphocyte whose membrane bound antibody molecules are specific for that antigen. Antigen stimulates this B cell to divide and produce a clone of cells with the same antigen specificity as the parent cell. Some of these cells become memory cells while others become effector B cells or plasma cells to produce antibodies. Same thing happens to T lymphocytes which produce clones of memory T cells and effector T cells **(Fig. 1.5).** This binding of antigen to a particular B lymphocyte and activating it to proliferate is called ***clonal selection theory***.

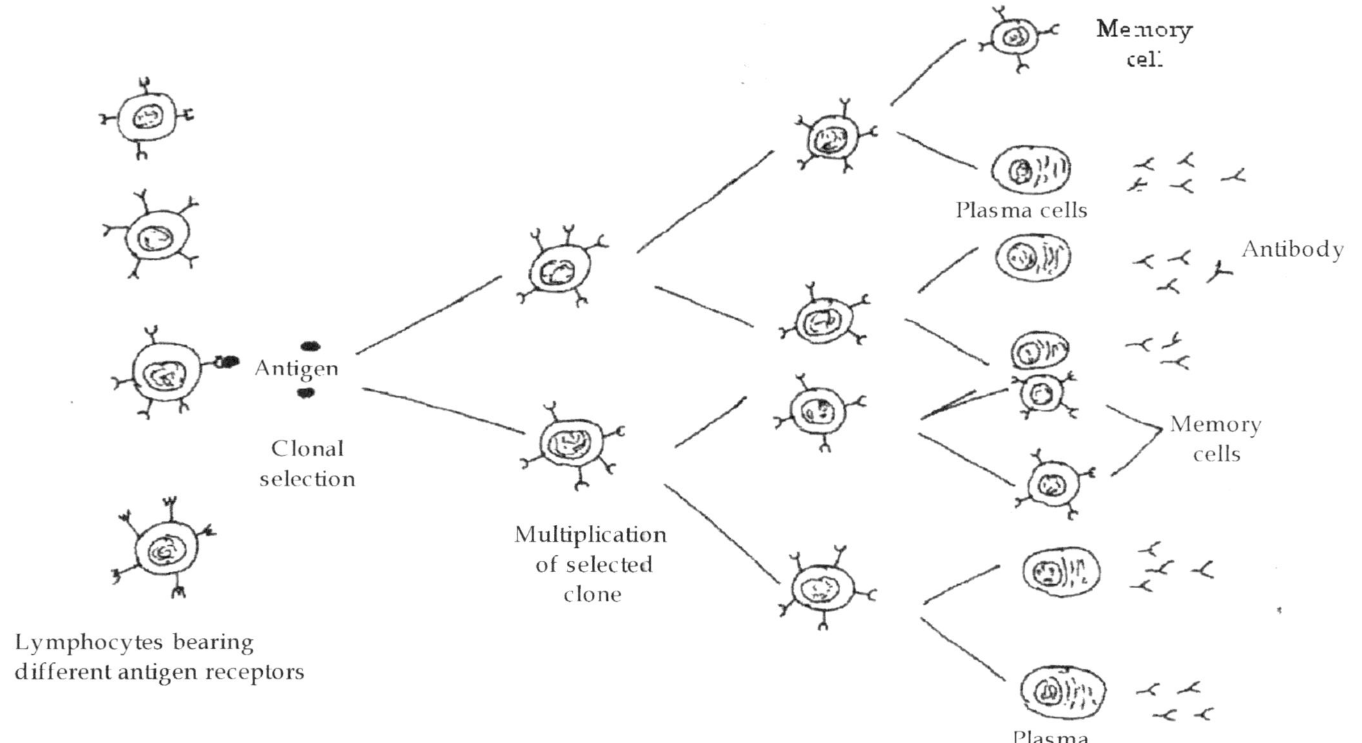

Fig. 1.5 : Clonal Selection model of antibody formation

RESPONSE TO AN ANTIGEN

(Collaboration between humoral and cell-mediated immune response)

The two components of immune system-humoral and cell-mediated immune response are not independent to each other. Both act in collaborative manner to produce more effective immune response.

When the body is exposed to an antigen, the antigen stimulates both B and T lymphocytes. On recognizing an antigen B cells start dividing and undergo differentiation into memory B cells and plasma cells or effector cells. Memory cells remain in the body for a long time. Plasma cells secrete antibodies. These antibodies are the major effector molecules of humoral immune response and neutralize the antigens in several ways.

T cells also recognize the antigen but they recognize it only when it is processed and bound to a cell membrane protein, called *major histocompatibility complex (MHC)*. When T_H cells recognize antigen and interact with it they get activated and become effector cells. Activated T_H cells secrete different kinds of cytokines. These cytokines act on various types of cells of the immune system. They activate B cells to produce antibodies. Cytokines also activate phagocytic cells to phagocytose the antigens. They also stimulate T_C cells to mediate direct killing of infected cells **(Fig. 1.6).**

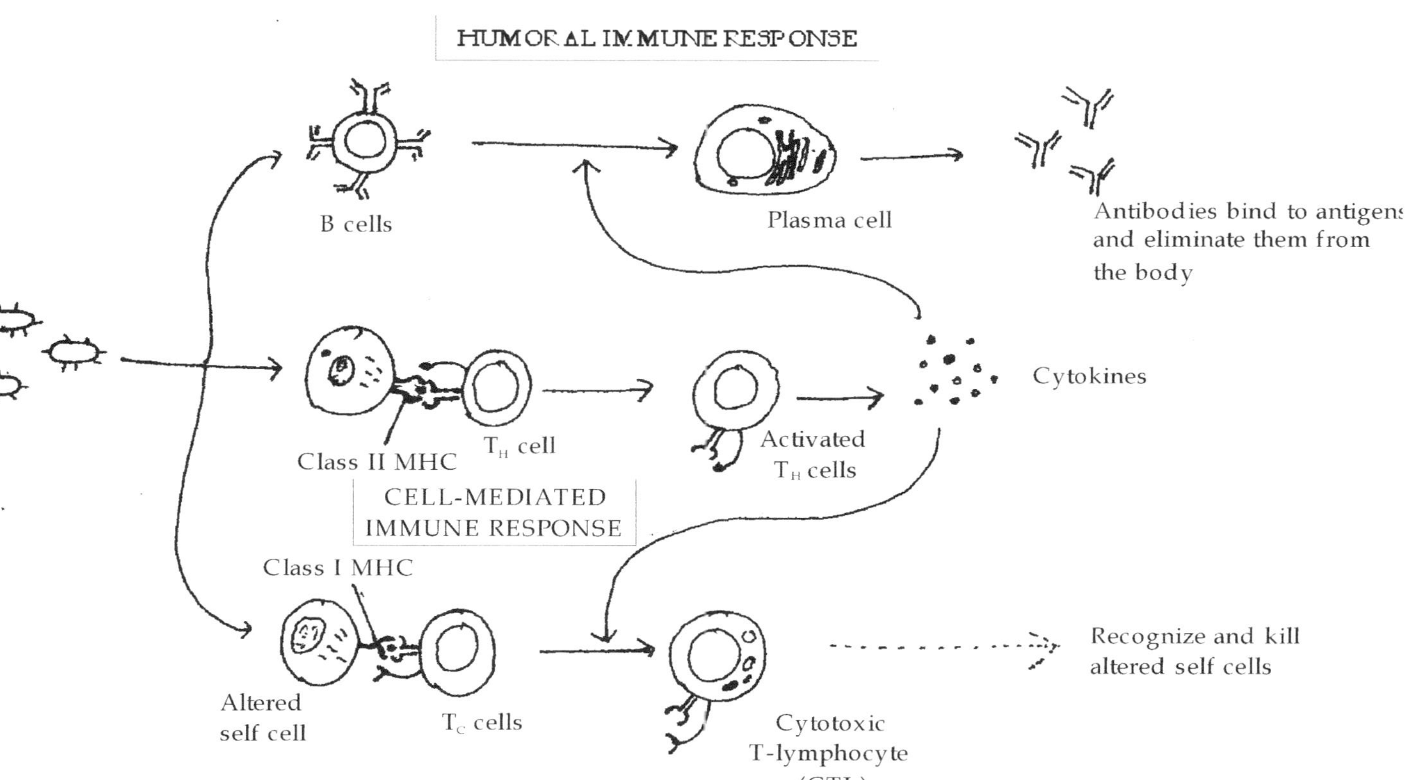

Fig. 1.6 : Collaboration between humoral and cell-mediated immune response

MONOCLONAL ANTIBODIES

Monoclonal antibodies are homogeneous preparations of antibody molecules all of which exhibit the same antigenic specificities. They are represented by myeloma proteins. Kohlar and Milstein and Margulies *et. al,* in 1970s, introduced hybridoma technique for the production of monoclonal antibodies. The technique involves the somatic fusion of two cell types-antibody producing cells, obtained from an immunized animal and myeloma cells **(Fig.1.7)**. Successful hybridoma production is influenced by the characteristics of each of the cell populations, the fusion conditions and the subsequent selection and screening of the hybrids.

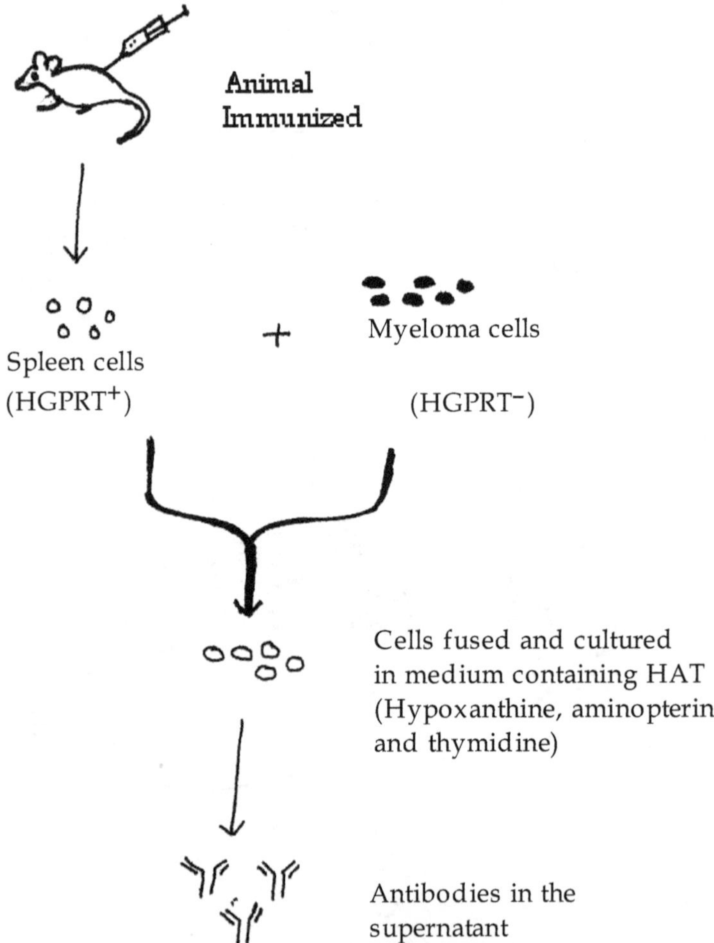

Fig. 1.7 : Steps involved in production of monoclonal antibodies.

They play enormous role in biological research and applications. Monoclonal antibodies are widely used in the diagnosis and therapy of many diseases. They are extensively used in RIA and ELISA techniques for measurements of various substances such as hormones and toxins, in biological fluids. Monoclonal antibodies can also be used as a catalyst. They can be used as an alternative to enzymes. Such antibodies can be used to perform various chemical reactions for which there are no catalytic enzymes. These catalytic antibodies would likely to have many more applications.

POINTS TO REMEMBER

- Immunity is the state of protection from infectious diseases.
- The study of the immune system in its various aspects is called immunology.
- Many important contributions have been made in the field of immunology right from the beginning.
- Innate immunity is a non-specific component of the immunity.
- There are many components of innate immunity like physical barriers, chemical barriers, phagocytosis, inflammation and fever.
- Skin and mucosa are the physical barriers which prevent the entry of the microorganisms.
- Many antimicrobial substances are secreted by different parts of the body.
- Fever and inflammation are the defense mechanisms against infections.
- Adaptive immunity is highly specific in nature which recognizes a specific microorganism.
- The two components of adaptive immunity are-humoral and cell-mediated immunity.
- Humoral and cell-mediated immune responses are not independent to each other. They act in a collaborative manner to produce more effective immune response.

REVIEW QUESTIONS

1. What is immunity? Differentiate between innate and adaptive immunity.

2. Write a brief account of historical development in immunology.
3. What are the different components of innate immunity? How do they contribute in innate immunity?
4. How are innate and adaptive immunity affect each other in their functions?
5. Differentiate between:
   i. Humoral and cell-mediated immune response
   ii. B and T lymphocytes
6. Write notes on:
   i. Clonal selection theory
   ii. Acute Phase proteins

CHAPTER - 2

Anatomy of the Immune System

Like any other organ system in the body, immune system comprises of various organs and cells. The major component of the immune system is the lymphocytes which constitute lymphoid system. The lymphoid organs are classified into two categories: primary and secondary lymphoid organs.

Primary lymphoid organs: These are the organs where B and T lymphocytes undergo maturation. Bone marrow and thymus are the primary lymphoid organs. These are the organs where lymphocytes differentiate, proliferate, get selected and mature into functional cells.

Secondary lymphoid organs: These are the organs where lymphocytes interact with antigens. They provide site for the interaction of lymphocytes with antigens. Secondary lymphoid organs include spleen, lymph nodes and mucosa associated lymphoid tissues (MALT). These may be well organized encapsulated organs such as spleen and lymph nodes or non-encapsulated accumulation of lymphoid tissues, as the lymphoid tissues found in association with mucosal surfaces (MALT). Mucosa associated lymphoid tissues interact with antigens which enter the body via gastrointestinal, respiratory or urinogenital tracts. The lymphoid tissues in the intestinal tract are called gut-associated lymphoid tissues (GALT) and those associated with respiratory tract are called bronchus-associated lymphoid tissues (BALT).

PRIMARY LYMPHOID ORGANS

Thymus and bone marrow are the primary lymphoid organs.

Thymus

Thymus is a bilobed organ, found in the thoracic cavity below the sternum and above the heart. In thymus, each lobe is organized into lobules which are separated from each other by connective tissue strands, called *trabeculae.* The lobules are divided into two compartments- the outer portion, called cortex and the inner medulla **(Fig.2.1, a, b).**

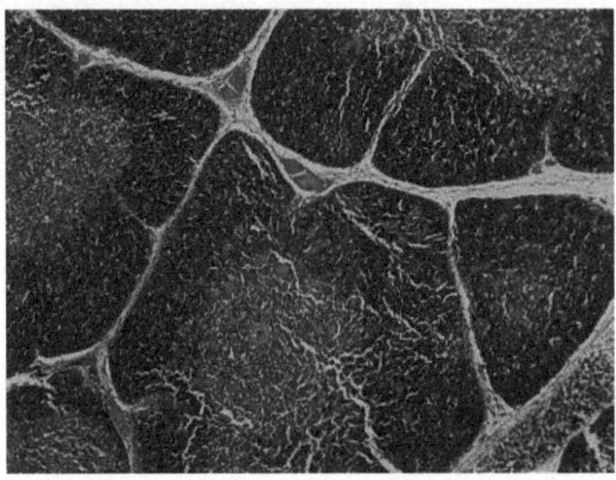

Fig.2.1 : (a) Histological section of thymus showing lobules with cortex and medulla regions

The outer cortex is densely packed with relatively immature proliferating T cells, called ***thrombocytes.*** Thrombocytes mature as they migrate to the inner part, called medulla. The medulla contains mature cells. From the thymus, the mature T cells are carried to the secondary lymphoid organs. Several types of cells such as stromal cells, epithelial cells, dendritic cells and macrophages are found both in the cortex and the medulla. Interaction with these cells is critical for maturation of thrombocytes. A large portion of cells that undergo proliferation in the cortex undergo apoptosis (programmed cell death) and only a small portion of cells migrate to medulla and undergo maturation.

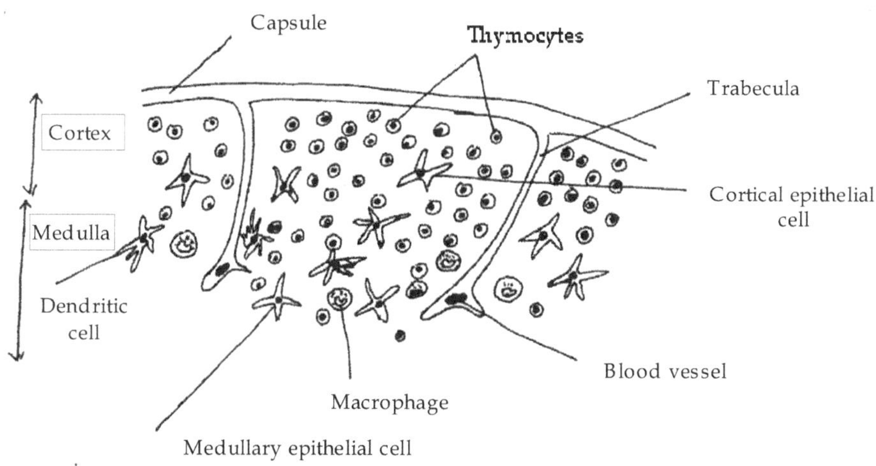

Fig. 2.1 (b) : Structure of thymus, showing lobules, separated by trabecular

In the thymus, three types of epithelial cells are found which have different roles in proliferation, maturation and selection of thrombocytes. The cortex contains *nurse cells* which sustain the proliferation of progenitor T cells. *Cortical thymic epithelial cells* are associated with positive selection of maturing thrombocytes. *Medullary thymic epithelial cells* are mostly organized into clusters and display a large variety of organ specific peptides.

Bone Marrow

Bone marrow is the site of the origin of B and T lymphocytes in mammals. During fetal development, hematopoiesis occurs, first in the yolk sac and then in the liver and spleen. The function is taken up by bone marrow in the adult. B cells originate, proliferate and differentiate in the bone marrow. In bone marrow, the stromal cells interact with B cells and secrete cytokines which are critical for maturation of B cells **(Fig.2.2).** After their maturation, B cells leave the bone marrow and accumulate in the peripheral lymphoid organs. In birds, B cells maturation occurs in gut associated lymphoid organ called ***bursa of Fabricius***. In case of T cells, after their origin in bone marrow, T cells leave the bone marrow and migrate to thymus for maturation.

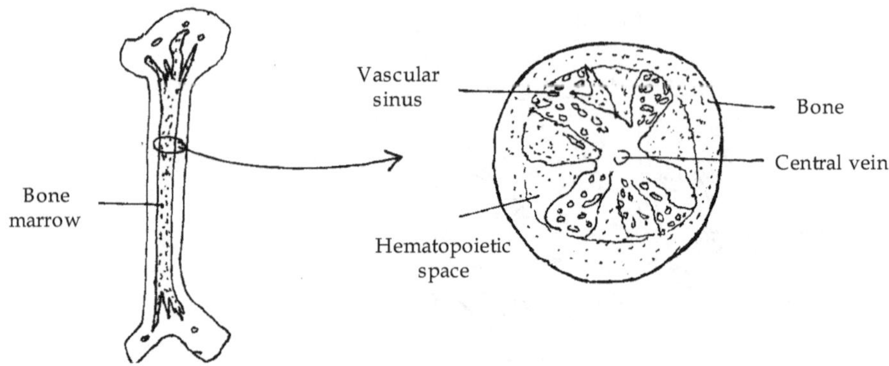

Fig. 2.2 : Bone marrow

SECONDARY LYMPHOID ORGANS

Secondary lymphoid organs have both organized and unorganized lymphoid tissues. Spleen and lymph nodes are highly organized lymphoid organs while MALT is a less organized lymphoid tissue.

Spleen

Spleen is an important lymphoid organ which efficiently traps antigens and provides site to the lymphocytes to act on antigens. It is an important site for phagocytosis. Spleen is an indispensable organ of immune system. It is a large and ovoid organ which lies in the left side of abdomen, below the diaphragm. Spleen is surrounded by a capsule of collageneous fibers. The organ is subdivided into compartments by capsular connective tissue called trabeculae. There are two types of splenic tissues: white pulp and red pulp. The white pulp consists of bulk of lymphoid tissue which is arranged around a central artery and forms periarterial lymphoid sheaths (PALS)**(Fig.2.3).** PALS consist of both T and B cells. T cells are found around the central arteriole while B cells are present in primary or secondary lymphoid follicles. Secondary lymphoid follicles contain germinal centers. Besides having B cells, germinal center also contains follicular dendritic cells and macrophages. These cells function in presenting antigens to B cells. The red pulp is formed of spongy cellular masses called the *splenic cords* or *pulp cords*. The venous sinuses and pulp cords contain macrophages and many blood cells such as erythrocytes, granulocytes, platelets, lymphocytes and plasma cells. The red pulp is concerned with the destruction of old erythrocytes. At the outer limit of the white pulp is present a third region, called the marginal zone. It is rich in macrophages, B cells and dendritic cells. Marginal zone strongly responds to the thymus independent antigens including capsular polysaccharides of bacteria.

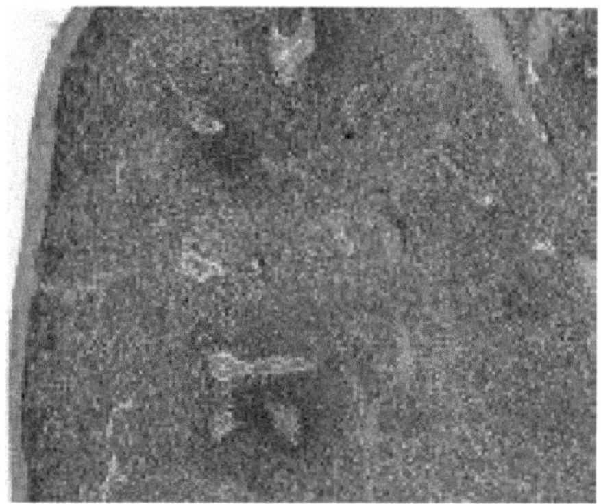

Fig. 2.3 : (a) Histological section of spleen showing red and white pulp

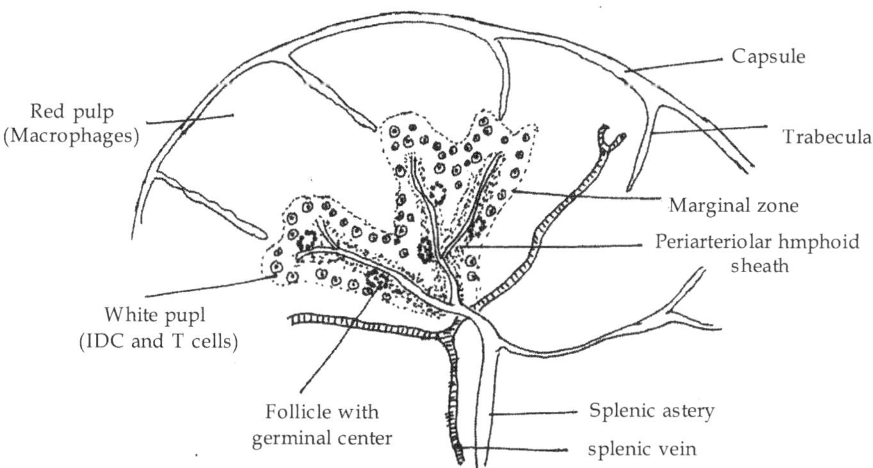

Fig. 2.3 : (b) Structure of spleen

Spleen is not supplied by lymphatic vessels. The antigens are carried through the blood which passes through the spleen. In the spleen interdigitating cells trap the antigens and carry them to periarterial lymphoid sheath.

Spleen is a vital organ of the immune system. It not only provides site for the interaction of antigen and lymphocytes but also helps in the clearance of worn-out red blood cells and also immune complexes. Spleen also serves as a reservoir for platelets, erythrocytes and granulocytes. Splenectomy has many serious and long-term consequences. It can lead to a compromised immune system. Certain bacteria, including *Streptococcus pneumoniae* (Pneumococcus) and *Haemophilus influenzae* that are usually confined to local infections may become blood-borne (septic) and wide spread in splenectomized persons.

Lymph Nodes

Lymph nodes are small, flattened, round or ovoid structures that occur at the branches of the lymphatic vessels. They are the sites where B and T lymphocytes interact with antigens. They are usually found in groups in various regions of the body such as neck, armpit, groin, chest and abdominal cavity.

In humans, lymph nodes are 2 to 10 mm in diameter. A lymph node is enclosed by a capsule of collageneous fibers from which connective tissue trabeculae extend into the parenchyma of the node. Internally, a lymph node can be divided into three regions- cortex, paracortex and medulla **(Fig.2.4).** Cortex is the outer area which is rich in B cells, macrophages and follicular dendritic cells. B cells are organized into primary follicles. On stimulation by antigens, primary follicles enlarge to form secondary follicles, each containing a germinal center. B cells proliferate profusely and undergo affinity maturation i.e. B cells develop affinity receptors for antigens. Germinal centers are also the sites for the development of memory cells. In addition to B cells, secondary follicles also contain Follicular Dendritic Cells (FDC), macrophages and a few CD4+ T cells. All these cells help in generation of matured B cells and memory cells. These cells produce certain cytokines which activate B cells. The activated B cells in the germinal centers divide and undergo differentiation to form plasma cells or memory cells. The plasma cells move to the medulla and start secreting antibodies.

The region, between the cortex and medulla is called paracortex. Paracortex is rich in interdigitating cells and T cells. Interdigitating cells are rich in class II MHC molecules and present antigenic peptides to T cells. T cells are activated and differentiated into effector cells and memory cells.

The innermost part is called medulla. Medulla consists of cellular cords, separated by lymph (medullary) sinuses. This region is rich in lymphocytes and plasma cells which are transported from cortex via lymphatic vessels. Many of the plasma cells secrete antibodies.

As the lymph percolates through the lymph node, the particulate antigens which come along with the lymph are trapped by phagocytic cells, follicular cells and interdigitating dendritic cells.

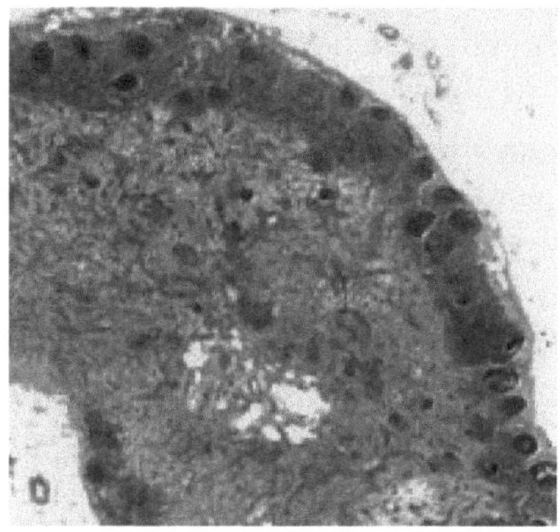

Fig.2.4 : (a) Histological section of a lymph node

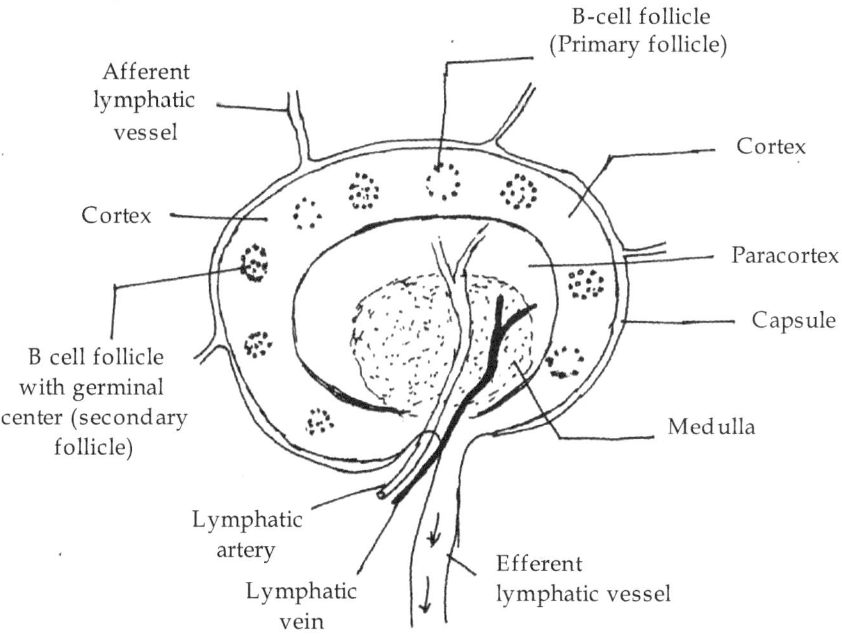

Fig.2.4 : (b) Structure of a lymph node

Mucosa Associated Lymphoid Tissues

Gastrointestinal, respiratory and urinogenital tracts are the common sites for the entry of microbes and other antigens in the body.

In order to provide protection from foreign antigens at these levels, aggregates of lymphoid tissues are found in the mucosal surfaces of these tracts. These are collectively called Mucosa Associated Lymphoid Tissues (MALT). In the intestines mucosa associated lymphoid tissues are represented by Peyer's patches. They are also called gut-associated lymphoid tissues (GALT).

Peyer's patches are found in the lower ileum and are organized in follicles, as in case of lymph nodes and have germinal centers **(Fig.2.5).** They are primarily composed of B cells but small numbers of T and APC cells are also present. Peyer's patches function to trap antigens entering the gastrointestinal tract and are the sites for the interaction of B and T cells with antigens. The interstitial epithelium, overlying the Peyer's patches, contains some specialized epithelial cells, called M cells. These cells have many microfolds on their surface. M cells contain deep invaginations on the internal side which form pockets. These pockets contain lymphocytes, macrophages and dendritic cells. Antigens are transported into these pockets where these antigens are processed and presented to T cells. This results in the activation of T cells.

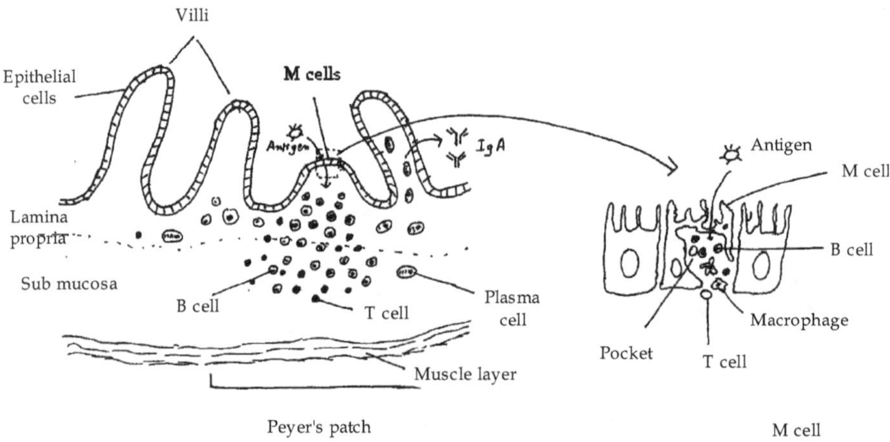

Fig.2.5 : Gut-associated lymphoid tissue (Peyer's patches) and M cell

Tonsils also represent highly organized lymphoid tissues. Tonsils are present in the mucous membrane of pharyngeal cavity and provide defense against the antigens entering through mouth and nasal passage. They have a nodular structure with a network of reticular cells and connective tissue fibers. They are rich in lymphocytes, granulocytes and macrophages. There are three kinds of tonsils: palatine, pharyngeal and lingual.

In addition to Peyer's patches and tonsils, numerous lymphocytes and plasma cells are found in the mucosa of the stomach, intestine, respiratory organs and several other organs in the body. They provide defense at mucosal lining.

Cutaneous Associated Lymphoid Tissues (CALT)

Skin is an important part of the immune system. It is an important component of innate immunity which acts as a surface barrier and prevents the entry of harmful microorganisms in the body. But in case of skin injury, many pathogens may enter through the skin. In such situations, skin counteracts these antigens with its specialized cells. The epidermal layer of the skin contains some specialized cells called keratinocytes. These keratinocytes act as antigen presenting cells (APCs). These cells also secrete a number of cytokines which help in local inflammatory reactions. Epidermis also contains lymphocytes and dendritic cells. The dendritic cells, called Langerhan's cells phagocytose the antigens which enter the skin. After encountering antigens, these cells migrate to the lymph node and differentiate into interdigitating dendritic cells. The dermal layer of the skin contains T cells, dendritic cells and macrophages.

CELLS OF THE IMMUNE SYSTEM

Many cells are involved in immune responses. Most of these cells originate from hematopoietic stem cell. Cells of immune system include lymphocytes, natural killer cells, dendritic cells, mononuclear phagocytic cells, mast cells, neutrophils, basophils and eosinophils.

Lymphocytes

Lymphocytes are non-phagocytic motile cells of the immune system which continuously circulate in the blood and lymph. They are capable of migrating into the tissue spaces and lymphoid organs. They are the only cells that produce high affinity receptors for antigen. Lymphocytes constitute about 20% of total leucocytes in the body. They are small, round and nucleated cells **(Fig.2.6).**

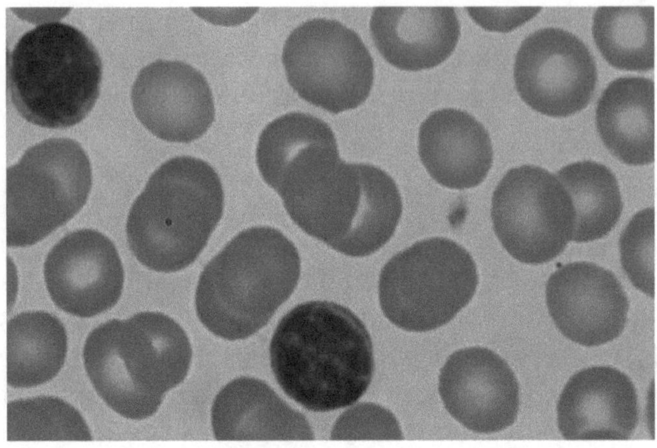

Fig. 2.6 : Lymphocyte

A newly formed lymphocyte (naïve lymphocyte) has a large nucleus with a small amount of cytoplasm. Naive cells have short life span. As they encounter an antigen, these cells get activated and triggered by certain cytokines to undergo proliferation and differentiation. Lymphocytes are non-granular type of leucocytes. Lymphocytes express many types of cell surface markers which are involved in recognition of antigen, cell-cell interaction and adhesion and transduction of signals. These cell surface markers are called cluster of differentiation (CD). Different types of lymphocytes can be distinguished on the basis of their cell surface markers. On the basis of their functions, lymphocytes are of two types-T-lymphocytes and B lymphocytes.

B lymphocytes

B lymphocytes are called so as they are formed and mature in the bone marrow, in mammals and in bursa of Fabricius, in birds. After their maturation in bone marrow, B cells leave the bone marrow and are directly released into circulation. They constitute about 5- 15% of total circulating lymphoid cells. B lymphocytes express a unique antigen binding receptor, mIgM which is a membrane bound antibody molecule. This mIgM along with Ig-α /Ig-β forms the B cell receptor for an antigen. IgD is also expressed on the surface of mature B cells. In addition to its unique antibody receptor molecule, B cells also carry some other receptors which have several important functions.

When B cells encounter an antigen through their receptors, they get activated and proliferate. They are differentiated into memory B cells and effector cells or plasma cells. The plasma cells do not possess membrane bound receptors. They start secreting antibodies. A single plasma cell can secrete hundreds to thousands of antibodies per second. Memory cells express the membrane bound receptors with the same specificity as expressed by the parent cell.

Most B cells carry MHC class II molecules which present antigen peptides to CD4+ T cells during interaction with T cells. B cells can be further divided into B1 and B2 cells. B1 cells appear earlier in the fetal development. They carry CD5 marker, originally found in T cells. They represent a very less amount of fetal cell population and are predominantly found in peritoneal and pleural cavities. B2 ells develop later than B1 cells and are the major part of the B cell population. They are the conventional B cells.

T lymphocytes

T lymphocytes are the lymphocytes which after their formation in the bone marrow migrate to thymus for maturation. As their maturation occurs in thymus, these cells are called T lymphocytes. Like B cells, T lymphocytes also express a membrane bound unique antigen receptor. This is called TCR (T cell receptor). TCR are structurally different from immunoglobulin but have some similar features. TCR can recognize an antigen only when it is presented along with MHC molecules. TCR is associated with several polypeptides collectively referred to as CD3. CD3 is a unique molecule which is identical on all T cells and is expressed only on T cells. It is a signaling molecule. In addition to TCR, T cells express several other surface molecules with specific functions.There are two different types of TCR. Most T cells express $\alpha\beta$ TCR, consisting of α and β chains. Some T cells express $\alpha\beta$ TCR with γ and δ chains.

T cells are further divided into two sub populations: T_H cells and T_C cells.

T_H cells: In addition to TCR and CD3, T_H cells also express CD4 glycoprotein on their surface. They are also called CD4+ $T_H$cells. These are the cells which have crucial role in immune response. They help B cells to produce antibodies. They also activate Tc cells. T_H cells secrete certain cytokines which act on B and Tc cells. CD4+ T_H cells recognize their specific antigen in association with classII MHC molecules. CD4+ T_H cells can be further divided into two subsets T_{H1} and T_{H2} cells. These subsets of T_H cells differ in the cytokines they secrete. T_{H1} cells secrete

IL-2 and IFNγ and mediate several functions associated with cytotoxicity and local inflammatory reactions. On the other hand T_{H2} cells which secrete IL-4,IL-5,IL-6 and IL-10 are more effective in stimulating B cells to produce antibodies.

Tc cells:Tc cells display CD8 molecules on their surface and are called CD8+ T_H cells. CD8+ T_H cells recognize antigen in association with class I MHC molecules. They are the effector cells which eliminate particularly viral infected cells, tumor cells and cells of a foreign graft.

Besides CD4+ T_H cells and CD8+ T_H cells, there is another category of T_H cells, called ***NKT cells***. These are the cells which express T markers as well as some NK cell markers. NKT cells recognize glycolipid antigens and are capable of producing large amount of IFNγ and IL-4.

Natural Killer Cells

Natural Killer Cells (NK cells)constitute about 5% to 10%of the peripheral blood lymphocytes in human. They are large granulated lymphocytes. They lack T cell receptors or membrane bound immunoglobulin molecules that are expressed on T or B lymphocyte surfaces. Their function is to recognize and kill the virus infected cells and certain tumor cells. The mechanism of their recognition is not fully understood. They may interact with tumor cells directly in a non-specific antibody independent manner. In some cases, natural killer cells express CD16 molecules which actas Fc receptors for IgG molecules. These NK cells attach to the antibody through their receptors and subsequently destroy the target cells.

Dendritic Cells

Dendritic cells arise from hematopoietic stem cells and are present in different forms. After their origin in the bone marrow, dendritic cell precursors circulate in the blood to specific sites in the body where they mature and reside. Their primary function is in antigen presentation. Structurally, they contain long membranous extensions, similar to the dendrites of nerve cells **(Fig.2.7).** It is due to the presence of these membranous extensions that they are called dendritic cells. There are four types of dendritic cells, found in different locations in the body : *Langerhans cells* are found in the epidermis of the skin, *interstitial cells* are found in the interstitial spaces of virtually of all organs except brain, *monocyte derived dendritic cells* are derived from the monocytes and *plasma cytoid derived dendritic cells* are derived from plasmacytoid cells.

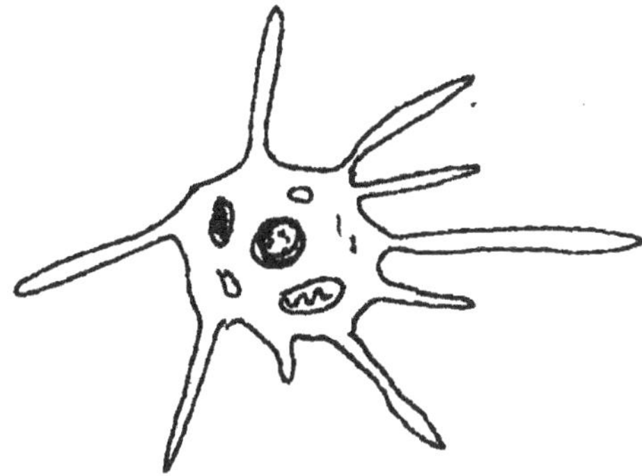

Fig. 2.7 : A dendritic cell

In immature form dendritic cell express less number of MHC molecules and function as phagocytic cells. When mature, their phagocytic activity is decreased. They express large quantity of class II MHC molecules and co-stimulatory molecules such as CD80, CD86 and CD40. Presence of these molecules make them more potent antigen presenting cells as compared to B cells and macrophages. They are 10 to 100 times more potent than B cells and macrophages which need to be activated by antigen to express these molecules. Dendritic cells process the antigens and present it to T_H cells along with class II MHC molecules.

Follicular Dendritic Cells

Follicular dendritic cells are a distinct group of dendritic cells which help in the maturation and diversification of B cells. They are exclusively present in lymph follicles within the lymph nodes. They do not display class II MHC molecules. Although they do not display class II MHC molecules, they express high levels of membrane receptors for antibody and complement through which they bind to circulating antigen-antibody complexes and help in activation of B cells.

Mast Cells

Mast cells are important cells of the immune system which are involved in allergic reactions. They are present in almost all parts of the body along with endothelial cells of the blood vessels as well as the mucosal epithelial tissues. Their precursors are produced in the bone marrow from where they are released in the blood circulation. They do not differentiate into mature cells in the blood. Even in blood they remain as undifferentiated cells and differentiate only when they enter the tissues. They contain large number of cytoplasmic granules that contain histamines and other pharmacologically active substances **(Fig.2.8).** Mast cells play a major role in allergic reactions.

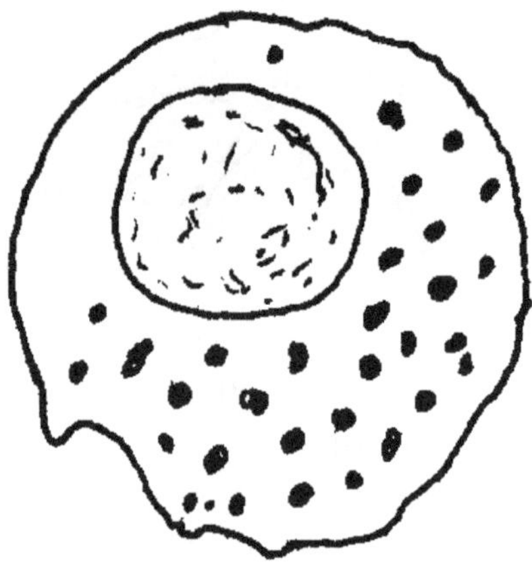

Fig. 2.8 : A mast cell

Neutrophils

Neutrophils, also known as polymorphonuclear phagocytic cells, are the most predominant type of granulocytes and are an essential part of the innate immune system.They are produced in the bone marrow. Their cytoplasm contain granules which stain with neutral dyes. They are short lived and highly motile. Neutrophils constitute about 50-70% of the total leucocytes in the body. They have characteristic multilobed nuclei **(Fig.2.9).** Neutrophils have a variety of specific receptors, including complement receptors, cytokine receptors for interleukins and interferon gamma (IFNγ), receptors for chemokines, receptors to detect and adhere to endothelium, and Fc receptors for opsonin. They are the first cells to arrive at the site of inflammation. They are important phagocytic cells capable of ingesting micro organisms or particles. They contain a number of lytic enzymes and bactericidal substances which are present in primary and secondary granules. Primary granules are a type of lysosome which contains hydrolytic enzymes. Secondary granules contain lactoferrin, collage nase and lysozyme. During phagocytosis, both these types of granules fuse with phagosome and digest its content.

During an infection, a transient increase is observed in the number of neutrophils. This is called *neutrophilia*.

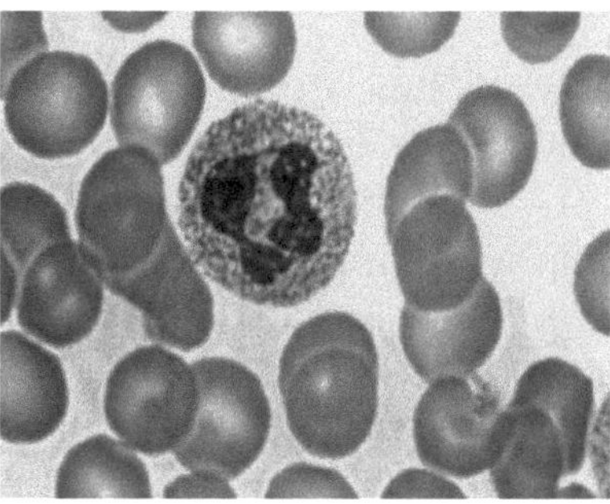

Fig. 2.9 : Neutrophil

Eosinophils

These are the granulocytes that develop during hematopoiesis in the bone marrow.They stain with acidic dyes.In a normal healthy individual, eosinophils constitute about 1-3% of total leucocytes. They are about 12-17 micrometers in size. Eosinophils have bilobed nuclei and granular cytoplasm **(Fig.2.10).**They also act as phagocytic cells but their major role is in combating multicellular parasites and certain infections in vertebrates. An increase in the number of eosinophils is observed in the people with a parasitic infestation. During infection they release the contents of their granules which cause damage to the parasitic membrane. Eosinophils play a prominent role in the pathogenesis due to allergy.

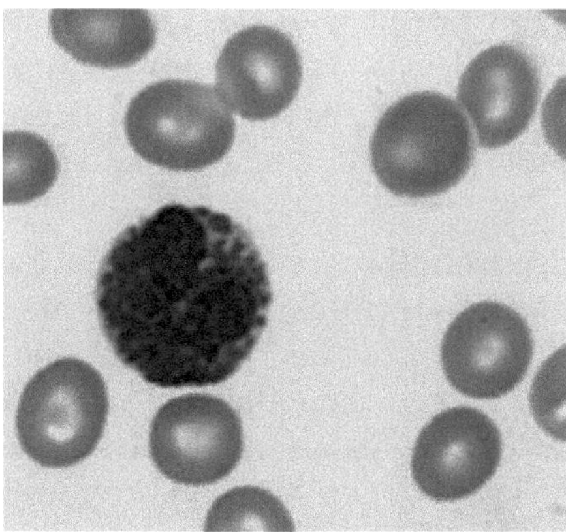

Fig. 2.10 : Eosinophil

Basophils

Basophils are the least common granulocytes in the body. They constitute <1% of total leucocytes in the body. Basophils contain large cytoplasmic granules with lobed nuclei **(Fig.2.11).** They stain with basic dyes. They are non-phagocytic cells which release pharmacologically active substances which lead to certain allergic responses in some people. Basophils appear in many specific kinds of inflammatory reactions. On activation, basophils release histamines, proteoglycans and other proteolytic enzymes which contribute to inflammation, particularly those that cause allergic symptoms. They play a major role in hypersensitivity reactions. Basophils are also release the cytokine interleukin-4.

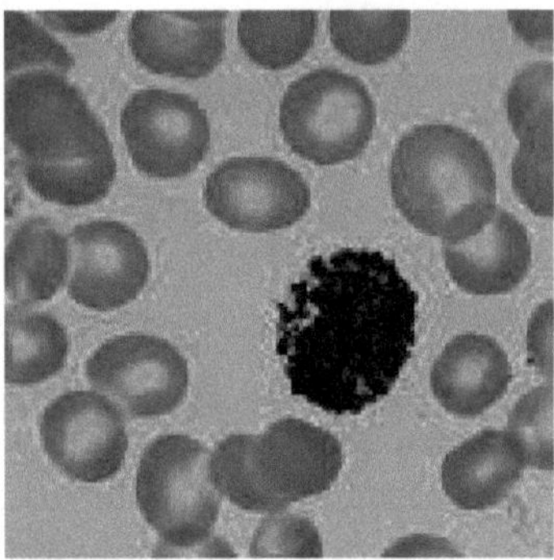

Fig. 2.11 : Basophil

Mononuclear Phagocytic Cells

There are two types of phagocytic cells: mononuclear phagocytic cells and polymorphonuclear granulocytes. Mononuclear phagocytic cells include monocytes and macrophages. These cells are widely distributed throughout the body and are found in virtually all organs of the body.

Monocytes arise from the monocyte progenitor cells in the bone marrow. During hematopoiesis, monocyte progenitor cells differentiate into promonocytes. Promonocytes leave the bone marrow and enter the blood where they are further differentiated into mature monocytes. Monocytes are relatively larger cells with horse shoe-shaped nuclei. They contain azurophilic granules and many intra cytoplasmic lysosomes. They have many specialized receptors like scavenger receptors, toll-like receptors and mannose receptors which help them to attach to the microbial surfaces. The monocyte which migrate to different tissues, differentiate into macrophages **(Fig.2.12).** Microphages are of two types: those that remain in the specific tissues become *fixed macrophages* and the others which remain mobile are known as *wandering or free macrophages*. Macrophages are highly phagocytic cells.When activated, they become very effective phagocytic cells. Phagocytosis of particulate antigen itself serves as a stimulus for the activation of macrophages. Certain cytokines such as IFNγ , secreted by T_H cells are also potent activators of macrophages. Activated macrophages secrete many cytotoxic proteins which help them to eliminate a broad range of microbes. Activated macrophages express higher levels of class II MHC molecules and act as antigen presenting cells for T_H cells. They can phagocytose the whole micro organism, cell debris and dead and injured cells.

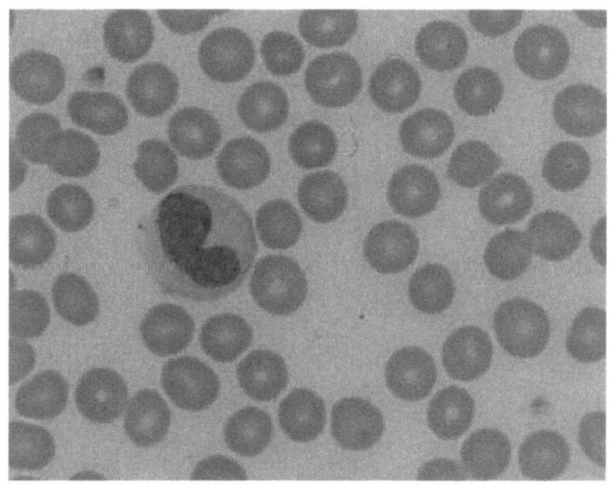

(a)

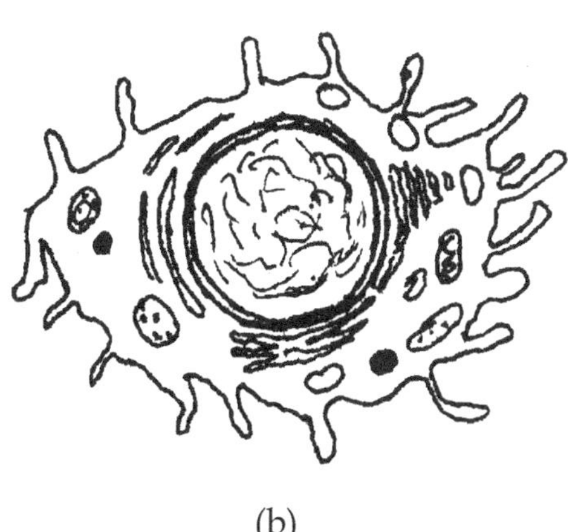

(b)

Fig. 2.12 : (a) Monocyte and (b) Macrophage

During phagocytosis, as mentioned in chapter 1, the antigen molecule attaches itself to the cell membrane of macrophages. Macro phages develop membrane protrusions, called pseudopodia, around the attached antigen. The pseudopodia completely surround the membrane bound material and fuse to form a vacuole called *phagosome.* Phagosome is taken inside the cytoplasm where it fuses with the lysosomes to form phagolysosome. The lysosomes contain many hydrolytic enzymes which digest the ingested material. Finally, the digested material is thrownout **(Fig.1.2).**

POINTS TO REMEMBER

- The lymphoid organs are classified into two categories: primary and secondary lymphoid organs.
- Primary lymphoid organs are those where B and T lymphocytes undergo maturation e.g. bone marrow and thymus.
- Secondary lymphoid organs provide site for the interaction of lymphocytes with antigens such as spleen, lymph nodes and MALT.
- Many kinds of cells play various important roles in immune responses. These include lymphocytes, natural killer cells, dendritic cells, mononuclear phagocytic cells, mast cells, neutrophils, basophils and eosinophils.
- Lymphocytes are the major cells of the immune system which are present in blood, lymph and lymphoid tissues.
- Lymphocytes are of two types- T lymphocytes and B lymphocytes.
- B lymphocytes form antibodies.
- T lymphocytes are of two types- T_H and Tc. T_H lymphocytes help B cells to form antibodies.
- Neutrophils are the most predominant type of granulocytes.
- Thymus is a bilobed organ where T cells mature.
- Bone marrow is the site of the origin of B and T lymphocytes in mammals.
- Spleen is a large and ovoid organ, situated in the left side of abdomen, below the diaphragm.

- Spleen is a vital organ of the immune system which not only provides site for the interaction of antigen and lymphocytes but also helps in the clearance of wornout red blood cells.
- Lymph nodes are small, round or ovoid structures where B and T lymphocytes interact with antigens.
- MALT refers to the groups of lymphoid tissues found in the mucosal surfaces of gastrointestinal, respiratory and urinogenital tracts which include Peyer's patches and tonsils.
- Skin is an important component of immune system which has some specialized cells with specific immune functions.

REVIEW QUESTIONS

1. Discuss, "Thymus is an important organ of immune system."
2. Describe the role of spleen in immune response.
3. Describe the structure and functions of a lymph node.
4. Describe the role of neutrophils in immune system.
5. Write short notes on-
   i. MALT
   ii. Mononuclearphagocytic cells
   iii. Natural killer cells
   iv. Dendritic cells
6. Differentiate between:
   i. Dendritic cells and follicular dendritic cells
   ii. T_H cell and Tc cell
   iii. Primary and secondary lymphoid organs
   iv. B lymphocyte and T lymphocyte

CHAPTER - 3

Antigens

The main purpose of the immune system is to protect the body from foreign substances, diseases or infections. When the body is exposed to a microbe or a foreign substance (antigen), the body makes some response which is called *immune response*. The purpose of immune response is to recognize, neutralize and eliminate the antigens from the body. Generally, the term immunogen is also used for antigen. Both, antigen and immunogen are used as synonyms but specifically, they are different from each other.

IMMUNOGENS *VS* ANTIGENS

An *immunogen* is a substance which can induce an immune response in the body. The immune response may be cell mediated or humoral. The ability of an immunogen to induce a specific immune response in the body is called ***immunogenicity***. *Antigens*, on the other hand, are the molecules that specifically interact with B or T cell receptors. These may be components of foreign substances such as pathogens or some self-molecules may also serve as antigens, as in case of autoimmune reactions.The ability of an antigen to combine with antibody or T cell receptor is called ***antigenicity***. Most antigens are either proteins or polysaccharides in nature, though lipids and nucleic acids may also act as antigens. The immunogenicity of these substances is increased when they are associated with proteins. Antigens may not be capable of inducing immune response but it can specifically react to the components of immune system (antibodies). Thus, all immunogens are antigens i.e. they can elicit an immune response and can also bind to the antibodies, while all antigens are not immunogens i.e. they are not capable of inducing an immune response. For example, there are substances, called **haptens,** which can bind to antibodies but cannot elicit an immune response at their own. They are, thus, antigens but not immunogens. But if they bind to a carrier molecule, they can bring out an immune response and act like immunogens. Haptens show the property of antigenicity and not immunogenicity.

FACTORS AFFECTING IMMUNOGENICITY

There are several factors which determine the immunogenicity of an immunogen. Both, the properties of the immunogen and the properties of biological system contribute in the immunogenicity.

Properties of Immunogens

Foreignness

Generally, the body elicits immune response only against non-self (foreign) molecules. The degree of the immunogenicity against a substance depends on the degree of foreignness. In general, greater is the phylogenetic distance between two species, greater the antigenic disparity between these constituent molecules. For example, if Bovine serum albumin (BSA) is injected into mice, cow and goat, it will induce a stronger response in mouse as compared to goat. It is because cow is more closely related to the goat than mouse. When injected in cow, BSA will not elicit any immune response because the animal will recognize it as self and makes no immune response. Sometimes, body can also elicit immune response against self-components, as in case of autoimmune diseases.

Molecular mass

Normally, only the large molecules with molecular mass of > 100,000 Da are immunogenic in nature. Molecules with molecular mass of <10000 Da are poorly immunogenic,with a few exceptions.

Chemical complexity

Chemical composition and complexity of the molecules also determine its immunogenicity. Synthetic homopolymers (molecules containing multiple copies of a single amino acid or sugar) are not immunogenic though they may be large in size. Heteropolymers (molecules containing different amino acids or sugars) are more immunogenic than homopolymers.

In addition to its composition, the conformation of a molecule could also be important to its antigenicity. For example, lysozyme molecules which consist of several amino acids, folded into a loop with aid of disulphide bond, are good antigenic. If the conformation is disturbed, the antigenicity of the molecule is reduced.

Antigen processing and presentation

When an antigen enters the body, it is processed (degraded) by antigen presenting cells (APC) and presented to the T cells along with MHC molecules. T cells recognize this processed antigen and help in developing immune response. Hence, large insoluble molecules which are easily phagocytosed and processed are strongly immunogenic. Molecules that are resistant to enzymatic degradation in APC and cannot be presented with MHC molecules are poorly immunogenic.

Properties of Biological System

Genotype of the host

Several experiments have shown that the nature and degree of immunogenicity also depend on the genetic constituent of the host animal. When an antigen is injected into two different strains of an animal, both strains give different levels of serum antibodies. MHC molecules, B and T cell receptors play important role in recognition of an antigen and initiation of immune response. Thus, the genes encoding these molecules and various other protein molecules, involved in immune system, influence immune response against an antigen.

Dose and route of administration of antigen

Degree of immunogenicity of an immunogen also depends on the dose and route of administration of an immunogen. A certain dose of immunogen is required for an optimal immune response. If the dose is too small, it may not be capable of stimulating the immune response. It is because too small amount of an immunogen may not be capable of activating sufficient number of lymphocytes. Too small dose, in some cases, may also induce immunological tolerance. A too high dose can also induce tolerance. For many immunogens, a single dose may not be capable of inducing a strong immune response and repeated administration of immunogen is required for a proper immune response.

An immunogen can be administered in the body by several routes such as intravenous, intradermal, subcutaneous, intramuscular or intraperitoneal. Antigens, entering the body by various routes, are carried to different lymphoid organs. The type and extent of immune response depend on the organ and cell population. Antigens administered via intravenously are first transported to spleen whereas a subcutaneously injected antigen is taken up by Langerhans' cells in the skin and is carried to regional lymph nodes. A subcutaneously injected antigen provokes a strong immune response.

Adjuvants

There are substances, called *adjuvants,* which also influence the immunogenicity of an antigen. These are the substances which, when mixed and injected with antigen, enhance its immunogenicity. They act non-specifically and are generally used along with soluble antigens. Aluminum potassium sulfate (alum) is a very common adjuvant for general human use.

There are several ways by which an adjuvant enhances the immunogenicity of an antigen. It may prolong the persistence of an antigen so that the effective time to the exposure to the antigen increases from a few days without adjuvants to several weeks with adjuvant, as in case of alum. Alum, when mixed with antigen, precipitates the antigen and increases its persistence. It also increases the size of the antigen which increases the probability of the antigen to be phagocytosed. Some adjuvants such as Freud's complete adjuvant act by enhancing the co-stimulatory signals. Muramyl dipeptide, a component of mycobacterial cell wall, in Freud's complete adjuvant, activates macrophage and dendritic cells. Activated macrophages and dendritic cells are more phagocytic than inactivated ones. The activated cells express higher levels of molecules that trigger co-stimulation and enhancement of T cell immune response. Some adjuvants also increase local inflammation and attract phagocytes and lymphocytes. In some cases, adjuvants stimulate non-specific proliferation of lymphocytes.

EPITOPES

When an antigen interacts with an antibody, it is not the entire molecule which interacts with the antibody. But there are certain parts of the antigen that are recognized and bound by antibodies or T cell receptors. These discrete sites on the antigen molecules are called *epitopes* or *antigenic determinants.* Even in an epitope there are certain specific regions of the epitope constituted by a few amino acids which bind with greater affinity to specific areas of the antibody-binding site, and thus are primarily responsible for the specificity of antigen-antibody interaction. Macro molecules possess several epitopes on their surface. A polypeptide with 100 amino acids may have as many as 14 to 20 non-overlapping epitopes.The epitopes may be continuous (linear) or discontinuous (conformational) as shown in the **(Fig. 3.1**.) A linear epitope is a conformation on the antigen, formed by a continuous

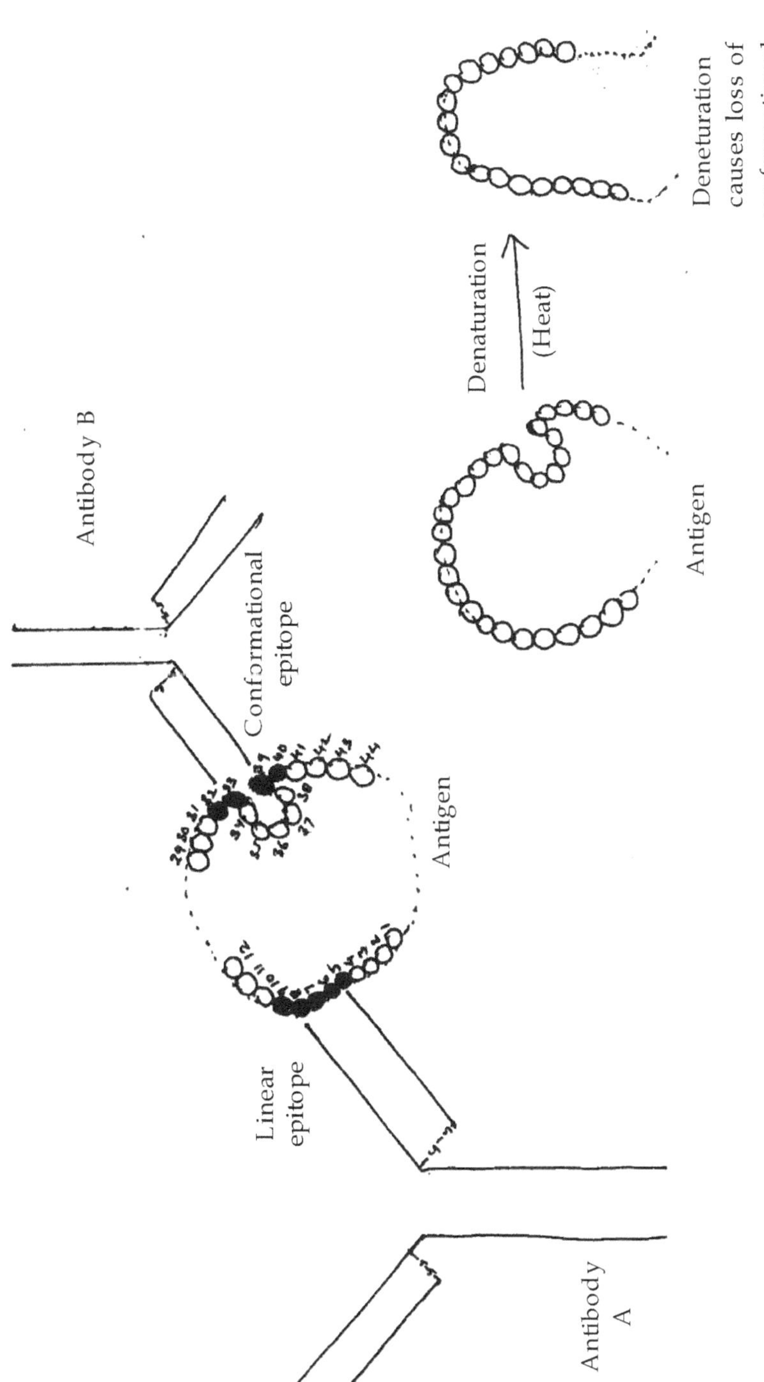

Fig. 3.1 : Linear and Conformational Epitopes

sequence of amino acids, while conformational epitopes are formed by folding of the antibodies i.e. when amino acids which are separated in primary sequence are brought close together in the tertiary structure due to folding of the polypeptide chain. Generally, in an antigen molecule, each epitope is different from other, though in some polypeptide chains, there are many repeats of the same epitope. B and T cells recognize different epitopes which have certain specific features.

B cell Epitopes

Epitopes which are recognized by membrane bound antibodies on B cells and secreted antibodies are called B cell epitopes. B cell epitopes are typically hydrophilic sequences, present on the surface of antigens. Amino acid sequences that are hidden inside the protein molecule cannot function as B cell epitope. B cell epitopes may be linear sequential residues or conformational determinants. If a protein is denatured, denaturation causes loss of conformational determinants and hence antibodies raised against native protein do not react with denatured protein. B cell epitopes are located in the flexible region of an immunogen. Studies have shown that most of the surface of the globular proteins is immunogenic and it consists of multiple overlapping B cell epitopes.

T cell Epitopes

T cells recognize an antigen (protein) molecule only when it is degraded into small peptides and displayed along with MHC molecules by APCs. The T cell epitopes are linear epitopes. Denaturation of the molecule by high temperature, which destroys the conformation of the protein, does not affect T cell molecules, as it affects B cell epitopes. The maximal size of T cell epitope depends on the peptide binding cleft of an MHC molecule.

CROSS REACTIVITY

Antigen-antibody reactions are very specific i.e. an antibody binds with a particular antigen. But if two antigens share structurally similar epitopes, cross-reactivity occurs **(Fig. 3.2).** This is because antibody, directed against a particular epitope, can also bind to another epitope which is located on another antigen and is structurally related to it. In case of some bacteria and viruses, they have epitopes which are similar to some of the normal host components. When antibodies are produced against these microbial antigens, the antibodies react with self-components of the host. This results in autoimmune reactions.

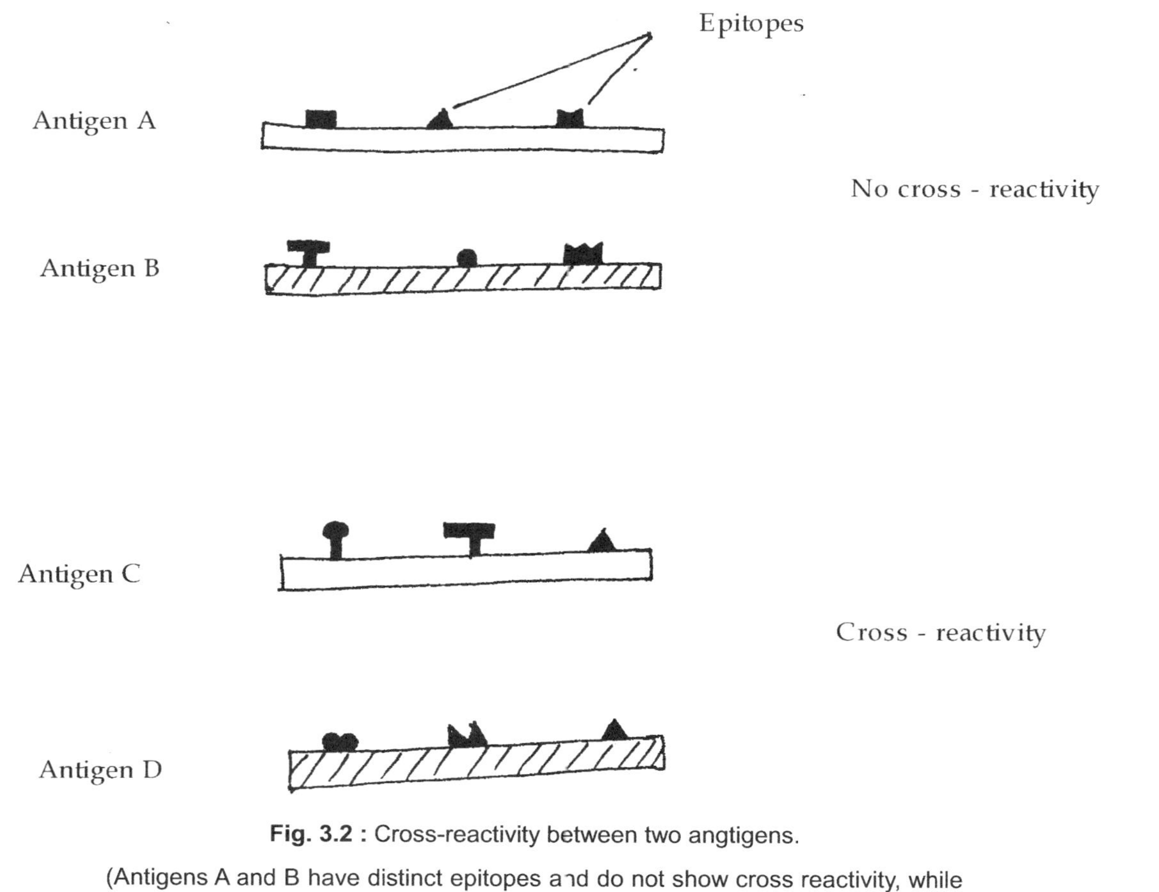

Fig. 3.2 : Cross-reactivity between two angtigens.

(Antigens A and B have distinct epitopes and do not show cross reactivity, while antigens C and D show cross-reactivity due to the presence of similar epitope)

POINTS TO REMEMBER

- Antigens are the molecules that specifically interact with B or T cell receptors.
- An immunogen is a substance which can induce an immune response in the body.
- The ability of an antigen to combine with antibodies or T cell receptors is called antigenicity.
- The ability of an immunogen to induce a specific immune response in the body is called immunogenicity.
- Many factors affect immunogenicity of an immunogen. These are foreignness, complexity of the molecule, molecular mass, antigen processing and presentation, genetic constituent of the host animal, dose and route of administration of antigen and adjuvants.
- Adjuvants are the substances which enhance its immunogenicity, when they are mixed and injected with antigen.
- The discrete sites on the antigen molecules which bind with antibodies are called epitopes or antigenic determinants.
- Epitopes which are recognized by membrane bound antibodies on B cells and secreted antibodies are called B cell epitopes and are different from T cell epitopes.

REVIEW QUESTIONS

1. What are antigens? How are they different from immunogens?
2. Describe various factors which affect immunogenicity of an immunogen.
3. What are adjuvants? How do they affect the immunogenicity of an immunogen?
4. Discuss,"All immunogens are antigens but all antigens are not immunogens."
5. Differentiate between:

   i. Antigenicity and immunogenicity

   ii. B cell and T cell epitopes

CHAPTER - 4

Antibodies

Antibodies are a group of glycoproteins having a globular compact structure. They are known as immunoglobulins. On the basis of their experiments, Tiselius and Kabat, in 1939, demonstrated that most antibodies are found in the γ globulin fraction of serum proteins. When serum from rabbit, immunized with ova albumin, was subjected to electrophoresis, it showed a peak for γ globulin **(Fig. 4.1).** Studies have revealed that the major part of IgG is found in the γ globulin fraction of the serum proteins. Other fractions, α and β, too, have other immunoglobulins along with small part of IgG.

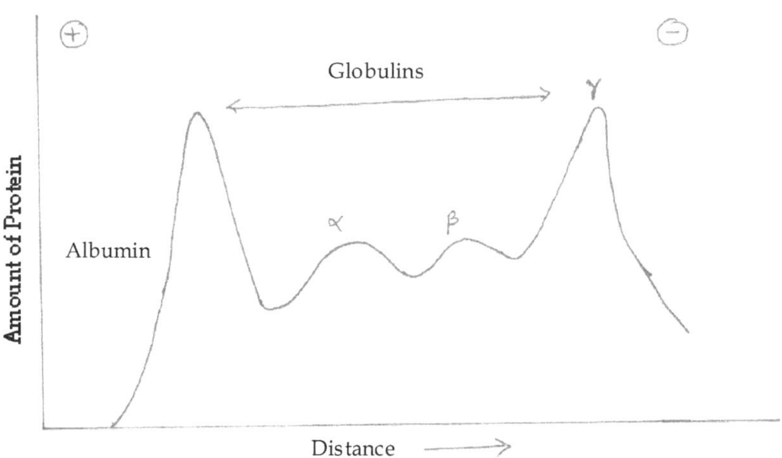

Fig. 4.1 (a) : Electrophortic mobility of serum from a rabbit immunized with ovalbumin.(Experiments by Tiselius and Kabat (1939) demonstrated that most antibodies are found in g glibulin fraction of serumm proteins : Tiselius and kabat (1939) : Journal of Experimental Medicine 69:119-131)

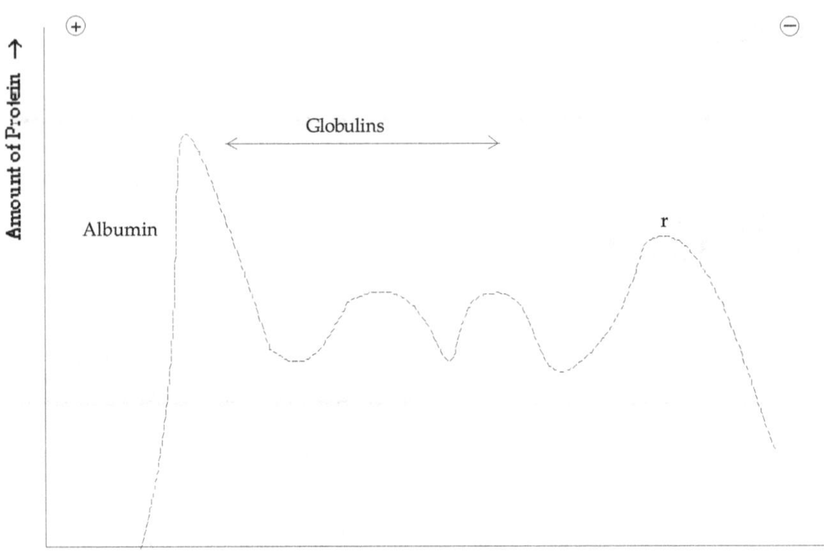

Fig. 4.1(b) : Electrophoretic mobility of serum from a rabbit immunized with ovalbumin. (The serum was first inclubated with ovealbumin to remove anti-ova antibody.)

Porter and Edelman, in 1950, revealed the basic structure of an immunoglobulin molecule. They showed that when immunoglobulins were subjected to proteolytic enzymes such as papain and pepsin, different types of fragments were produced. The fragments produced by such proteolytic digestion had contributed in elucidating the structure and functions of immunoglobulin. Porter showed that when rabbit IgG was digested with enzyme papain, at neutral pH, the enzyme cleaved the immunoglobulin into three fragments. Out of these, two fragments were identical. These fragments had specific antigen binding sites, called *Fab (fragment antigen binding)*. The third fragment did not have antigen binding site and could be easily crystallizable. It was termed as *Fc (fragment crystallizable)* **(Fig. 4.2).** When immunoglobulin was treated with enzyme pepsin, it cleaved the molecule at different site. It produced a bivalent fragment and small peptides. The bivalent fragment had two antigen binding (Fab) sites. The fragment was termed F(ab')2 because it had two antigen binding sites and could precipitate antigens. The F(ab') 2 could again be split into two univalent fragments, similar to Fab fragments as produced by papain, by reduction of S-S bonds. The remaining Fc fragment, as produced by papain, was not recovered in pepsin digestion because it was digested into small peptides. On the basis of their experiments Porter and Edelman proposed that immunoglobulin molecule is a Y shaped structure with two antigen binding sites.

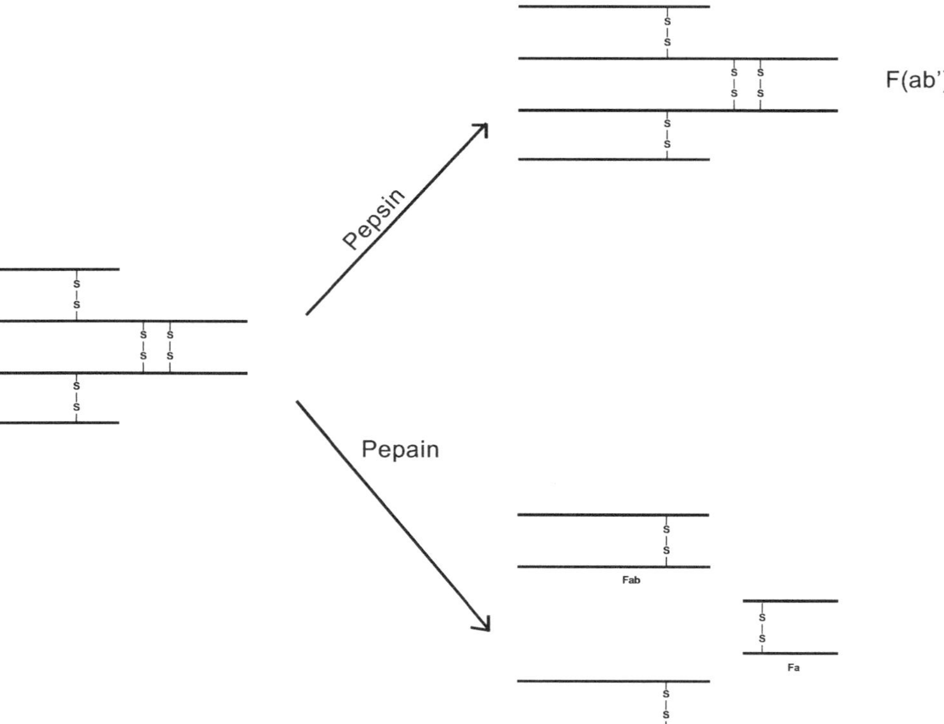

Fig. 4.2 : Degradation of IgG by Pepsin and Pepain Enzymes.

(Digestion by Pepsin results in a divalent F(ab')$_2$ fragment while digestion with Papain yields two univalent Fab fragments and one Fc fragment)

BASIC STRUCTURE

The basic structure of an immunoglobulin molecule is a Y-shaped consisting of four polypeptide chains – two identical heavy (H) chains and two identical light (L) chains **(Fig. 4.3)**. The molecular weight of a monomer Ig molecule is about 150000 Da . The molecular weight of each light chain is 25000 Da and that of each heavy chain is 50000 Da. Light chains are bound to heavy chains by interchain disulfide bridges and multiple non-covalent interactions such as salt linkages, hydrogen bonds and hydrophobic interactions. The heavy chains are also bound to each other by interchain disulfide bridges and multiple non-covalent interactions. Light and heavy chains can be divided into two parts - one variable region and one constant region. Both these regions have separate functions. The variable regions of light and heavy chains, known as V_L and V_H respectively form the antigen binding sites and the constant regions, known as C_L and C_H, determine the effector functions.

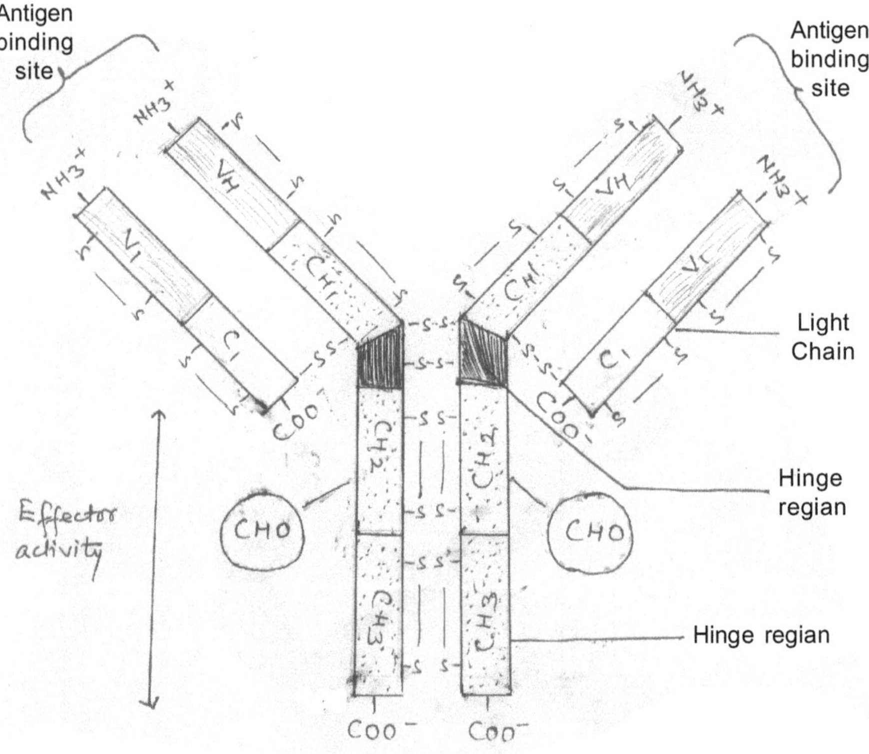

Fig. 4.3 : Structure of IgG.

The light chains are of two types: kappa (κ) and lambda (λ). These chains differ considerably in their amino acid sequences. A given immunoglobulin molecule can have either kappa or lambda chains but not both. A normal individual has a mixture of immunoglobulin molecules in his serum, some with kappa chains and others with lambda chains. The ratio of κ and λ chains varies in different species. In human beings, about 65% of Ig molecules have κ chains and 35 % have λ chains. In mice, κ chains form the major part of light chains and about 95 % of all the light chains are of κ type. In some patients of myeloma tumor free L-chains i.e. the light chains that are not associated with heavy chains, have been detected. These free L- chains are called *Bence Jones proteins*.

The amino terminal regions of each light and heavy chain consist of about 100-110 amino acids which show huge variability. These are known as *variable regions* (V_L in case of light chain and $V_{H,}$ in case of light chains). The carboxyl terminal of the light chain is called the *constant region* (C_L). The light-chain constant regions are found to be almost identical in light chains of the same type, but to differ markedly in κ and λ chains. The difference in antigenicity between the two types of light chains may be directly correlated with the structural differences in constant regions. In heavy chains, the constant region $C_{H'}$ show relatively constant sequences among different antibodies. There are five basic amino acid sequence patterns, corresponding to five different heavy chain constant regions. These are - γ(gamma), α(alpha), μ(mu), δ(delta) and ε (epsilon). In γ, α and ε chains, the constant regions are about 330 amino acids long and for μ and ε , they are of 440 amino acids. In an immunoglobulin, each segment of about 110 amino acids, of both light and heavy chains, is folded to form compact loops called, ***domains*** **(Fig. 4.4)**. Each domain is folded into two layers of pleated sheets containing antiparallel strands of amino acids. The structure is stabilized by intrachain disulfide bond. Therefore, a light chain consists of one variable domain (VL) and one constant domain (CL). In heavy chain, there is one variable domain (VH) and three or four constant domains ($CH_{1'}$ $CH_{2'}$ CH_3 and CH_4), depending on the type of heavy chain. In an immunoglobulin molecule each domain has a specific function. The variable region domains of both light and heavy chains form the *antigen binding sites*. The antigen binding sites, present on the antibody with which antibody binds with the epitope of the antigen, is called *paratope*.

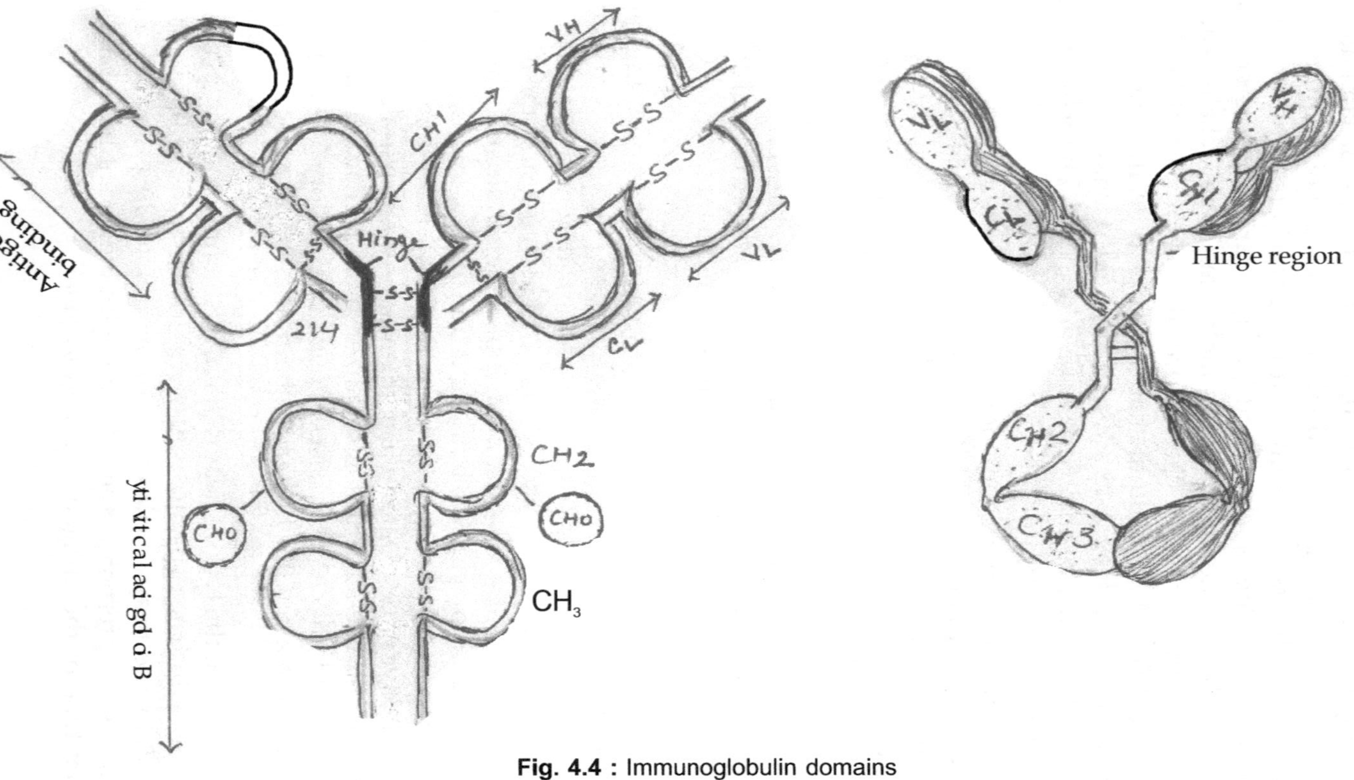

Fig. 4.4 : Immunoglobulin domains

The region between CH_1 and CH_2 domains of heavy chains is flexible region and is rich in proline. This region is called *hinge region.* The flexibility of this region enables antibody molecule to bind to antigen determinates that are widely spaced on cell surface.

In the light and heavy chains, within the variable regions, there are certain regions which show high variability. These regions are called *hyper variable regions.* These are the regions which are actually involved in antigen binding. These regions show compatibility with epitope of antigens. These regions are also called *complemenatrity determining regions* (CDRs). There are three hyper variable or CDR regions. These regions are located among the amino acid positions 30, 50 and 95. The intervening regions, in the variable regions, are called *framework regions (FRs).* Hence, there are three CDRs and four framework regions in the variable regions **(Fig. 4.5)**.

IMMUNOGLOBULIN CLASSES

On the basis of the type of heavy chains there are five types of immunoglobulins, in human beings. These are: IgG, IgM, IgA, IgD and IgE. Each of these classes is distinguished by its unique heavy chain which confers its specific structural and functional properties. Immunoglobulins of each of these classes differ in their size, charge, amino acid sequence and carbohydrate content. Some of these classes are further divided into subclasses.

IgG

IgG is the most abundant immunoglobulin in human serum and in the serum of most mammalian species. It comprises about 80% of the total serum immunoglobulin. IgG is a monomer and represents the typical structure of immunoglobulin. It consists of two γ heavy chains and two κ or two λ light chains. It is the major immunoglobulin, found in the blood, during secondary immune response. There are four subclasses of human IgG. These are designated as IgG1, IgG2, IgG3 and IgG4. These subclasses differ in the length of hinge region and in number and position of interchain disulphide bonds **(Fig.4.6 Next Page)**. IgG1 is the major IgG, comprising about 70% of total IgG. Normal serum concentration of these subclasses is IgG1- 9 mg/ml, IgG2- 3 mg/ml, IgG3- 1 mg/ml and IgG4- 0.5mg/ml. The subclasses of IgG molecule also differ in their biological activities. IgG1, IgG3 and IgG4 readily cross the placenta and provide protection to the developing fetus. IgG1 and IgG3 are very effective complement activators. IgG3 is the most effective activator followed by IgG1. IgG2 is least effective. But IgG4 is not at all able to activate complement.

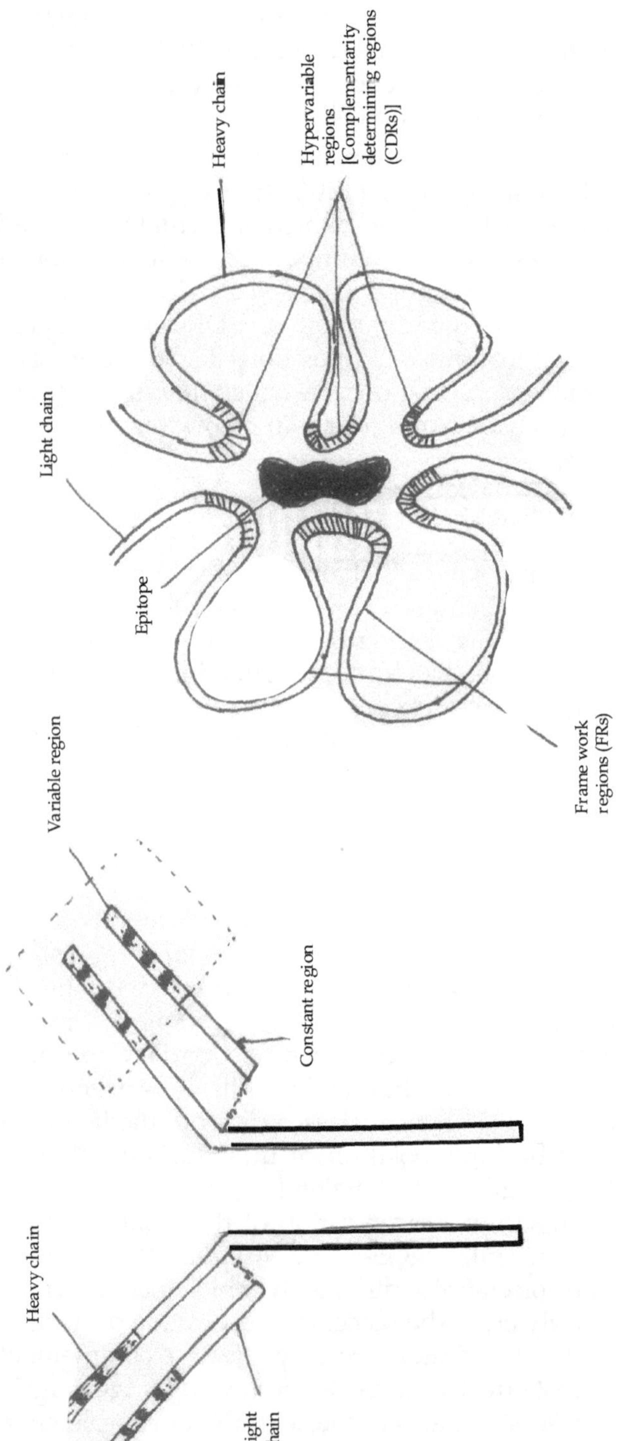

Fig. 4.5 : Hypervariable regions in variable regions of immunoglobulin.

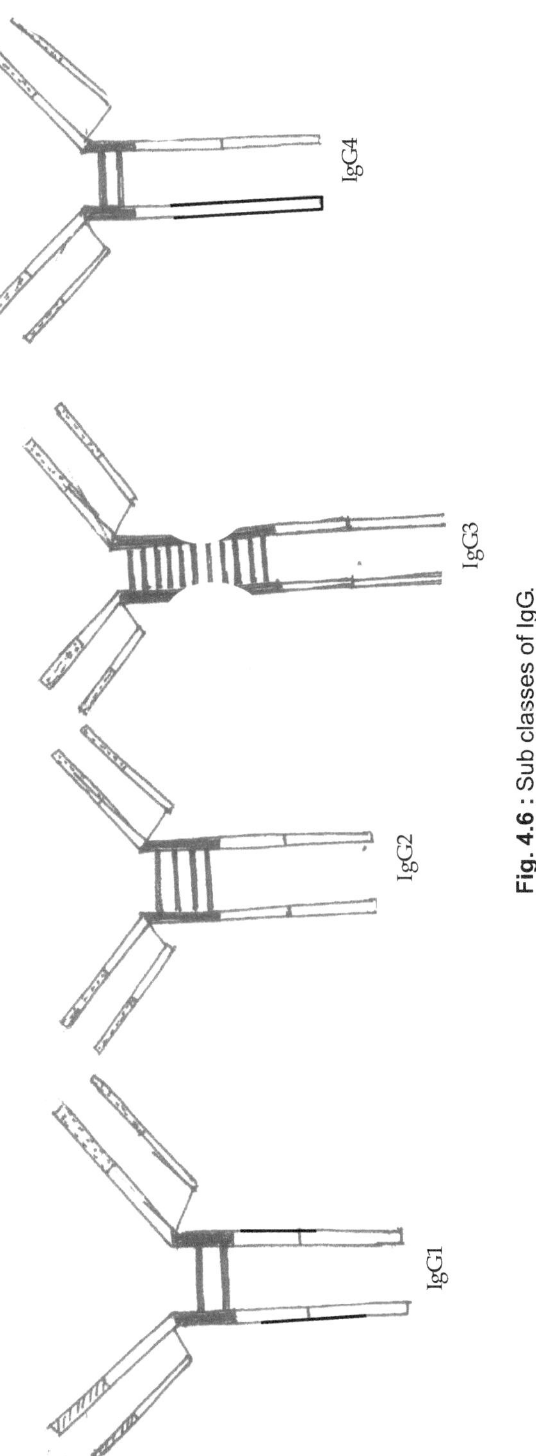

Fig. 4.6 : Sub classes of IgG.

These subclasses also differ in their binding affinity to Fc receptors on phagocytic cells. IgG1 and IgG3 have high affinity to Fc receptors, followed by IgG4. IgG2 has very less affinity to Fc receptors.

IgM

IgM represents about 5-10 % of the total serum immunoglobulin. The average serum concentration of IgM is 1.5 mg/ml. The half life is 5 days. It is the first immunoglobulin to appear, after initial exposure to an antigen i.e. during primary immune response. It is a macromolecule which exists in the blood as a pentamer, consisting of five subunits, cross linked to each other by disulphide bonds between their Fc regions and by a polypeptide chain, called J chain (joining chain). The Fc region of each unit is directed towards the center of the molecule and Fab sites are directed towards the periphery of the molecule **(Fig. 4.7)**. J-chain is believed to be required for the polymerization of the monomers to form the pentamer. In addition, the J-chain also helps IgM antibodies to bind to the receptors on the secretary cells. IgM molecules have an extra domain at hinge region. Thus, unlike IgG which has three constant domains, IgM has 4 constant domains. The molecular weight of the IgM is 900 000 Da. Each monomer unit has a molecular weight of 180 000 Da. Theoretically, an IgM can bind with 10 antigen molecules but practically, due to steric hindrance, it binds only with 5 large antigen molecules.

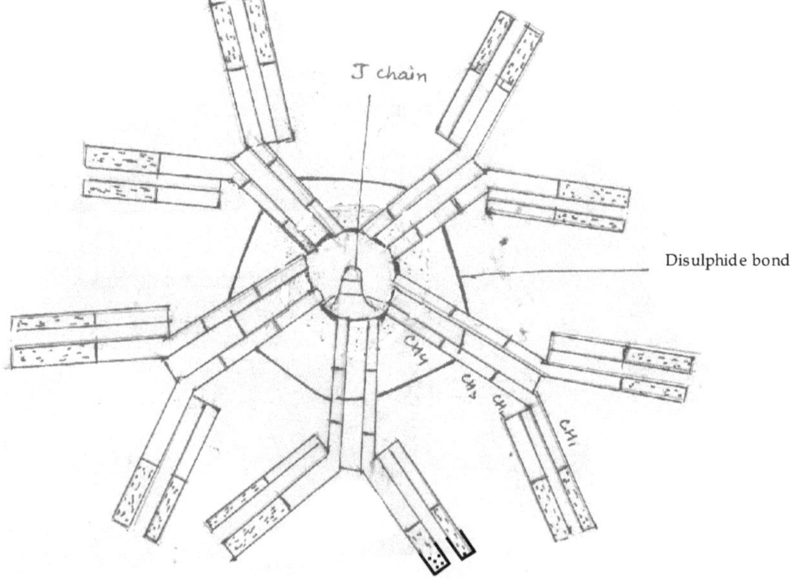

Fig. 4.7 : Structure of IgM.

IgM is more effective agglutinating antibody than IgG. This property of IgM is due to its high valency. It is also very efficient in complement activation and bacteriolysis.

IgD

IgD is a membrane bound immunoglobulin which is co-expressed on B cells' membrane along with IgM. Certain antigens such as insulin, penicillin, milk proteins, and diphtheria and thyroid antigens are known to stimulate IgD production. It is present in serum in very low concentration and constitutes about 0.2 % of the total serum immunoglobulin. Total serum concentration of IgD is about 0.03 mg/ ml. IgD has a molecular weight of 150000 Da. The half life is 3 days. IgD is a monomer with an extended hinge region which is susceptible to proteolysis **(Fig.4.8)**. The hinge region bears multiple O-linked oligosaccharides. The heavy chain bears three N-linked oligosaccharides. IgD functions as transmembrane antigen receptors on mature B cells. It does not have any known effector function as a serum protein.

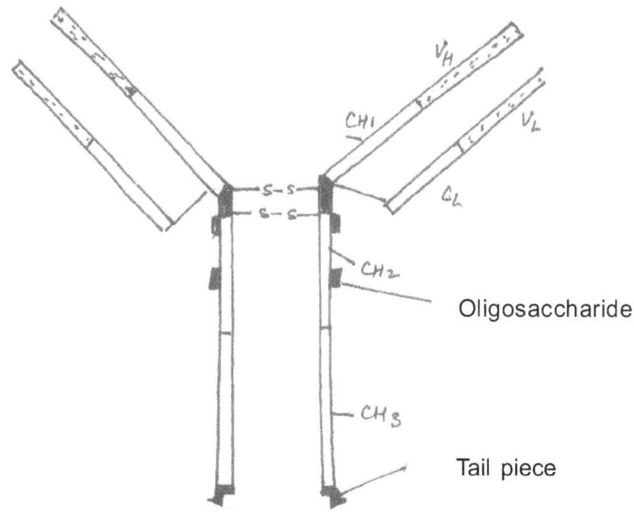

Fig. 4.8 : Structure of IgD.

IgA

IgA constitutes about 10-15 % of the total immunoglobulins. It is present in external body secretions such as saliva, tears, sweat, nasal fluid and colostrum and also in the secretions of lung, urinogenital tract and gastro-intestinal tract. In serum, IgA primarily exists as a monomer with molecular weight 150000 Da. In addition to its monomeric form, IgA also exists in polymeric forms (dimmer, trimer and tetramer) in secretory fluids. In the secretory fluids it is called as secertory IgA (sIgA). Secretory IgA mainly exists in dimeric form. The molecular weight of dimer IgA is 600000 Da. The dimeric form has a polypeptide called J chain and a secretory component **(Fig. 4.9)**. The J-chain is identical to the J-chain of IgM and helps in the polymerization of both serum IgA and secretory IgA. Secretory component is made up of five immunoglobulin like domains and binds to the Fc region of IgA dimmer. Secretory component is made in the epithelial cells and helps in the transport of IgA from submucosa across epithelial cells to mucosal surfaces in the alimentary and respiratory tracts. Here, it provides local protection by binding to the bacterial and viral surface antigens and prevents their attachment to the mucosal cells. IgA cannot cross the placenta but is abundantly present in colostrum. IgA, present in colostrum, provides immunity to the newborn during the earlier months of his life. There are two subclasses of IgA in humans- IgA1 and IgA2. IgA1 predominates in human serum and secretions such as tears, saliva, milk and nasal mucus. IgA2 predominates in colon. IgA1 and IgA2 differ significantly in the structure of their hinge region. IgA2 has a shorter hinge region than IgA1 and is more resistant to attack by bacterial proteases.

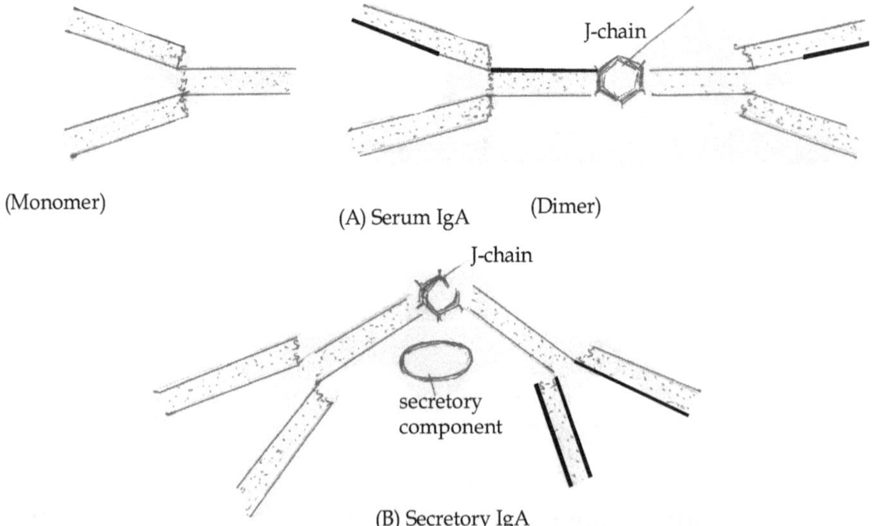

Fig. 4.9 : IgA antibody-Structure of serum and secretory IgA.

IgE

IgE is present in serum in extremely low concentration. Its concentration is about 0.0003 mg/ml. It is a cell bound molecule which is mostly present on the surface of basophils and mast cells. It binds to Fc receptors on mast cells and basophils and then complexes with antigens. The molecular weight of IgE molecule is 190000 Da. The half life is 2.5 days. Like IgM, IgE also has four constant domains and lacks hinge region **(Fig. 4.10)**. IgE antibodies are involved in hypersensitivity reaction and lead to the symptoms of hay fever, asthma, hives and anaphylactic shock. They function to protect the external mucosal surface by recruitment of the effector and plasma cells by stimulating acute inflammatory reactions.

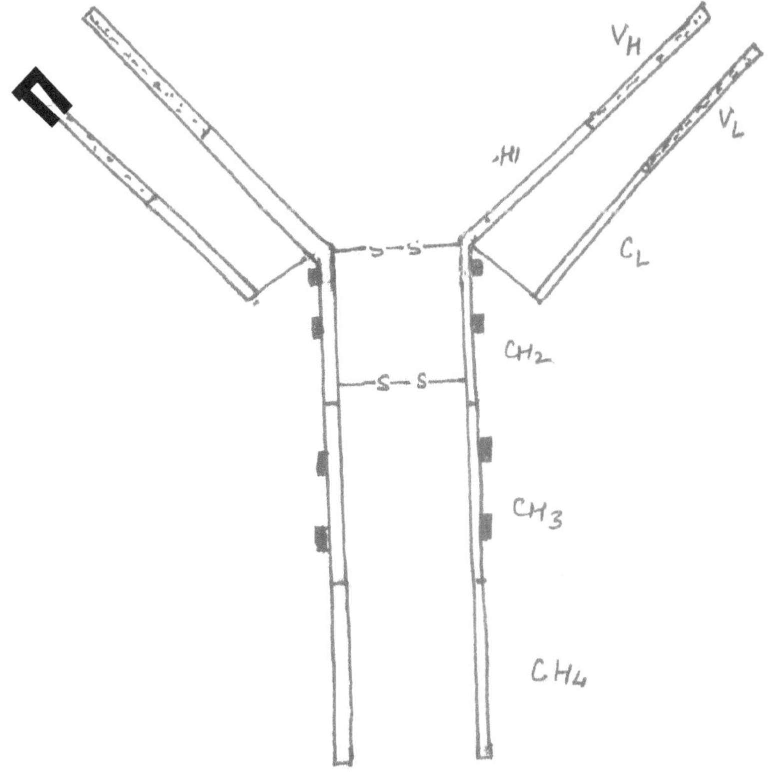

Fig. 4.10 : Structure of IgE.

Important features of different classes and subclasses of immunoglobulins are given in table 4.1.

Table No. 4.1 : Important characteristics of different classes and subclasses of immunoglobulins

CLASS/ SUB-CLASS	% of total IgG in normal serum	MOLECULAR WEIGHT (Dalton)	NO. OF HEAVY CHAIN DOMAINS	NORMAL SERUM CONCENTR ATION (mg/ml)	TYPE OF HEAVY CHAIN	TYPE OF LIGHT CHAIN	HALF LIFE (in Days)	IMPORTANT BIOLOGICAL FUNCTIONS
IgG1	75	150000	4	9	γ1	κ or λ	23	• Activation of classical complement pathway • Transplacental transfer • Pathogen neutralization in tissues • Opsonization
IgG2		150000	4	3	γ2	κ or λ	23	• Transplacental transfer
IgG3		150000	4	1	γ3	κ or λ	8	• Activation of classical complement pathway • Transplacental transfer • Pathogen neutralization in tissues
IgG4		150000	4	0.5	γ4	κ or λ	23	• Activation of classical complement pathway • Transplacental transfer
IgM	5-10	900000	5	1.5	μ	κ or λ	5	• Activation of classical complement pathway • Membrane B-cell receptor
IgA1	10-15	150000-600000	4	3	α1	κ or λ	6	• Mucosal transport
IgA2		150000-600000	4	0.5	α2	κ or λ	6	• Mucosal transport
IgD	0.2	150000	4	0.03	δ	κ or λ	3	• Membrane B-cell receptor
IgE		190000	5	0.0003	ε	κ or λ	2.5	• Mast cell degranulation

Functions of Ig

Except IgD, immunoglobulins are bifunctional in nature. They perform two types of functions: first, they recognize and bind with the antigens with their antigen binding sites and subsequently promote its killing and removal; second, antibodies act as adapter molecules for different effector immune functions: they activate complement classical pathway and lead to lysis of the antigens; bind to the surface of the pathogens, opsonize them and cause their phagocytosis ; antibodies which are bound to cells promote their recognition and killing by natural killer cells ; and finally, antibody bound to Fc receptors sensitize the cells so that they can recognize antigens and cell becomes activated when antigens bind to surface antibodies **(Fig. 4.11).**

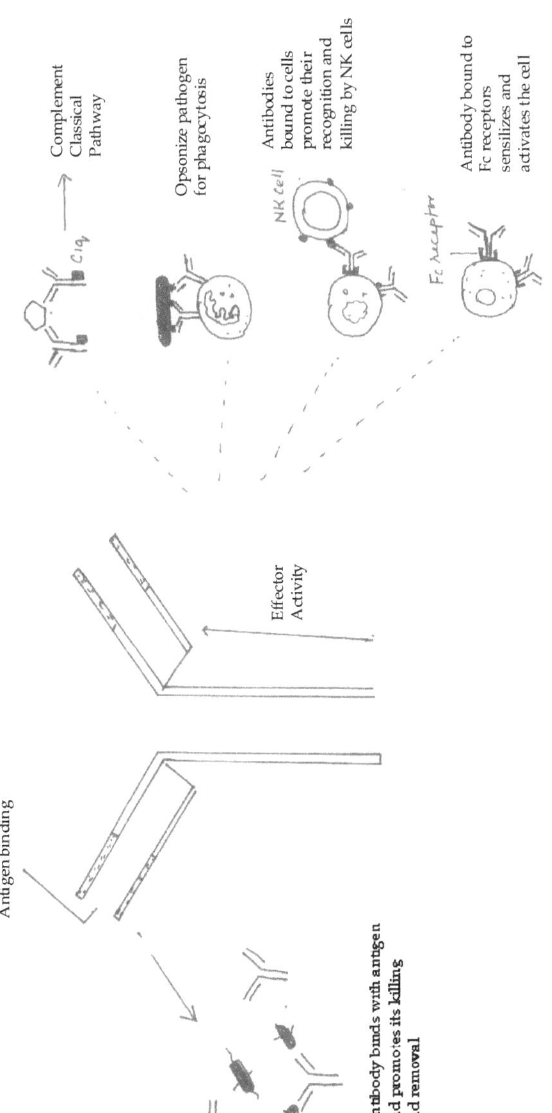

Fig. 4.11 : Functions of immunoglobulins

ISOTYPES, ALLOTYPES AND IDIOTYPES

When an antibody, from one species is injected into another species, it induces an immune response and the animal produces anti-antibodies. These anti-antibodies are produced against specific antigenic determinants on immunoglobulins. These are three types of antigenic determinants : isotypes, allotypes and idiotypes **(Fig.4.12).**

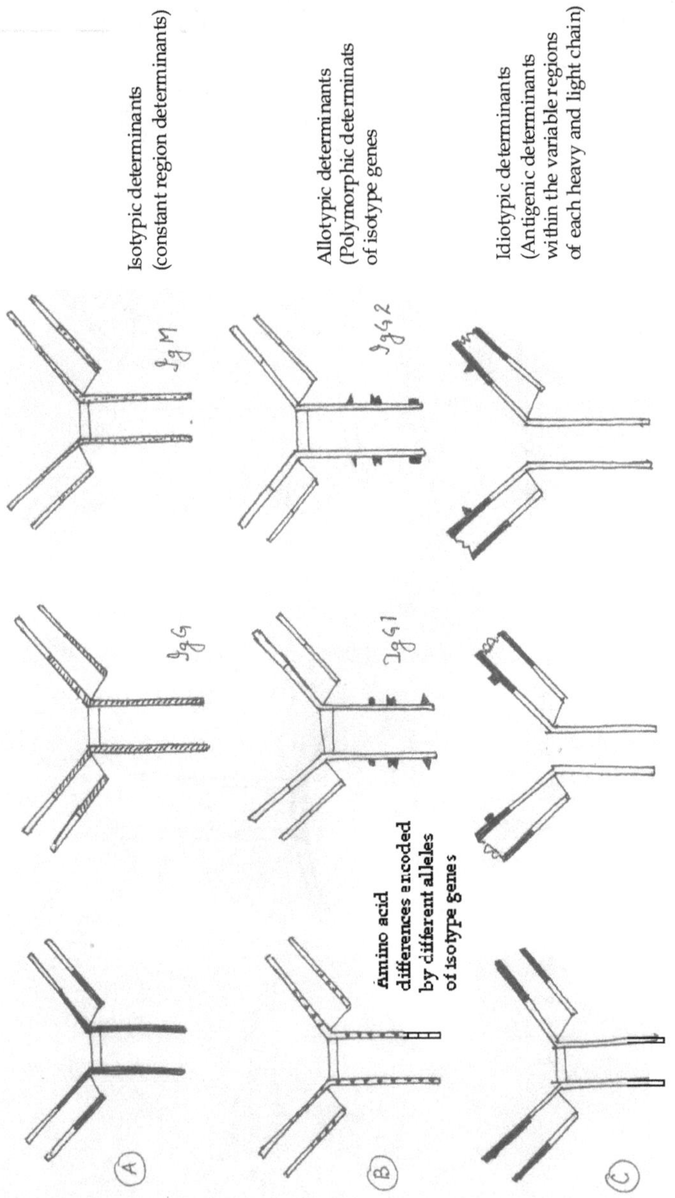

Fig. 4.12 : Antigenic determinants of immunoglobulins.

Isotypes

Isotypes are constant region determinants which distinguish each Ig class and subclasses within a species. For example, rabbit IgG1 and IgM represent two different isotypes. Each isotype is encoded by a separate constant region gene. In humans, immunoglobulin heavy chain isotypes are encoded by Cγ1(IgG1), Cγ2(IgG2), Cγ3(IgG3), Cγ4(IgG4), Cμ (IgM), Cδ(IgD), Cε(IgE) , Cα1(IgA1) and Cα2(IgA2) genes. Light chain κ and λ isotypes are encoded by Cκ and Cλ genes. All members of a species have different constant region genes and thus express different isotypes.

Allotypes

Allotypes are polymorphic variants of the isotype genes. The different allelic forms of the genes produce subtle amino acid differences in some of the members of a species. If antibodies from an individual of a species are injected into another individual of the same species, carrying different allotypic determinant, the body will produce anti-allotypic antibodies. Sometimes, pregnant mother produces such anti-allotypic antibodies in fetus, if the father has different type of allotypic determinants. In humans, 25 different γ chain allotypes called Gm allotypes or Gm markers, have been identified. Allotypes are also encountered with IgA2 and κ light chain.

Idiotypes

Within the variable regions of each heavy and light chain, each individual antigenic determinant is called idiotypic determinant. These idiotypes may be a part of antigen binding site or may be a part outside of the antigen binding site, in the variable region. The term idiotype refers to the sum of all individual idiotypic determinants of an antibody. These antigenic determinants represent unique amino acid sequences of heavy and light chain variable regions, specific for each antigen. Sometimes, when two variable regions are homologous, cross reactivity is observed. The term *public / cross reactive* or *recurrent idiotype* is used for such idiotypes.

Antibodies, produced by B cells or plasma cells of an individual, have identical variable region sequences and have same type of idiotypes, also referred to as *private idiotypes*.

Soluble secreted Ig and membrane bound Ig

Antibodies exist in two forms: as a soluble form and an integral cell membrane protein on the surface of B cells. Both types of antibodies have different biological functions. Presence of membrane bound antibodies enables B cells to recognize a specific antigen and to induce a specific immune response. IgM and IgD antibodies are expressed on B cell surface as integral proteins. Each B cell has specific antigen specificity and therefore all the Ig (IgM and IgD) on a B cell have same variable regions as they have same type of heavy and light chain variable regions.

Both surface and secreted Ig are identical in structure except for the C terminal end of their H chains **(Fig.4.13)**. The C terminal end of membrane Ig contains many hydrophobic amino acids that enable it to anchor in the cell membrane.

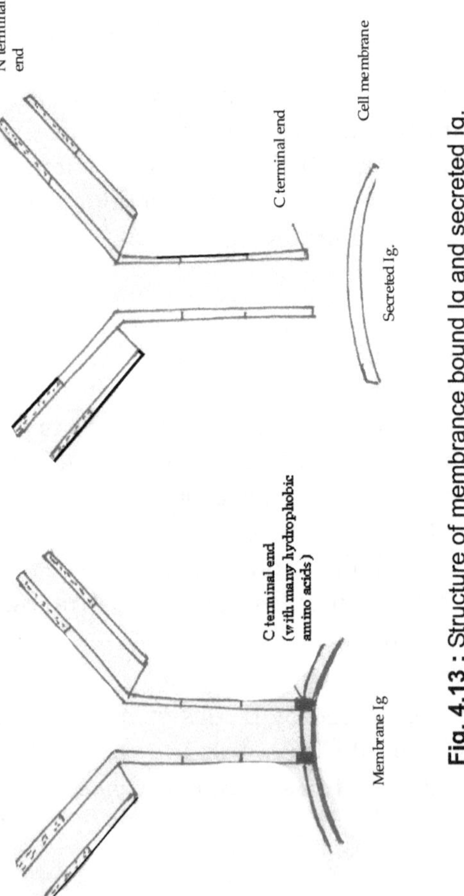

Fig. 4.13 : Structure of membrance bound Ig and secreted Ig.

The membrane Ig on B cells are associated with two other proteins called Igα and Igβ. Ig, along with these proteins, form a signaling complex. Igα and Igβ help in initiating intracellular signaling while membrane Ig binds to antigen.

GENERATION OF ANTIBODY DIVERSITY

The main characteristic of the immune system is its specificity. Antibodies are very specific in nature. Each antibody recognizes a specific antigen. But as the number of antigens is very vast (although it seems to be impossible to know all potential epitopes, it is estimated that the potential number of epitopes is in excess of 10^{11}), the question arises how it is possible to produce a wide diversity of antibodies with large number of specificities. Two different sets of theories were proposed to explain the antibody diversity. These were *germ line theories* and *somatic variation theories*. According to *germ line theories,* the genome in the germ line contains a large repertoire of Ig genes that encode diverse antibody molecules. The somatic variation theories argue that genome contains a relatively small number of Ig genes and mutations or recombination lead to antibody diversity. But none of these theories alone could explain the mechanism of generation of antibody diversity as well as the mechanism of maintaining the constancy in antibody molecules. In 1965, W. Dreyer and J. Bennett proposed a two gene-one polypeptide model. According to them, two separate genes encode a single Ig heavy or light chain, one for constant region and another for variable region. These two genes come together at the DNA level to be transcribed and translated into a single Ig heavy or light chain. In 1976, S. Tonegawa and N. Hozumi, experimentally found that separate genes encode for variable and constant regions of immunoglobulin. During differentiation of B cells, variable and constant region genes undergo rearrangement and produce antibodies with diverse specificities.

Genes for Immunoglobulins

With cloning and sequencing of light and heavy chain DNA, it is now believed that both light and heavy chains are encoded by separate multigene families, situated on different chromosomes. The genes or gene segments for κ light chain are located on chromosome 2, in man and chromosome 6, in mouse. For λ light chain, genes are located on chromosome 22, in man and on chromosome 16, in mouse. Heavy chain genes are situated on chromosome 14, in man and chromosome 12, in mouse.

Structure of κ light chain genes

There are different gene arrangements for k and λ light chains. The κ light chain consists of two regions variable region and constant region. Similar to other genes in eukaryotes, gene coding for κ light chain consists of exons (DNA segments that code for the proteins) and intron (non-coding DNA sequences). The constant region is coded by a single gene called C_K gene. For variable region, there are two exons (gene). Most of the part of the variable region is coded by the gene called Vκ gene. A separate gene called J gene or Jκ gene codes for the part of the variable region closest to the constant region **(Fig.4.14).** In human there are 41 Vκ genes and 5 Jκ genes. In mice, κ light chain contains 75 Vκ and five Jκ genes and one C_K gene. During development of B cells one of the V exons joins to one of the five J_K genes to form VJ DNA sequence. Then, this VJ DNA sequence is transcribed into RNA. Different B cells choose different Vκ and Jκ chains at random. In an individual, the population of B cell has all different V and J segments. Both Vκ and J_K gene influence antigen specificity of the immunoglobulin, made by the B cell.

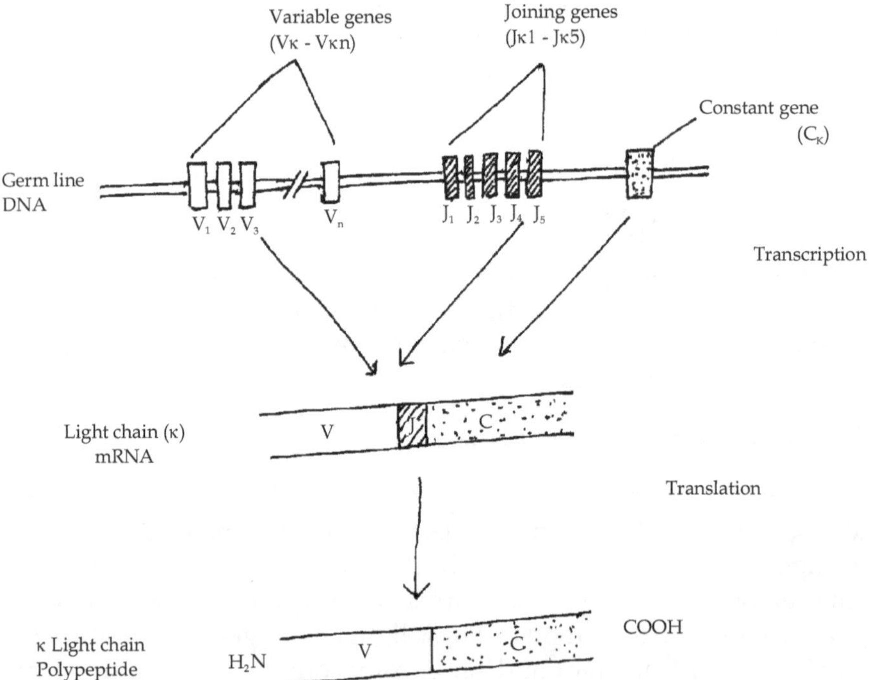

Fig. 4.14 : Structure of genes of Kappa (κ) light chain

As mentioned earlier, in Ig within the variable regions, there are three regions which are highly variable. These regions are called hyper variable regions or CDRs. Out of these three CDRs, CDR1 and CDR2 are coded by Vκ gene. CDR3 is coded partially by Vκ gene and partially by Jκ gene.

Structure of λ light chain genes

λ light chain also has V and J genes which code for different parts of the variable region, as in case of κ light chain. But in case of λ light chain, there are 30 V λ genes and four repeats of functional J and C genes **(Fig. 4.15)**. Any V gene can join at the four repeats of C genes. Each of the Cλ gene produces structurally same type of λ chain.

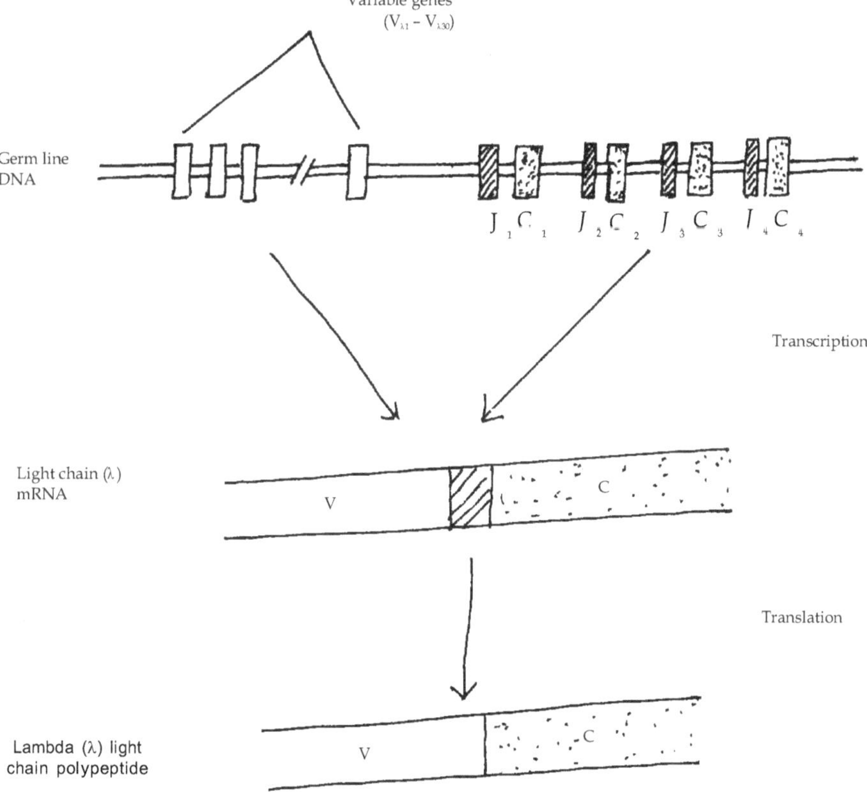

Fig. 4.15 : Structure of genes of Lambda (λ) chain

Structure of Ig heavy chain genes

Heavy chains too have variable and constant regions. As in light chain, in heavy chain too, the variable region is coded by V and J genes. But there is an additional set of exons, called D segment, present between V and J genes. During its development, each B cell chooses one V, one J and one D gene. This segment of VDJ is then transcribed and translated to generate H chain. Along with V and J, D is also involved in coding CDR3 of heavy chain variable region. The constant region is coded by C gene, present downstream of the J gene. There are total nine constant region genes (Cμ, Cδ, Cγ3, Cγ1, Cα1, Cγ2, Cγ4, Cε and Cα2) in humans, coding for each class and subclasses of antibody.

Rearrangement of Genes during B Cell Development

Which of the V gene would join with which of the D or J gene, is a random process. The whole process occurs during B cell development. One B cell may choose, for instance V1, D1 and J1 genes while another B cell may choose V2, D2 and J2 genes **(Fig.4.16)**. In case of light chains also, this selection occurs randomly. Different B cells choose different V and J segments but in an individual all gene segments are chosen by different lymphocytes.

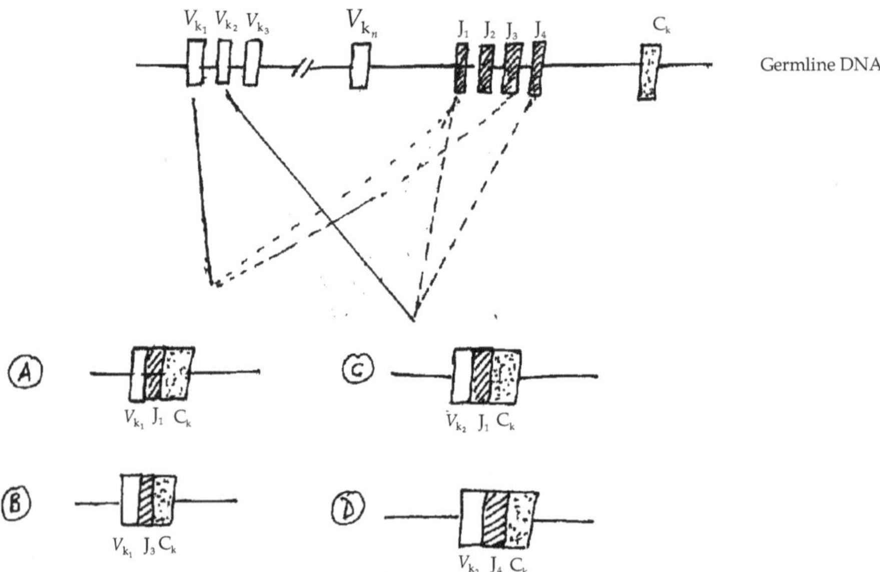

Fig. 4.16 : Rearrangement of genes during B-cell development. (In the figure are shown four combinations, for example, B-cell (A) has $V_{\kappa 1}$ J_1 C_κ genes, Cell (B) has $V_{\kappa 1}$ J_3 C_κ gene cell (c) has $V_{\kappa 2}$ J_1 C_κ and (D) $V_{\kappa 1}$ J_1 C_κ gene)

As shown in the **(Fig. 4.17)**, V genes are located at 5′ end of D genes, followed by J genes. During rearrangement, V genes move and joins to 5′ end of D segment with its 3′ end. In the same way 3′ end of the D segment joins with the 5′ end of the J segment. The process involves specific recognition sequences and enzymes. These unique DNA sequences, flanking each germ line V, D and J gene segments are called *recombination signal sequence* (RSSs) which are located to 3′ to each V gene segment, 5′ to each J gene and 5′ and 3′ to D gene segment. The enzymes which catalyze the recombination are collectively called *V(D)J recombinases.*

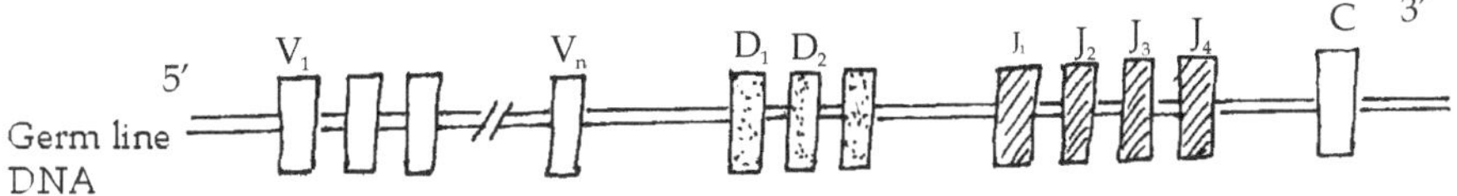

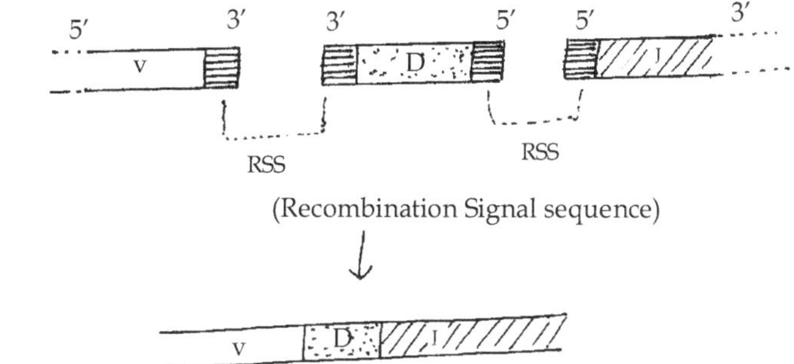

Fig. 4.17 : Presence of RSS (Recombination Signal Sequence) at the V, D and J gene segments

Mechanisms of Generation of Antibody Diversity

It is clear from the above discussion that the rearrangement of V, D and J gene segments leads to generation of diversity in antibody. In addition to this, there are several other mechanisms which are responsible for antibody diversity in humans. Various mechanisms which produce antibody diversity are:

- Multigene family
- Combination of V, D and J gene segments
- Imprecise joining during gene segment
- P-addition
- N-addition
- Somatic hypermutation
- Association of light and heavy chains

Multigene family

As mentioned earlier, there are about 75 Vκ and 5 Jκ gene segments in mice and 41 Vκ and 5 Jκ genes in human κ light chain. In κ light chain there are 34 Vκ gene and 5 Jκ genes In heavy chain, there are 48 VH gene segments and 6 J segments and 23 D segments. Presence of multigene germ line V, D and J gene segments contribute to generate diversity of the antigen binding site in antibody.

Combination of V, D and J gene segments

The rearrangement of V, D and J segments is a random process. This random rearrangement of the gene segments, in somatic cells, greatly increases antibody diversity. For example, if we consider only V genes there are 48 different VH genes in heavy chains and 75 V genes for light chain. As any heavy chain can pair with any light chain, their combination will lead to 3570 different combinations. As there are 48 V, 23 D and 61 J genes and any of these can join with any of the V, D and J genes, there occur 48*23*6 i.e. 6624 combinations of VD J genes in heavy chain. Similarly, for light chain, 41*5 =205 combinations are there for κ chain and 34*5= 170 combinations for λ chain. Thus, there are total 375 combinations for light chain. Since any light chain can combine with any of the heavy chains, total number of combinations increase upto 2484000.

Imprecise joining during gene segment

During rearrangement of V, D and J gene segments 3′ end of DNA segment of one gene joins with 5′ end of DNA segment of other (say D gene). Sometimes, this joining is not precise i.e. there is not the exact joining of the nucleotides that are present at the end of each segment. A nucleotide present at the 3′ end of V gene segment may join with nucleotides downstream ends of 5′ end of D segment, as shown in the **Fig. 4.18.** Similarly a nucleotide which is upstream of the 3′ end of V segment may join to D or J region. This type of imprecise joining which may occur at different levels i.e. V to D, D to J and V to J contributes in generation of antibody diversity.

P-addition

Sometimes, during the joining of V D and J DNA segments, some nucleotides which are the part of the non-coding DNA segment (intron) do not get cleaved and remain left attached to V D and J DNA segments. Such nucleotides are called P-nucleotides. They form part of the DNA

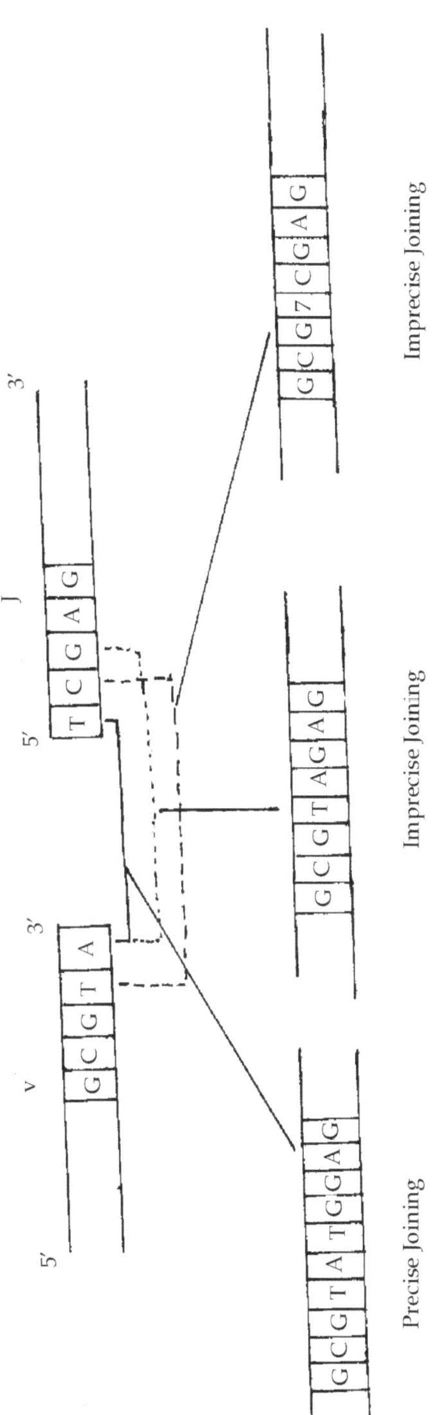

Fig. 4.18 : Imprecise joining during gene rearrangement

segment which codes for protein. The presence of such P-nucleotides further contributes in antibody diversity.

N-nucleotides

During joining of V D and J segments, sometimes, some nucleotides are lost. There are mechanisms to counter such problem. An enzyme called TdT (terminal deoxynucleotide transferase) is expressed by B cells during their development. This enzyme has the ability to add the nucleotides to a DNA segment. Though the nucleotides that are added are not the same one that are lost during joining of V D and J segments. The addition of nucleotides is random. These nucleotides which are added, in place of lost nucleotides, are called N- nucleotides and addition of N-nucleotides occurs at H chain genes. These N-nucleotides code for new amino acids and it is another mechanism to increase the antibody diversity.

Somatic hypermutation

During differentiation of B cells into plasma cells, some mutations may occur. Such mutations are thought to occur in Ig variable regions. These mutations may replace individual nucleotides on V J or VDJ segments. Replacement of nucleotides affects the antigen binding affinity of the encoded antibody. Such mutations are called ***somatic hypermutations*** because the rate of these mutations is very high, at least a hundred-thousand folds higher than the spontaneous mutation rate for other genes.

Combinatorial association between heavy and light chain

As a result of rearrangement of variable region genes, there occur a large number of variations in heavy and light chains. Association of these heavy and light chains occurs randomly and these associations further increase the combinatorial diversity. For example, if we assume variable region gene rearrangements leads to 6624 heavy chains and 375 light chains, random association of these will lead to 2484000 combinations.

POINTS TO REMEMBER

- Antibodies, also known as immunoglobulins, are a group of glycoproteins having a globular compact structure.
- Most antibodies are found in the γ globulin fraction of serum proteins.

- When immunoglobulins are subjected to proteolytic enzymes such as papain and pepsin, different types of fragments are produced.
- The basic structure of an immunoglobulin molecule is a Y-shaped, consisting of four polypeptide chains - two heavy chains and two light chains.
- Light and heavy chains are divided into two parts - one variable region and one constant region.
- The variable regions of light and heavy chains form the antigen binding sites.
- Constant regions determine the effector functions.
- The light chains are of two types: kappa (κ) and lambda (λ).
- In an immunoglobulin, each segment of about 110 amino acids, of both light and heavy chains, forms a compact loop called domain.
- The region between C_{H1} and C_{H2} domains is called hinge region.
- Hyper variable regions are the certain regions in the light and heavy chains which show high variability.
- On the basis of the type of heavy chains, there are five types of immunoglobulins in human beings- IgG, IgM, IgA, IgD and IgE.
- IgG is the most abundant immunoglobulin in human serum which comprises of about 80% of the total serum immunoglobulin.
- There are four subclasses of human IgG - IgG1, IgG2, IgG3 and IgG4.
- IgM is the first immunoglobulin to appear during primary immune response. It is a pentamer, consisting of five subunits,
- IgA exists both in monomeric and dimeric form.
- IgD is a membrane bound immunoglobulin, co-expressed on B-cells' membrane along with IgM.
- IgE is present in serum in extremely low concentration and is involved in hypersensitivity reactions.
- Antibodies can exist either in a soluble form or in a membrane bound form.

- There are three types of antigenic determinants on immunoglobulins- isotypes, allotypes and idiotypes.
- Germ line and somatic variation theories were proposed to explain the antibody diversity.
- Both light and heavy chains are encoded by separate multigene families, situated on different chromosomes.
- There are different gene arrangements for κ and κ light chains.
- Heavy chain variable region is coded by V_H D_H and J_H genes. For constant region, there is C_H gene.
- Various mechanisms which produce antibody diversity are: multigene family, combination of V, D and J gene segments, imprecise joining during gene segment, P-addition , N-addition, somatic hypermutation and association of light and heavy chains.

REVIEW QUESTIONS

1. What are antibodies? Describe the basic structure of an immunoglobulin.
2. Describe the structure of IgM.
3. What is secretory IgA? How is different from serum IgA?
4. Describe various mechanisms of generation of antibody diversity.
5. Write notes on :
   i. Immunological domains
   ii. J chain
   iii. Subclasses of IgG
6. Differentiate between:
   i. CDRs and FRs
   ii. Epitopes and paratope
   iii. Isotype and allotypes
   iv. Secreted and membrane bound Ig

CHAPTER - 5

Antigen-Antibody Interaction

Antigen-antibody interaction is very specific one in which an antibody recognizes an antigen through complementarities of shapes on paratope and epitope. It is like lock and key, as occurs in case of enzyme-substrate reaction. It does not involve covalent linkage rather a spatial complementarity occurs between antigen and antibody. For example, when antibodies, raised against m-amino benzene sulphonate, are tested for their ability to bind with ortho-, meta- and para-isomers of the hapten, the antibodies show highest strength with meta than ortho- isomer. Antibodies show poor reactivity with para isomer. This is because though chemically all these molecules are same but structurally they are different.

The force between antigen-antibody is of noncovalent type. Four types of noncovalent bonds are involved in the interaction between antigen and antibody. These are: hydrogen bond, electrostatic bond, van der Waal's forces and hydrophobic forces. Hydrogen bonds are relatively weak bonds that can be formed between hydrophilic groups. Electrostatic bonds result from the attraction between oppositely charged ionic groups of two protein side chains. Van der Waal's forces depend upon interactions between the "electron clouds" that surround the antigen and antibody molecules and hydrophobic groups tend to associate due to van der Waals bonding and combine in an aqueous environment, excluding water molecules from their surroundings. All these forces are relatively weak forces but together they can generate high affinity antigen-antibody interaction. As the strength of the noncovalent bonds depend upon the distance between the interacting groups, the affinity of antigen-antibody reaction depends on the complementarity of the shape between epitope and paratope. If there is a high degree of complementarity between epitope and paratope the interacting groups will be in quite intimate contact of each other resulting in a high affinity antigen-antibody interaction.

AFFINITY

Antigen-antibody (Ag-Ab) interaction is a reversible process. The strength of the interaction between an epitope and antibody is called *affinity*. Affinity is the total strength of all non-covalent interactions occurring between a single epitope on an antigen and a single paratope on an antibody. The reaction can be represented as:

$$Ag + Ab \underset{k_{-1}}{\overset{k_1}{\longleftrightarrow}} Ag\text{-}Ab$$

where, k_1 is the association constant and k_{-1} is the dissociation constant.

As the above reaction is reversible, so at equilibrium law of Mass action is applied and equilibrium constant (ka) can be determined.

$$ka = \frac{[Ag\text{-}Ab]}{[Ag][Ab]}$$

Where [Ag] is the concentration of free antigen, [Ab] is the concentration of free antibody and [Ag-Ab] is the concentration of Ag bound to Ab. The ratio between k_1 and k_{-1} is *intrinsic association constant* or *equilibrium constant*. It is also designated as *affinity constant*(ka).

$$ka \text{ (Affinity constant)} = k_1/k_{-1}$$

The value of ka varies for different antigen-antibody reactions.For example, ka is high for small haptens. High affinity antibody binds with antigen relatively more tightly than low affinity antibody and remains bound for longer time.

AVIDITY

When we talk about affinity, we consider the interaction between a single antigen site and a single epitope. But under physiological situations, since antibody is at least bivalent or multivalent and interaction occurs between multivalent Ab with multivalent antigen i.e.

$$mAg + nAb \xrightarrow{k} Ag_m - Ab_n$$

In such situations, the strength of such multiple interactions between multivalent antigen and multivalent antibody is termed as *avidity*. Avidity depends on the affinity of each of the binding sites for the epitopes present on an antigen. Avidity is always greater,usually many folds greater, than arithmetic sum of the individual antibody links. For example, if there are two determinants A and B and if the affinity for determinant A is K_1 and for B is K_2, the avidity of the mixture of antibodies for antigen would be

$$K_{(av)} = K_1 * K_2$$

If $K_1 = 10^4$ and $K_2 = 10^6$ then ,$K_{(av)} = 10^{10}$ l/mol. This is called *bonus effect*. This is because all Ag-Ab bonds would have to be broken simultaneously for the complex to dissociate which would require more amount of energy.

TYPES OF ANTIGEN-ANTIBODY INTERACTIONS

The binding of antigen and antibody results in various types of reactions, depending on the nature or form of the antigen. These reactions may be of the following types:

- Precipitation reaction
- Agglutination reaction

Precipitation Reaction

When a soluble antigen reacts with an antibody it leads to the formation of a visible precipitate. This is called a precipitation reaction. The antibodies that bring about precipitation formation are called *precipitins*. The reaction occurs due to lattice formation between antigen and antibody. For precipitate to be formed, the antibody should be bivalent or multivalent i.e. more than one antigen binding sites should be available for binding. The antigen should also be multivalent i.e. it should have two or more epitopes. When multivalent antigen reacts with antibody lattice formation occurs. The lattice formed is too large to remain in the solution and hence precipitate is formed. The formation of the precipitate of antigen-antibody complex depends on the concentration of both antigen and antibody. The reaction occurs within a narrow range of concentration when the concentrations of antigen and antibody are equal. This optimal range when the ratio of antigen and antibody is optimal is called *equivalence zone*(**Fig. 5.1**). If concentration of one exceeds than that of the other, no visible precipitate is formed. This is because when antibodies are in excess, unreacted antibodies are present in the supernatant. One molecule of antigen is bound by multiple molecules of antibodies and form small soluble complex. In case of excess of antigens, one or more molecules of antigen bind to a single molecule of antibody and unreacted antigens are found in the supernatant **(Fig. 5.2).**

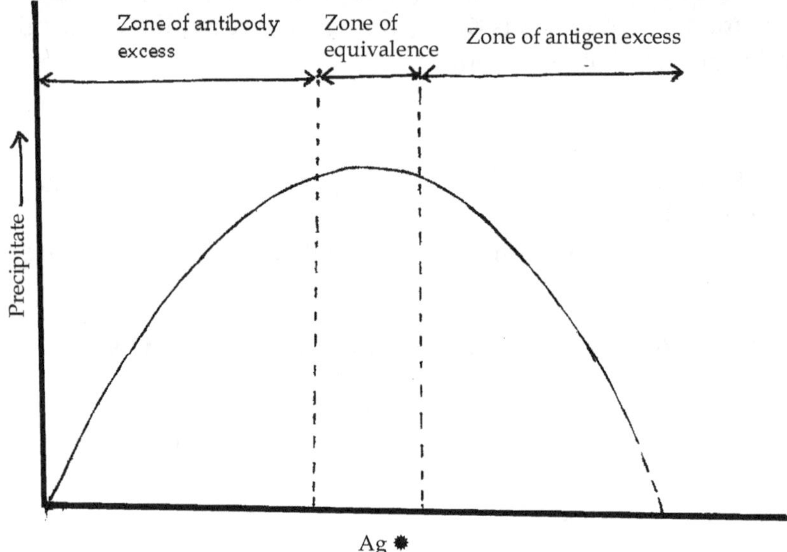

Fig. 5.1 : Quantitative precipitation reaction

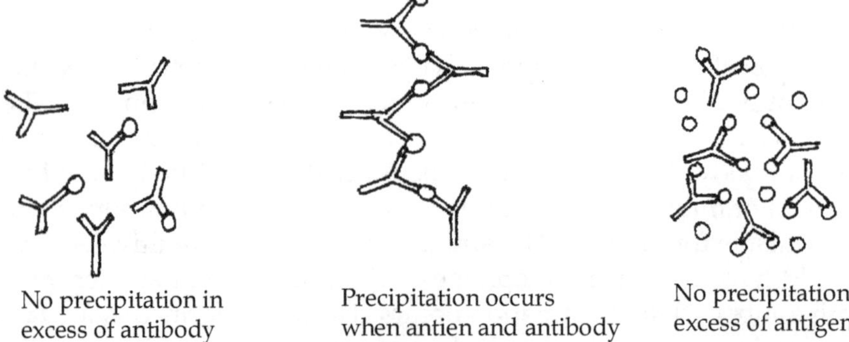

Fig. 5.2 : Precipitation reaction in presence of excess of antibody or antigen

The precipitation reaction can be demonstrated in agar gel either in a test tube or in a petridish. In agar gel method agar solution is poured into the petridish or on a glass side. When the agar solidifies wells are cut into the gel and antigen and antibody are put in the wells. The plates are left for some time and antigen and antibody are allowed to diffuse into the gel. A precipitation line is formed at the zone of equivalence. This is called immunodiffusion. There are many variations of immunodiffusion such as, *double immunodiffusion* (Ochterlony method), *radial immunodiffusion* (Mancini method) and *immunoelectrophoresis*. These techniques have been described in chapter 20.

Agglutination Reaction

Agglutination is the aggregation of insoluble particles or cells by antibody into large clumps. The reaction takes place as a result of particles coming together and producing visible clumping. The particles can be bacteria, cells of higher animals or plants or microfungi. The antibody that brings out the agglutination is called *agglutinin*. The agglutination can occur only if the particle size is more than 200-250 nm. Agglutination occurs due to cross linking of cells or large particles by antibody, directed against surface antigen.

The first step in the agglutination is the cross-linking of different particles or cells by antibody molecules that specifically attach to the antigenic determinants on the surface of the particles or cells. This results in the formation of large lattice which are generally visible with the naked eyes as the clumps **(Fig.5.3).** The strength of the agglutination reaction depends on the nature of antigens. Multivalent IgM antibodies with higher avidity are more effective agglutinating agent than IgG antibodies. Agglutination reactions have a higher sensitivity than precipitation reaction and can be performed easily. The tests are routinely carried out in the laboratory for blood group typing. Agglutination reaction can be carried out in a test tube or on the slides. When the reaction is carried out in the test tube and visualized by naked eyes, they are categorized as *macroscopic tests*. Tests carried out on slides and visualized under microscope are considered as *microscopic tests*. The method of agglutination on slideis faster and requires less amount of reagents than agglutination in test tubes.

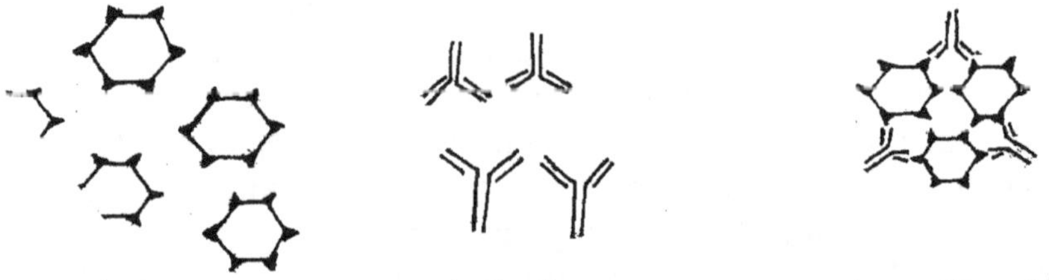

Antigen Antibodies Aggregation of antigens antibodies.

Fig. 5.3 : Agglutination reaction

In most agglutination tests, agglutination is carried out in tubes. For titration of antibody, a series of test tubes, containing serial dilutions of antiserum, is prepared. A constant amount of antigen is added in each tube and mixed properly. The tubes are left for incubation at 37°C. Agglutination is observed as clumps or flakes in a clear surrounding medium. No agglutination is observed in the first few tubes which contain higher concentration of antibody. This is because at higher concentration most of the antibodies bind to the antigen only univalently instead of multivalently. This prevents the formation of lattice. This inhibition of agglutination, in the excess of antibody, is called *prozone phenomenon* or *prozone effect*. In the excess of antigen also no agglutination is observed. This is called *postzone effect*.

Types of Agglutination Reaction

Agglutination reactions are of two types-*direct agglutination reaction* and *indirect agglutination reaction*.

When an insoluble particle or cell is directly agglutinated by an antibody it is called *direct agglutination*. Blood group typing, called hemagglutination, involves direct agglutination.

Hemagglutination

Blood group typing in ABO or other system is carried out by mixing a drop of test blood with a drop of specific antiserum and examining the visible agglutination. Red blood cells carry specific antigens on their surface which determine type of blood group in an individual. When, for example, blood from group A is mixed with antiserum A, clumping is observed. Similarly, blood group B agglutinates with antiserum B. In case of AB blood group, the red blood cells have both A and B antigens and clumping occurs with either of these antiserums. Red cells, in case of O blood group, do not carry any antigen on their surface and hence no agglutination occurs when blood from O group is either mixed with antiserum A or antiserum B. Agglutination is also observed in Rh antigen. When Rh+ blood is mixed with anti D, clumping is observed.

Indirect or Passive Agglutination

This test is used for the detection of antibody against soluble antigens. The test is used in the diagnosis of syphilis and to detect rheumatoid factors. The soluble antigens are coated onto various types of inert particulate surfaces. Particulate surface used may consist of tanned and killed RBCs of O blood group or sheep or chicken RBCs. The RBCs are first tanned by treating them with tannic acid or chromium chloride. Treatment with these compounds enhances the adsorption of antigen to the surface of RBCs. Serum from the patient is added to the antigen coated RBCs. The antibodies react to the soluble antigen, adhering to the particles, and agglutinate them.

Hemagglutination Inhibition Test

Certain viruses can spontaneously cause agglutination of red blood cells. This viral hemagglutination can be specifically inhibited in the presence of antiviral antibodies. If the patient has antiviral antibodies, these antiviral antibodies react with the viral proteins which are no longer available to cause agglutination. This is called hemagglutination inhibition reaction and can be used to detect the presence of specific antiviral antibodies in a patient's serum, in certain viral infections like influenza, mumps or measles.

Bacterial Agglutination

Bacterial agglutination reactions are carried out in the laboratories for identification of bacterial species. The tests are also used for determining the intensity of infection by measurement of antibody in the serum. Bacterial infection induces formation of antibodies, directed against surface antigens. When the serum of the patient(infected with the bacteria) is incubated with bacteria, bacteria agglutinate the antibody and a visible clump is formed. The formation of agglutination indicates the presence of bacterial antibody in the body. *Widal test,* used for diagnosis of typhoid, and *Weil Felix test* and *Brucella agglutination tests* used for the detection of Rickettsiosis and Brucellosis,respectively, are based on agglutination reaction.

CONSEQUENCES OF ANTIGEN-ANTIBODY REACTION

Antigen-antibody reactions involve different types of biological consequences such as *neutralization of the antigen, activation of complement system leading to lysis of the offending microbial cell* and *opsonization* and *phagocytosis of the microorganisms.* Antigen-antibody reactions can also lead to allergic reactions. Antibody-mediated allergic reactions are called *immediate hypersensitivity reactions.*

POINTS TO REMEMBER

- Antigen-antibody interaction is very specific one which occurs due to complementarities of shapes on paratope and epitope.
- Four types of noncovalent bonds-hydrogen bond, electrostatic bond, van der Waal's forces and hydrophobic forces are involved in the interaction between antigen and antibody.
- The strength of the interaction between an epitope and antibody is called affinity.
- The strength of multiple interactions between multivalent antigen and multivalent antibody is called avidity. Avidity is always greater than arithmetic sum of the individual antibody links.
- The binding of antigen and antibody results in two types of reactions: precipitation reaction and agglutination reaction.
- When a soluble antigen reacts with antibody it leads to precipitation reaction.
- The formation of the precipitate of antigen-antibody complex depends on the concentration of both antigen and antibody.
- Agglutination reaction occurs as a result of particles coming together and producing visible clumping.
- Agglutination reactions are of two types-direct agglutination reaction and indirect agglutination reaction.
- Indirect agglutination is used for the detection of antibody against soluble antigens.
- Antigen-antibody reactions results into three types of effects: neutralization of the antigen, activation of complement system and opsonization and phagocytosis of the microorganisms.

REVIEW QUESTIONS

1. What is affinity? How is it different from avidity?
2. What is precipitation reaction? What are different methods to demonstrate precipitation reaction?
3. What is agglutination? Describe various types of agglutination reactions.
4. Write notes on :
   i. Bonus effect
   ii. Prozone effect

CHAPTER - 6

T Lymphocytes and B Lymphocytes

All blood cells arise from a basic cell called the hematopoietic stem cell. Stem cells are undifferentiated embryonic cells.They are self renewing and their population by cell division. Stem cells have the ability to develop into any type of blood cell. The formation of blood cells begins in the embryonic yolk sac during the first week of development. Every mature cell is derived from the same type of stem cell. In contrast to an unipotent cell which differentiates into a single cell type, a hematopoietic stem cell is multipotent or pluripotent. In the mammalian embryo, the blood corpuscles are normally formed in yolk sac, liver, spleen and thymus gland. In the adults RBCs, blood platelets and granulocytes are produced in red bone marrow, found at the ends of long bones. Lymphocytes are formed in the spleen, lymph glands and thymus gland while monocytes are formed in the reticulo-endothelial connective tissue. Formation of RBCs and WBCs is known as *erythropoiesis* and *leucopoiesis,* respectively. The stem cells develop in three lines, namely *lymphoid lineage, myeloid lineage* and *erythroid lineage.* In lymphoid lineage, the stem cells develop into lymphoid progenitor. The lymphoid progenitor cell develops into three groups of immune cells, namely T lymphocytes, B lymphocytes and null cells **(Figure 6.1).**

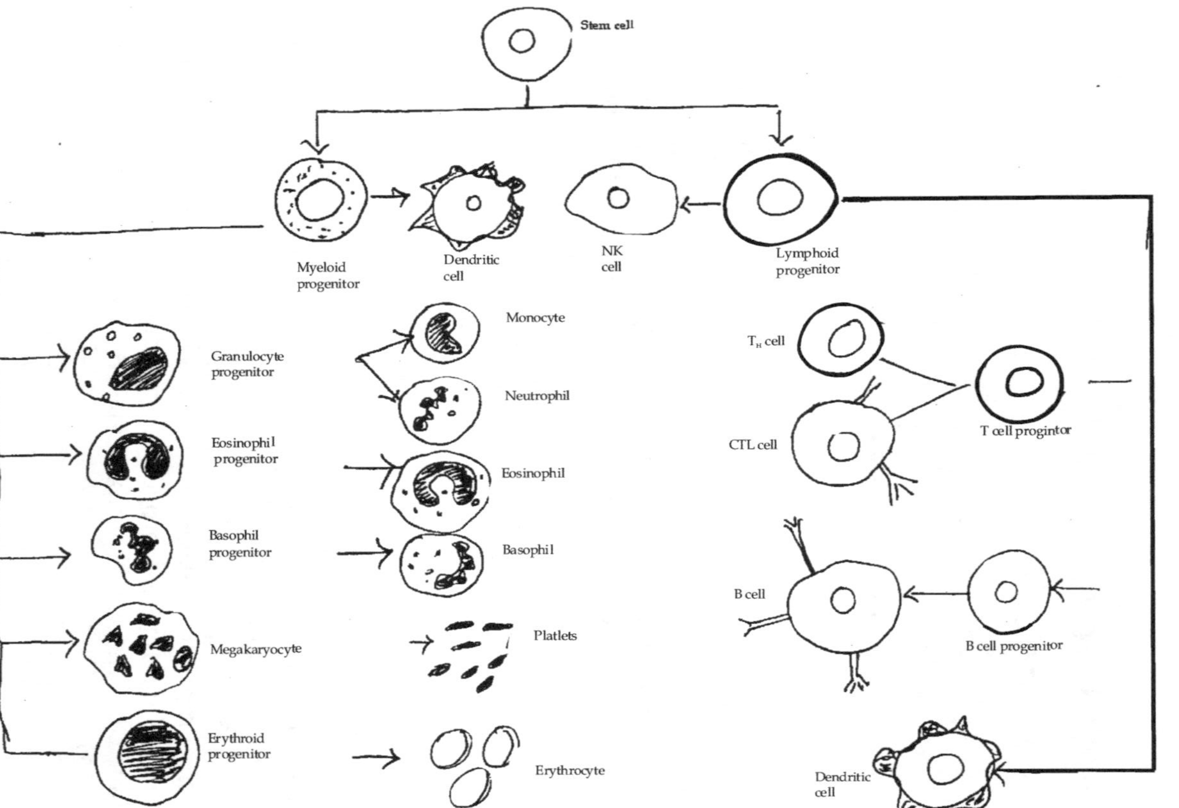

Fig. 6.1 : Hematopoiesis : Showing generation of T, B and other blood cells

Lymphocytes constitute 20% to 40% of the white blood cells and 99% of the cells in the lymph. They are found in blood, lymph and lymphoid tissues such as spleen, lymph nodes and tonsils. Lymphocytes are circulated in blood and later on transmitted into peripheral lymphoid organs. The T lymphocytes mature in the thymus, under the influence of thymic hormones. In thymus, they are differentiated into mature cells.Then, they come out from the thymus. In the circulating blood they comprise 70 to 80% of the total lymphocytes. T cells are classified in two groups: one is regulatory cells and other is effector cells. *Regulatory cells* control the development of effector cells. These are of two types: T helper cells (T_H) and T suppressor cells (T_S). $T_H$cells secret various cytokines which play a central role in the activation of B cells. T suppressor cells directly attack specific target cells. T suppressor cells are divided into other cells like cytotoxic cells, delayed type hypersensitive T cells (T_{DTH}) and natural killer cells (NK). Stems cells in the bone marrow produce B cells that migrate to lymphoid tissue. Mature B cells circulate in the blood.They are distinguished from other lymphocytes by synthesis and display of membrane bound immunoglobulin molecules.

GENERATION OF T CELLS

T lymphocytes mature in the thymus. They have specific receptors for antigen. These antigen binding T cell receptors are structurally distinct from immunoglobulins. These T cell receptors (TCRs) share some common features with an antibody. These T cell receptors, in cooperation of MHC class I and II molecules are able to recognize antigens. T cells also express other specific surface markers. These are Thy1, CD3, CD28, CD45, CD8 and CD4.

Thy1 are acquired by T cells during maturation in the thymus. CD3 is a membrane complex attached to T cell receptor. CD 28, a receptor for B7 molecules, is found on antigen presenting cells. CD45 is a signal transduction molecule. CD8 and CD4 are membranous glycoprotein molecules. CD8 recognizes antigen associated to class I MHC molecules. CD4 recognizes antigen associated to class II MHC molecules.

T cell maturation involves rearrangement of the germ line. In the thymus, developing T cells are known as *thymocytes*. Thymocytes that undergo productive TCR gene rearrangement are put through selection process which involves positive selection and negative selection. *Positive selection* permits the survival of only those T cells whose TCRs are capable of recognizing self MHC molecule. Cells that fail positive selection are eliminated within the thymus by apoptosis. *Negative selection* eliminates T cells that react strongly with self MHC. Negative selection that eliminates thymocytes bearing high affinity receptors for self MHC molecules alone, results in self-tolerance. Both process are necessary to generate mature T cells, that are self MHC restricted and self-tolerant. After three weeks of development of T cells in the thymus, the differentiating T cells progress through a series of stages that are marked by characteristic changes in their cell surface phonotype. Thymic stromal cell, macrophages, epithelial cells and dendritic cells play an important role in positive and negative selection.

Positive Selection of T Cells

Positive selection takes place in the cortical region of thymus and involves the interaction of immature thymocytes with cortical epithelial cells: Thymocytes interact with epithelial cells through their T cell receptors (TCR) which establishes the contact with class I and class II MHC on thymic epithelial cells. T cells displaying very high or very low receptor affinities for self MHC undergo apoptosis and die in the cortex. Only those thymocytes whose αβ T cell receptor recognizes a self MHC molecule are selected for survival. If a thymocyte fails to express αβ T cell receptor, that fails to recognize the self MHC, then the cell will die by apoptosis. Due to TCR diversity of the developing thymocytes, a large array of self peptides are required for efficient positive selection. Peptide specificity is a crucial component of positive selection.

Negative Selection of T Cells

Some of the positive selected T cells may have T cell receptors that recognize self-components other that self MHC. These cells are deleted by a negative selection process. T cells interact with antigen presented by interdigitating cells, macrophages and medullary thymic epithelial cells (TECs). The role of medullary TECs for negative selection has been emphasized recently by the findings that these cells express genes for all tissue antigens in the body. In negative selection of T cells, thymocytes with high affinity T cell receptor will interact with dendritic cells presenting self MHC alone or self MHC with self antigen. In the negative selection, selected cells are observed to undergo death by apoptosis. Only those T cells that fail to recognize self antigens are allowed to proceed to their development. Rest of them undergo apoptosis and are destroyed. Tolerance to self antigen is achieved by eliminating self reactive T cells and only allowing maturation of T cells that are specific for foreign antigens and altered self molecules. In the thymic cortex, the medulla is densely packed with antigen presenting cell (APCs) and medullary epithelial cells which are capable of inducing thymocyte negative selection. All self reactive T cells are eliminated during intra-thymic development. The thymic epithelial barrier that surrounds blood vessels may also limit access of circulating antigens. It is thought that the stromal cell type, involved in negative selection, depends upon the activity of T cell receptor and major histocompatibility complex (TCR-MHC) interaction.

T CELL RECEPTOR COMPLEX

The domain structure of T cell receptor is almost similar to immunoglobulin. Each domain has 65 to 70 amino acids and disulfide bonds. TCR variable region has three hypervariable regions which are similar to immunoglobulin light and heavy chain. The two chains of TCR are joined together by a single disulfide bond at the lower end, followed by a transmembrane region of 21 to 22 amino acids. At the carboxy terminal end TCR chain has 5 12 amino acids which extend into cytoplasm **(Figure 6.2).**

CD3 is closely associated with TCR and is situated adjacent to TCR on membrane. It helps in transforming signals. CD3 has five invariant polypeptide chains. There are several other membrane molecules which are required for performing various accessory roles in antigen recognition and T cells activation. These are collectively known as *accessory molecules.*

GENERATION OF B CELLS

B lymphocytes or B cells, like all other blood cell types, originate from pluripotent stem cells of the bone marrow of the adults. The pluripotent stem cells differentiate within the bone marrow microenvironment. In the maturation, a population of stem cells develop lymphocytes lineage and give rise to B and T cells. B lymphocyte maturation proceeds within the bone marrow. This first B cell is called *progenitor B cell* (Pro-B cell). Pro-B cells proliferate within the bone marrow and differentiate into *precursor B cells* (Pre-B cells). The pre-B cells require microenvironment provided by the stromal cells. Stromal cells secrete cytokines. Development and maturation of B cell depend on Ig gene rearrangement. Pre-B cells are characterized by

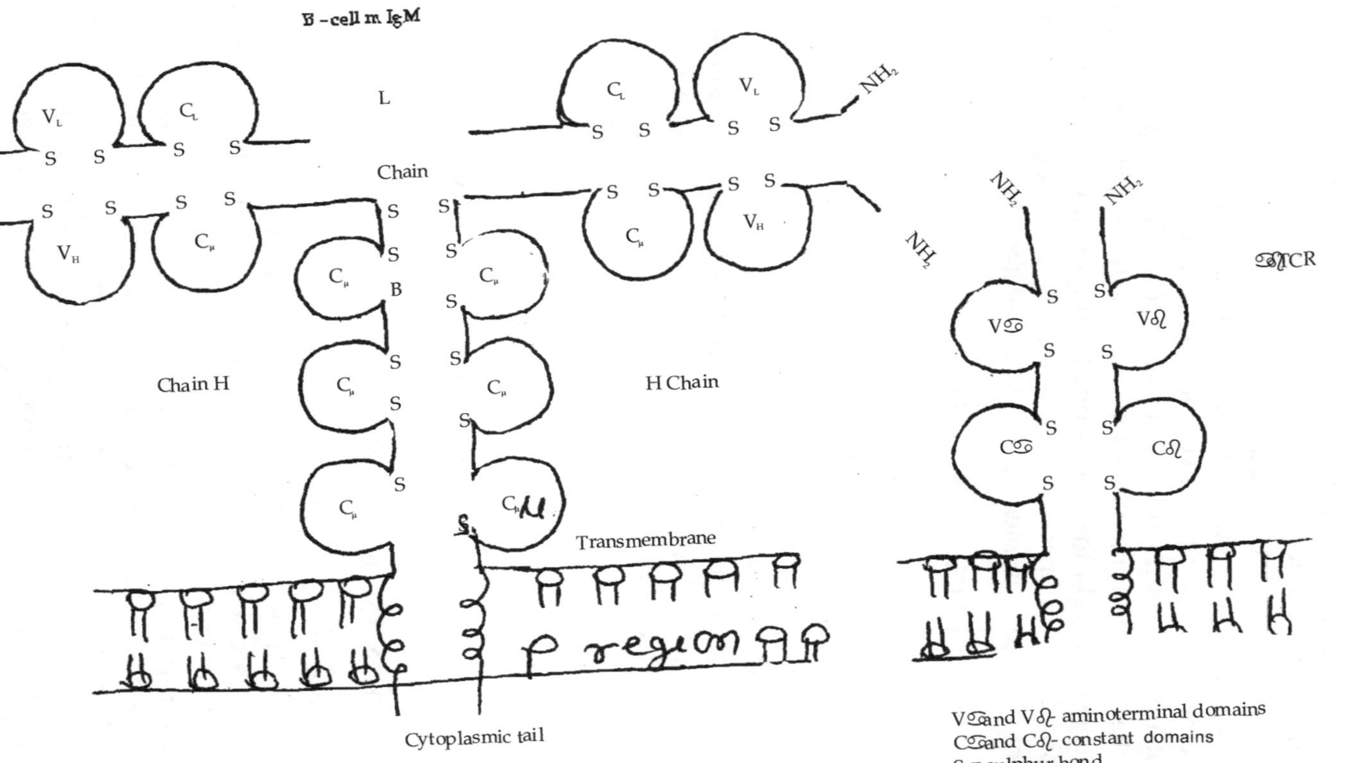

Fig. 6.2 : Structural similarity between αβTCR and Ig of B cell.

the cytoplasmic expression of Ig heavy chain. Pre-Bcells cannot recognize any antigen. The immature B cells express IgM on their cell surface, in association with Igα and Igβ. The mature B cells express both IgM and IgD and are called mature B cells. B cell stromal cell interactions enhance the survival of developing B cells and mediate a form of selection that rescues a minority of B cells with productive rearrangement of their immunoglobulin genes from cells death. Many immature B cells fail to complete maturation and are ultimately eliminated i.e. negative selection.

Positive Selection of B Cells

Gene expression in Pre-B cells is different from mature B cells. It is thought that a B cell requires the interaction between B cell receptors (BCR) and a specific ligand to survive and thus might be positively selected.

Negative Selection of B Cell

The cross linking of IgM on immature B cells leads to cell death. 90% of the B cells, produced each day, die without leaving the bone marrow. This is due to stringent selection procedures that operate, once the cell reach immature stage with high IgM and low IgD expression. This loss is called negative selection. Immature B cells migrate to the periphery, where high level expression of IgM may further ensure increased sensitivity to negative selection upon recognition of self antigens, found on the periphery.

B CELL RECEPTOR COMPLEX

Mature B cells co-express both membrane bound IgM and IgD immunoglobulins. Activation of IgM and IgD expressing mature B cell occur in a particular manner that requires B cell receptors (BCR) to bind to foreign antigen in the lymphoid organ. The BCR is a transmembrane protein complex composed of mIg and disulphide linked heterodimer which is Igα/Igβ. The Ig α chain has long cytoplasmic tail containing 61 amino acid and Igβ tail having 48 amino acids. Igα and Igβ tails interact with intracellular molecules like tyrosin kinases and G protein. Igα and Igβ perform same function in B cell. Igα and Igβ cytoplasmic tails contain 18 residue tyrosine rich motif or immunoreceptor tyrosin based activation motifs or (ITAMs). Antigen induced clustering of BCRs brings several ITAMs in close proximity and allows ITAMs to interact with tyrosin kinases. In B cells, a component of the B cell membrane, called *the B cell co- receptor,* provides stimulatory signals. The B cell co-receptor is a complex of three membrane bound transmembrane proteins CD19, CD21 and CD81. CD19 is a member of immunoglobulin and has a cytoplasmic tail and three extracellular domains. CD21 component is a receptor of C3d which is a product of complement system. CD21 is also a receptor for membrane molecule and transmembrane protein. CD21 protein and C3d binding is facilitated when a complex of C3d and antigen binds to mIgM of B cells. According to immunologists all the mIg isotypes are known to have very short cytoplasmic tails i.e. mIgM and mIgD. The mIgM and mIgD contain three amino acids. Ig recognizes the antigen and CD21 binds to C3d. This interections allow the CD19 to interact with Igα/Igβ component of BCR. CD19 possess three extracellular Ig like domain and six tyrosin residues.

POINTS TO REMEMBER

- All blood cells arise from the hematopoietic stem cell.
- Stem cells can be classified into three basic lineages: lymphoid lineage, myeloid lineage and erythroid lineage.
- Lymphoid lineage gives rise to B lymphocytes and T lymphocytes.
- Thymocytes that undergo productive TCR gene rearrangement are put through selection processes which involve positive and negative selection.
- Positive selection permits the survival of only those T cells whose TCRs are capable of recognizing self MHC molecules.
- Negative selection eliminates T cells that react strongly with self MHC.
- The B cell receptor (BCR) is a transmembrane protein complex composed of mIg and a disulfide linked heterodimer called Igα/Igβ.
- Two molecules of this heterodimer are in association with one mIg and they are present on either side of mIg molecule to form a single BCR.
- Development and maturation of B cell depends on Ig gene rearrangement.

- Pre-B cells are characterized by the cytoplasmic expression of Ig heavy chain.
- The immature B cells express IgM immunoglobulin on their cell surface in association with Igα and Igβ.

REVIEW QUESTIONS

1. Describe negative and positive selections of T cells.
2. Describe T cell receptor complex.
3. Explain : B cell receptor complex.
4. Explain negative and positive selections of B cells.

CHAPTER - 7

Humoral Immune Response

The immune system protects the body from harmful substances by recognizing and responding to antigens. Antigens are the protein molecules which are present on bacteria, virus, fungi or some other microorganisms.Some toxins can also be antigens. These antigens are recognized by immune system. The destruction of antigens by producing antibody is called ***antibody-mediated immune response.*** As antibodies are present in the body fluids i.e. humors,this type of response is called ***humoral immune response***. As humoral immune response is brought about by B cells , it is also called ***B cell- mediated immunity***. B cells are involved in antibody production. They are also called *effecter cells*. Many B cells are produced in the bone marrow throughout life but very few of these cells mature. The primary exposure to a microbe leads to the activation of native B cells and their subsequent differentiation into antibody producing cells and memory cells. Some of the antibody producing cells migrate to the bone marrow and continue to produce antibodies. Helper T lymphocytes (T_H) stimulate B lymphocytes for antibody production against antigen protein, so these antigens are classified as *T dependent(TD) antigens.* Non-protein antigens like polysaccharides and lipids are not dependent on T_H cell for antibody production. So these antigens are classified as *T independent (TI) antigens.* Humoral response against antigen generates different types of immunoglobulins.

B CELL MATURATION

The generation of B cell first occurs in embryo. B cells originate from pluripotent stem cells of the bone morrow. B cell development begins as lymphoid stem cell changes into earliest B cell which is a pro-B cell. This pro-B cell proliferates within the bone marrow and fills the large extra vascular space of the bone. Proliferation and differentiation of pro- B-cells into pre- B-cell requires the micro-environment provided by the bone marrow stromal cell. If the stromal cells are not present pro-B cell will not develop into mature B cell because stromal cell secrets various cytokines for maturation and development of B cells. Maturation involves a regulated gene expression which leads to over all change in the phenotype and functional aspects of lymphocytes for maturation. Pro-B cell requires direct contact with stromal cells in the bone marrow. It is mediated by several cell adhesion molecules. B cell maturation also depends upon rearrangement of the immunoglobulins. In the peripheral region the native B cells respond to antigen and continue proliferation and differentiation as activated B cells. Activated B cells undergo proliferation and differentiation into antibody secreting cells through heavy chain class switching. The developmental progression from progenitor to mature B cell is a changing pattern of surface molecules.

B cells contain surface immunoglobulins IgM and IgD. Each B cell may have as many as 10000 of specific antigenic receptors sites. In pre-B cells which are derived from stem cells immunoglobulins are absent in them. On the surface of immature B cells, membranous IgM monomers are present. IgM and IgD immunoglobulins are also present on the surface of matured B cells. Some cells becomes more differentiated because they express IgM, IgG and IgA immunoglobulins **(Figure 7.1).**

B CELL ACTIVATION AND PROLIFERATION

B cell activation depends upon the nature of antigen by two different routes. One depends upon T_H cell and other is independent to it. These are classified as thymus dependent (TD) and thymus independent (TI). B cell responses to TD antigens require contact with T_H cells whereas antigens that activate B cell in the absence of $T_H$cells are called TI antigens, as stated above. The most important TI antigens are bacterial cell wall components including lipopolysaccharides, glycolipids and nucleic acids. TI antigens are divided into two groups: TI antigen-1 (TI_1)is a bacterial lipopolysaccharides and TI antigen-2 (TI_2) are bacterial cell wall polysaccharides and polymeric proteins. TI antigen-1 are strong antigens. They are activators of B cells. At high concentration these antigens stimulate proliferation and antibody secretion by 1/3 of B cells. TI antigens-2 activate B cells by cross- linking mIg receptor. TI antigen-2differs from TI antigen-1 in several aspects:

- They are not B cell mitogens.
- TI_2 antigens activate B mature cells and inactivate immature B cells.

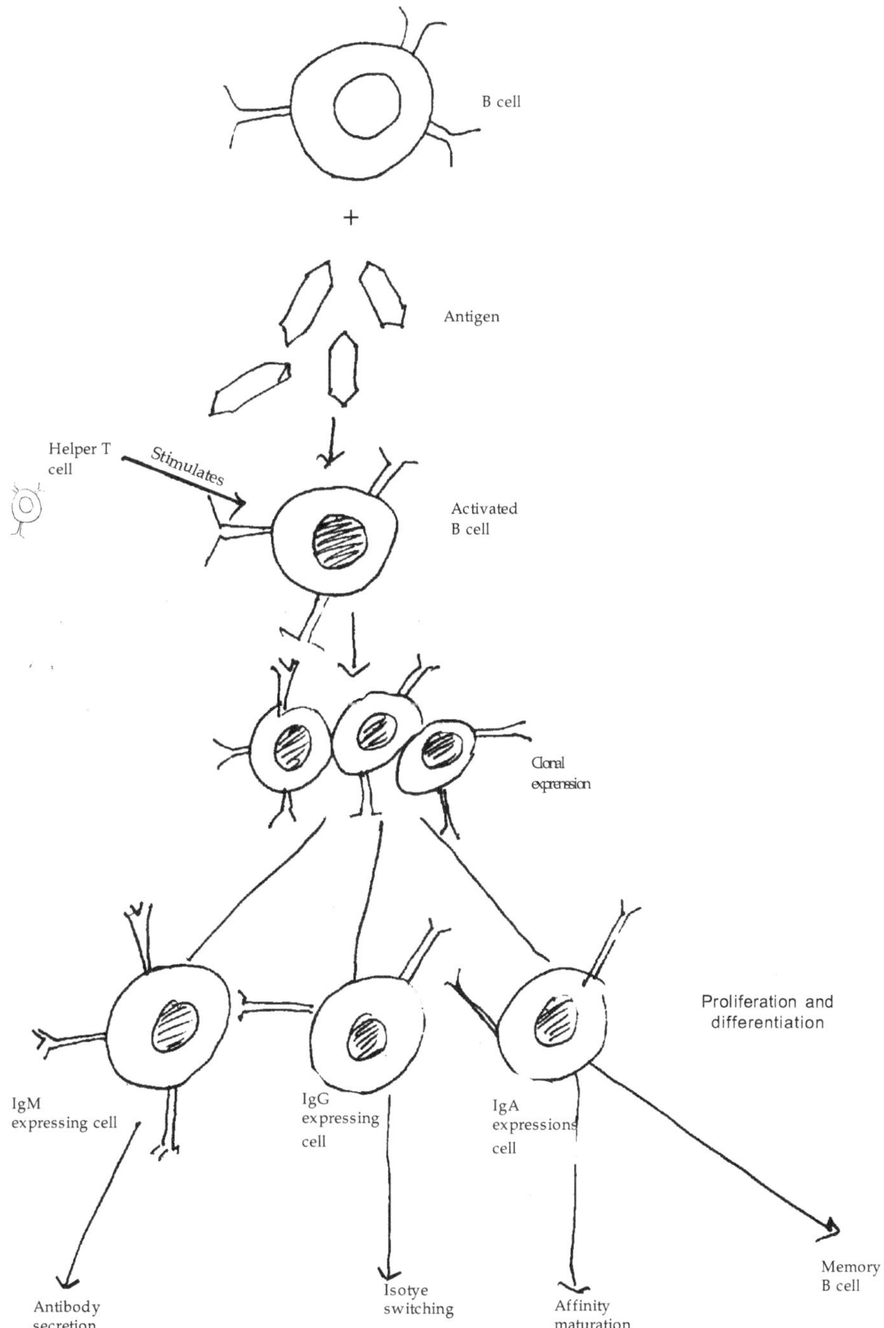

Fig. 7.1 : Humoral immune response

- TI_2 antigens do not require direct involvement of T_H cells.

The humoral response to thymus independent antigens is different from the response to thymus dependent antigens.

During resting stage, B cells are non-dividing cells.In the G_o phase of cell cycle,there are two signals for native B cell to enter into proliferative stage and differentiate into antibody producing cells. These two signals are competence signal and progression signals. *Competence signals* drive the native B cells from G_o into GI phase. *Progression signals* drive the cells from G1 to S phase and ultimately to cell division and differentiation.

Thymus dependent and thymus independent antigens involve two different ways of competence signaling. Proliferative expansion and differentiation are triggered through the pre- BCR (pre-B cell receptor) in pre-B cell to and BCR complex in mature B cell. BCR are also responsible for recognizing antigen processing compartments in B cells to promote presentation of antigen. BCR is functionally divided into the ligand binding immunoglobulin molecule and the signal transducing Ig-α/Ig-αβ hetrodimer. Ig-α contains 61 amino acids and Ig-β contains 48 amino acids. The cytoplasmic tails of both Ig-α and Ig-β contain 18 residue motif, termed the ITAM i.e. immunoreceptor tyrosine based activation motif. ITAM provides antigen receptor with specific binding sites. ITAM interacts with tyrosine kinase and plays an important role in signal transduction.

B cell activation requires a signal which is derived from another component of B cell membrane. That is called *B cell co-receptor*. It provides stimulatory modifying signals. The B cell receptor is a complex of three proteins CD 19, CD 21 and CD 81. This signal transduction process is mediated by complement protein C3. Complement system plays an important role in humoral immune response. Type 2 complements receptor binds to C3d which is generated as a photolytic product of C3.

HUMORAL RESPONSE

The effector functions of humoral branch depend upon the production of large number of antibodies against pathogens. Antibodies eliminate the pathogens by different way like antibodies initiate complement for lysis of pathogens, antibodies bind to pathogens for opsonization, antibodies bind to bacterial pathogens and neutralize it and antibodies help in killing of infected cells. The first contact between a pathogen and B cell gives rise to antibody secreting plasma cell and memory B cell. The first contact of pathogens with an individual antibody generates ***primary humoral response***. The primary response begins with a *lag phase*. During this phase the native B cells undergo clonal selection in response to antigen and differentiate into plasma cells and memory cells. The duration of lag phase varies with the nature of the antigen. The lag phase lasts for 3 to 4 days and during this period no antibody is detected in serum. The lag phase is followed by a *logarithmic increase* in serum antibody level. During primary humoral response IgM is secreted initially followed by IgG. Duration of primary humoral response depends upon the intensity of antigen. The memory cells formed during a primary response stop cell division and enter into the Go phase of cell division. These cells have variable life span with some persisting for the life of the individual. These cells may remain in the lymph nodes and relocate to other lymphoid tissue and survive for long period. Activation of these memory cells generates *secondary humoral response* which has shorter lag phase. In the secondary response the memory B cells which are formed during primary response and had entered into G_0 phase are now stimulated by the presence of same antigens. These B cells proliferate and differentiate into the plasma cells. Activation, proliferation and differentiation of memory B cells is a characteristic of secondary response. IgM produced during the secondary immune response is in much lower concentration than that which is produced during primary response but IgG level is higher during secondary response. The secondary response is also characterized by secretion of antibodies with a higher affinity for the antigens **(Figure 7.2).**

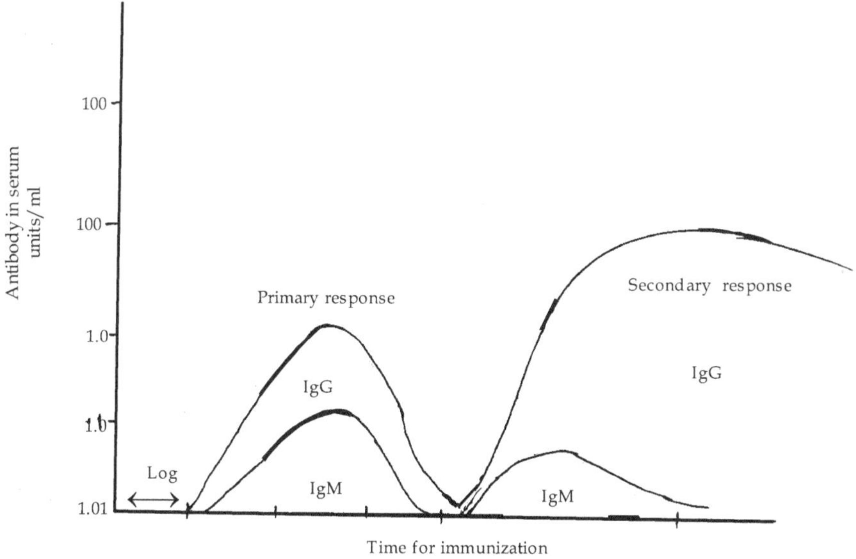

Fig. 7.2 : Primary and secondary immune response.

GERMINAL CENTERS AND ANTIGEN INDUCED B CELL DIFFERENTIATION

Initial B cell activation occurs outside the primary follicle. Germinal centers arise within 7 to 10 days after initial exposure to a thymus dependent antigen. Some activated B cells migrate deep inside the follicle and proliferate rapidly and appear in human germinal centers as a *dark zone*. This region is called *germinal center*. B cell proliferates and differentiates in the germinal center into memory B cell and antibody secreting cells. The germinal center is the site of selection. B cells that have high affinity receptors for the antigen are likely to be positively selected and leave the germinal center and those with low affinity are likely to undergo negative selection and die in the germinal center. The average affinity of the antibodies produced during the course of the humoral response increase remarkably during the process of affinity maturation. Monitoring of antibody genes during an immune response shows that extensive mutation of the immunoglobulin genes that respond to the infection takes place in B cell within germinal centers. Many mutations were found in the Ig genes obtained from B cells in germinal centers. Activation and differentiation of B cells occurs in a particular anatomical region. When an antigen enters into the blood it becomes concentrated in various peripheral lymphoid organs. Blood borne antigens are filtered by the spleen. Tissue borne antigens are filtered by the lymph node. A lymph node is an extremely efficient filter and at least 90 % antigens are trapped through the filtration. When the lymph is in contact with specific type of antigen, the B lymphocytes activate, divide and proliferate. Antigenic humoral immune response involves a complex reaction in a lymph node**(Figure 7.3).**

REGULATION OF B CELL DEVELOPMENT

A number of transcription factors regulate expression of various gene products at different stages of B cell development such as *B cell specific activator protein* (BSAP). Early B cell factors are important for B cell development. BSAP is a B cell regulator and it is expressed by B lineage cells and influences B cell maturation.

REGULATION OF THE IMMUNE EFFECTOR RESPONSE

Regulation of the immune response takes place both in the humoral and the cell-mediated immunity. In regulation of immune response antigens play important role. Cytokines also regulate the effector response. Immunological history too influences the quality and quantity of immune response.

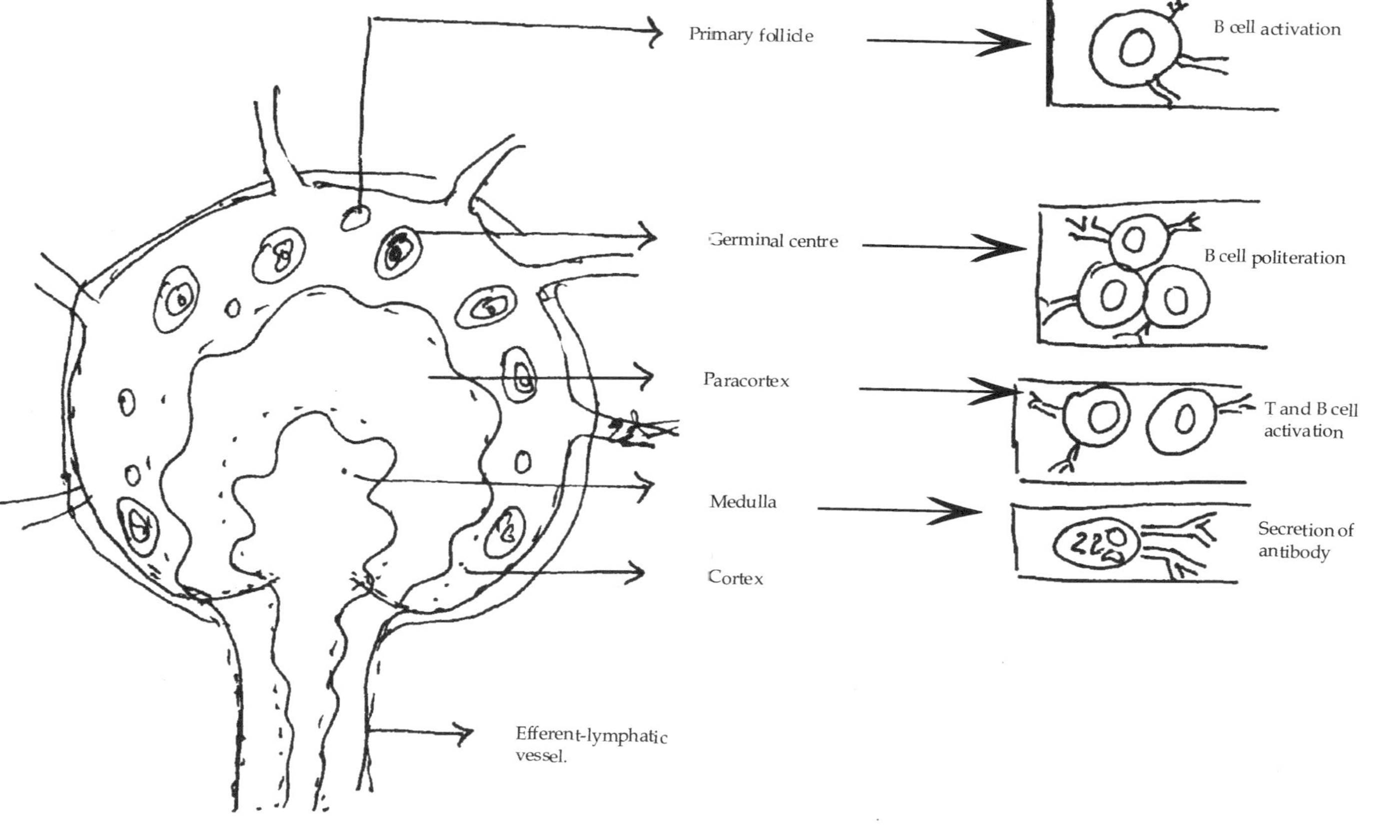

Fig. 7.3 : Humoral response in lymph mode

Cell Cooperation In Humoral Immunity

Humoral immunity is carried out by B cells. B cells secret antibodies but they need the cooperation of other cells, like macrophages, T_H cell, and killer cell. When B cells are cultured in the presence of antigen without macrophages, no antibodies are formed but when B cells are cultured in the presence of antigens with macrophages, antibodies are formed. The macrophages play role in processing and presentation of the antigen to the B cells. The processing makes the antigen, antigenic and the processed antigen is deposited on the surface of macrophage. The B cells recognize this antigen and it bind with the antigen with the help of immunoglobulin and produce antibody. These B cells are called thymus independent B cells.

In the T and B cell cooperation, B cells act as effector cell and T cell as a helper cell. B cells produce antibody with the help of T cells. These types of B cells are called thymus dependent B cells.

In the cooperation between killer cell and B cells, killer cells binds to cells coated with antibodies and produce cytotoxicity and kill the cell.

POINTS TO REMEMBER

- The immune system protects the body from various types of antigens.
- The destruction of antigens by producing antibody is called antibody mediated immune response.
- Humoral immune response is brought about by B cells. It is also called B cell-mediated immunity.
- B cell activation is the consequence of signal transduction process, triggered by engagement of the B cell.
- B cell activation depends on the nature of antigen by two different routes. One depends upon T_H cell and other is independent to it.
- The humoral response to thymus independent antigens is different from the response to thymus dependent antigens.
- The first contact of pathogens with an individual antibody generates primary humoral response.
- When the body encounters the same antigen to which it has been previously exposed, the response is called secondary humoral response.

- Many factors regulate the development of B cells.
- During humoral immune response, B cells require cooperation of many other cells like macrophages, T_H cell, and killer cell

REVIEW QUESTIONS

1. What is the role of humoral immunity?
2. Describe the structure and functions of B cell.
3. Explain : B cell activation and proliferation.
4. Describe antigen induced B cell differentiation.

CHAPTER - 8

Cell-Mediated Immunity

Cell mediated immune response performs different functions in protecting the immune system. When antigens are destroyed by cells, without producing antibodies, the immunity is called *cell-mediated immunity* (CMI) or *cellular immunity*. In this immunity T cells play a major role. The principal role of cell mediated immunity is to detect and eliminate the cells that have intracellular pathogens. When the antigen remains inside the cell the antibody cannot handle them. The intracellular pathogens are controlled by cell- mediated immunity. CMI is invited by the recognition of antigens by T cells. T cell lymphocytes recognize protein antigens of intracellular microbes that are displayed on the cell surface of infected cells. It is also important for elimination of cells which express foreign MHC molecules. Antigen specific and nonspecific cells can contribute to the cell-mediated immune response. The activities of both specific and non-specific components usually depend on effective local concentration of cytokines.

EFFECTOR RESPONSE

The adaptive immune response against microbes, within the phagosomes of phagocytes, is mediated by T lymphocytes. T-lymphocytes identify microbial antigens and produce chemical called cytokines that activate the phagocytes and stimulate inflammation. A person who is born without a thymus and lacks T cells is able to cope with infections of extracellular bacteria but can't eliminate intracellular pathogens.

Cell-mediated immune response can be divided into two major groups on the basis of *effector functions*, performed by cells. *One group* involves *effectors cells* that have direct cytotoxic activity to lyse virally infected cells but indirectly regulate their own growth and function.

The cytotoxic cells can be divided into two groups: one is cytotoxic T lymphocytes (CTL) and other is nonspecific natural killer (NK) cells. *The other group* involves special CD4$^+$ T cells showing delayed type hypersensitivity reaction **(Figure 8.1)**

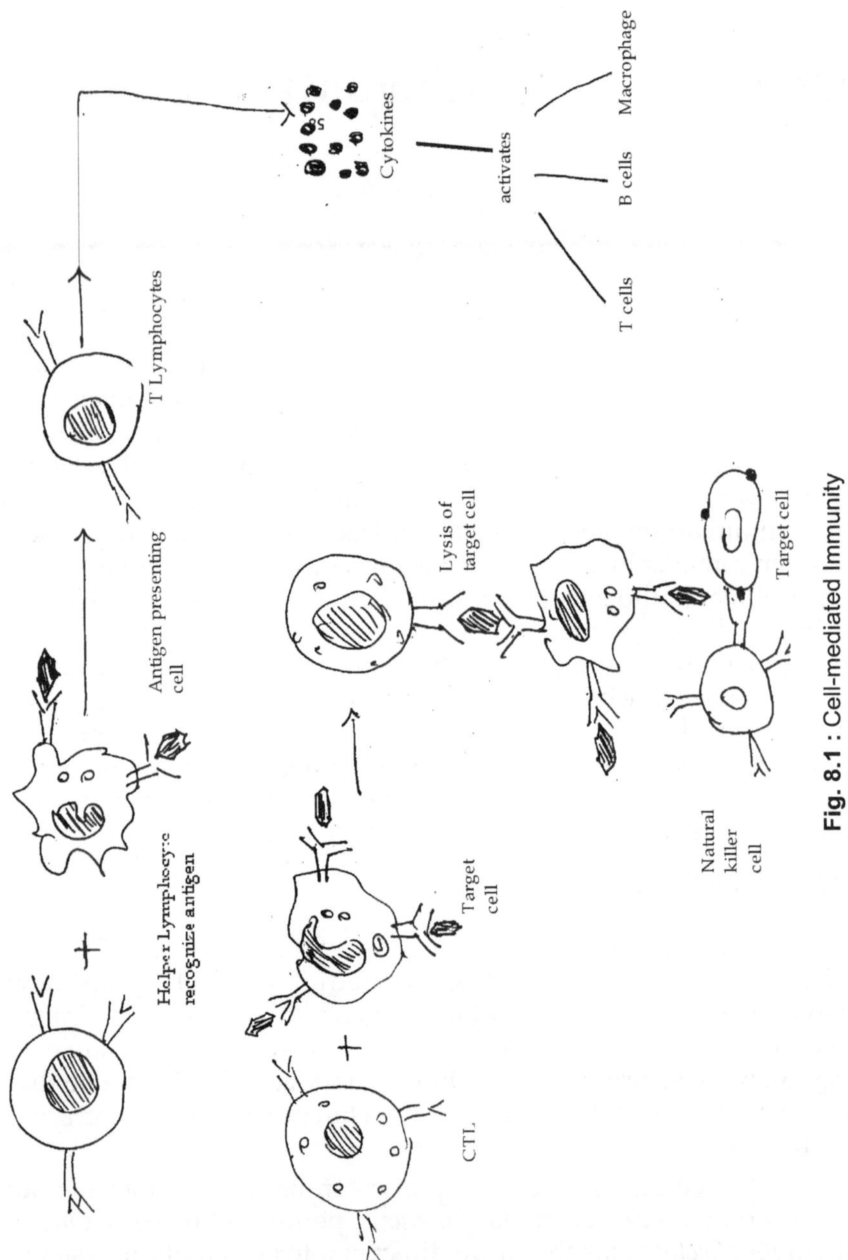

Fig. 8.1 : Cell-mediated Immunity

PROPERTIES OF EFFECTOR T CELLS

The effector functions in cell-mediated immunity depend on the effector functions of T lymphocytes. The three types of effector T cells include: $CD4^+$ T_{H1}, T_{H2} cells and $CD8^+$ cytotoxic T lymphocytes. The classification of these cells is on the basis of the production of the type of cytokine receptor and markers expressed by these cells. The adaptive immune response to intracellular microbes that reside within the phagosomes of phagocytes is mediated by T lymphocytes. T lymphocytes recognize microbial antigens and produce cytokines which increase phagocytosis. $CD8^+$cytotoxic T lymphocytes are the mediators of cell-mediated immunity against viruses that survive and replicate within various cell types. These T cells kill the infected cell. Effector T cells express certain effector molecules which may be membrane bound. The membrane bound molecules belong to tumor necrosis factor. The effector T cells produce adhesion molecules. These molecules play an important role in T cell adhesion to antigen presenting cell or target cells, infected with intracellular pathogen. These adhesion molecules act as signal transducers. Both the effector $CD4^+$ T cells and cytotoxic T lymphocytes (CTLs) are required for stimulation and proliferation of B lymphocytes which requires specific recognition by T cells. T lymphocytes recognize peptide antigens and foreign peptide antigens only when they are expressed on the cell surface of antigen presenting cell along with MHC molecules. Antigen presenting cells are capable of performing two important functions - endocytosis and processing of antigens and also expression of MHC molecules.

CYTOTOXIC T CELLS

Cytotoxic T cells or cytotoxic T lymphocytes(CTLs) are formed from activation of T cells. These effector cells have lytic capability and role in recognition and elimination of virus infected cells and cancerous cells.

The CTL mediated immune response is divided into two phases:one is activation and differentiation of naive T cells, and the second is recognition of antigens.

Activation and Differentiation of T Cells

B and T lymphocytes that have not interacted with antigen referred to as naïve or unprimed. These are resting cells in the Go phase of the cell cycle. The naïve lymphocytes have short life span. The naïve are small lymphocytes and their cytoplasm forms a scarcely visible rim around the nucleus, few mitochondria and poorly developed endoplasmic reticulum and golgibody. In this phase naive T cells are referred as cytotoxic T lymphocytes precursors (CTL- Ps) because naive T cells are incapable of killing target cells. For activation of CTLPs three signals are required:

(a) Antigen specific signal transmitted by TCR complex and TCR for recognition of the peptide class I MHC molecules.

(b) A co-stimulatory signal transmitted by the CD28 for interaction of CTL-P and antigen presenting cells (APC)

(c) Cytokine induced signal by the interaction of IL-2 with IL-2 R (receptor) resulting in proliferation and differentiation of antigen activation of CTL-P into effector CTLs.

Inactivated CTL-Ps do not express receptors and do not proliferate and show cytotoxic activity. On antigen mediated activation of CTL-Ps they begin to produce IL-2 receptor and IL-2.The cytokines are required for proliferation and differentiation of activated CTL-Ps into effector CTLs.

Recognition of Antigens

It involves interaction of CTLs membrane bound fas ligand (FasL) with fas receptor on the target cell surface. Fas ligand is found on the membrane of CTLs. The interaction of FasL with Fas ,on a target cell, triggers apoptosis. This interaction activates a particular cellular enzymes called *caspase* which ultimately induces apoptosis of the target cell. CTLs store cytotoxic proteins in the form of granules in their cytoplasm. These proteins belong to two categories. One is *perforins* and other is *fragmentins*. Perforins are involved in pore formation and fragmentins are involved in fragmentation of cellular products.

When antigen specific CTLs are incubated with target cells. Both these cells interact and conjugate. Conjugate formation is completed by Ca^{++} ions. The CTL dissociates from target cell and binds to another cell. After some time CTL dissociates and target cell dies by apoptosis **(Figure 8.2).**

CTLs and Target cell interaction → conjugation of CTLs and Target cell →cytoplasmic rearrangement of CTLs cell → CTLs granule exocytosis → dissociation → CTLs recycling → target cell membrane damage.

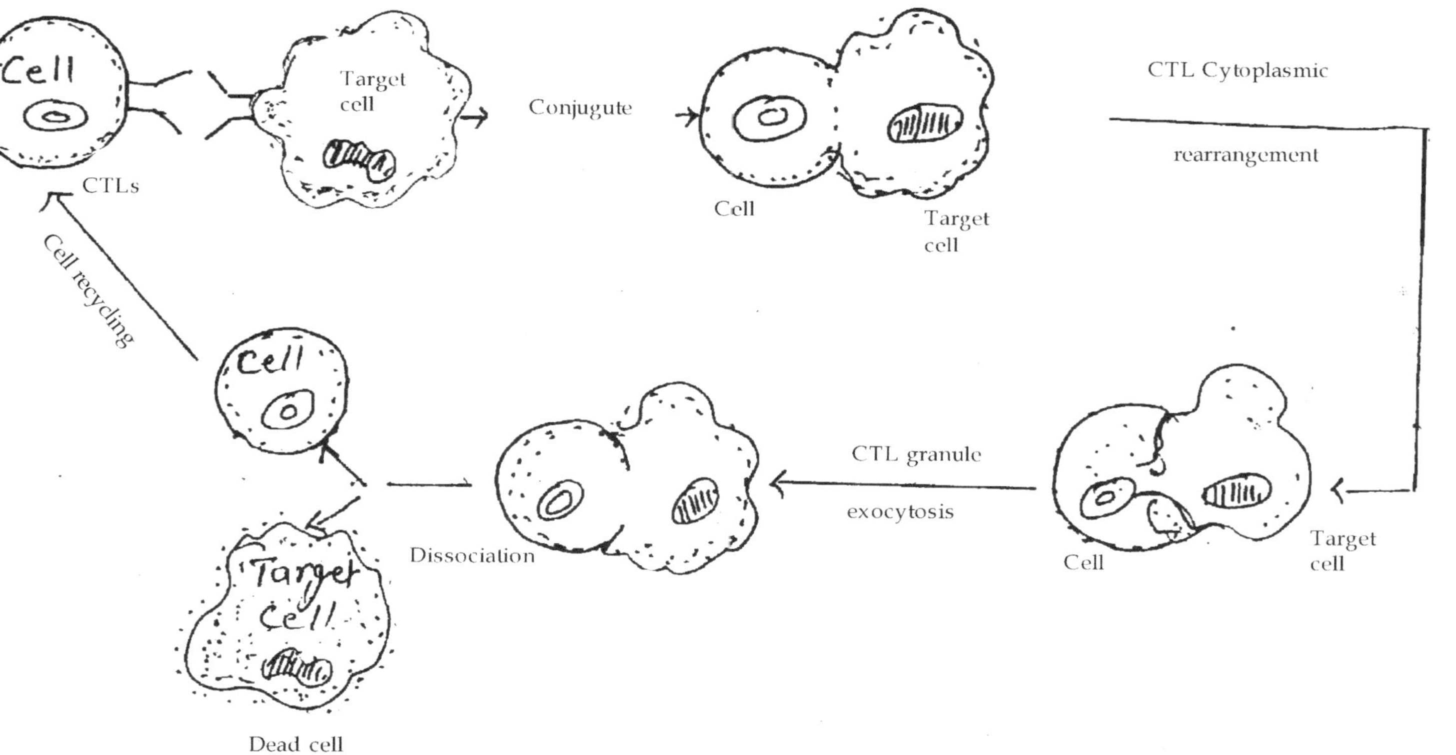

Fig. 8.2 : CTLs and Target cell interaction

Pore formation in the cell membrane of the target is the one way that perforin mediates granzyme. Many target cells have a molecule known as mannose $6PO_4$ receptor on their surface that binds to granzyme and forms complex and appear inside vesicles .When this enters the cytoplasm, granzyme initiates reaction and starts fragmentation.

NATURAL KILLER CELLS (NK CELLS)

NK cells are the bone marrow derived large lymphocytes which can kill or destroy various target cells without any need of additional activation. These cells are involved in immune defense against viruses and tumors. NK cells produce cytokines that play an important role in immune regulation and influence both adaptive and innate immunity. NK cells are large in size with granules in their cytoplasm. NK cell are also derived from pluripotent hematopoietic stem cells. These cells constitute 5 to 10% of lymphocytes and they do not express the membrane molecules and receptors that distinguish T and B cells lineages. They express some membrane markers that are found on monocytes and granulocytes. NK cells show early response in viral or bacterial infection. Nature NK cells are found in circulation and in several tissues.

Mechanism of NK Cell Killing

NK cells kill the virus infected cells or tumar cells by the same mechanism as employed by CTLs. The cytoplasm of NK cells also contains perforin and granzyme. NK cells bear FasL on their surface and readily induce death in fas bearing target cells. NK cells are cytotoxic and always have large granules in their cytoplasm. NK cells also show target cell destruction by apoptosis due to the expression of tumor necrosis. The role of perforin and granzyme in NK mediated killing of target cells by apoptosis are believed to be similar to their roles in the CTL mediated process but there are some differences between NK and CTL cells. First, NK cells do not express antigen specific T cell receptors. Secondly, target cell recognition by NK cells is not MHC restricted, and thirdly NK cells do not generate immunologic memory **(Figure 8.3).**

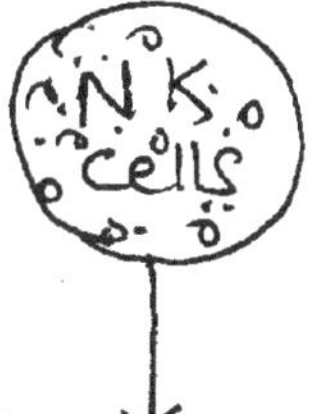

NK cell cytoplasm

NK cell attaches to a target cell

Degranulation with release of
perforin & granzyme B

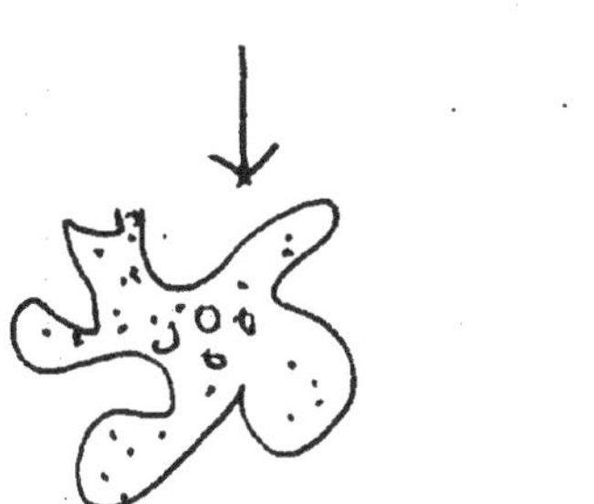

Apoptisis of target cell

Fig. 8.3 : NK cell mediated cytotoxicity

NK Cells Activation and Inhibition Receptors

The NK cells cytotoxic activity against target cells is easily triggered through involvement of many surface receptors. These cell surface receptors are important for activation of signals and for inhibition. The decision of killing a target is a balance between positive and negative signals received by NK cells. On the basis of their chemical nature, NK cell receptors are of two type I and II.

ANTIBODY DEPENDENT CELL-MEDIATED CYTOTOXICITY

The antibody molecule acts as a bridge between cytotoxic cell and target-cell and subsequently causes lysis of the target cell. Cytotoxic cells are nonspecific for antigen. The specificity of the antibodies directs them to specific target all. This cytotoxicity is referred to as antibody dependent cell-mediated cytotoxicity (ADCC). The cells that can mediate ADCC are NK cells, macrophages, neutrophils, monocytes, and eosinophils. Virally infected cells are killed through ADCC reaction, mediated by NK cells and macrophages.In the target cell killing mechanism by ADCC so many cytotoxic cells are involved. Neutrophils, eosinophils and macrophages, on activation through Fc receptor, release lytic enzymes at the site of Fc-mediated contact and damage the target cell. In this mechanism neighboring normal cells are also damaged. Monocytes, macrophages and NK cells secrete tumor necrosis factor (TNF) which has cytotoxic effects on bound target cell (**Figure 8.4**).

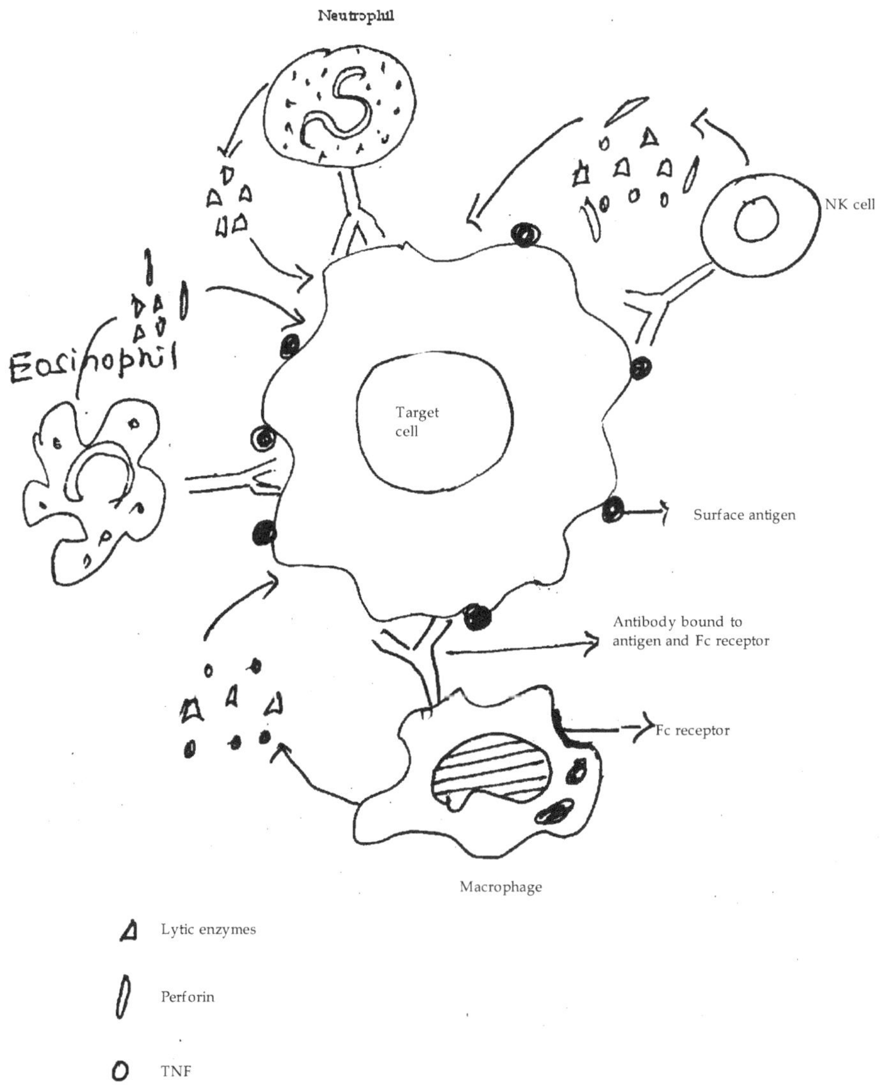

Fig. 8.4 : Antibody dependent cell mediated cytotoxicity

CELL CO-OPERATION IN CELL-MEDIATED IMMUNITY

In cell-mediated immunity, T cells directly destroy antigens. In this system, T cells act as effector cells. Some other T cells function as helper cells. In cell-mediated co-operation some other cells like macrophage and natural killer cells are also involved. The antigen is processed by the macrophage and antigen in deposited on the surface of macrophage. T helper cell recognizes the antigen along with the Ia antigen which is present on the surface of macrophage. Now these cells allow the macrophage to release IL- 1. IL- 1 and antigen Ia complex activates the T cell to release IL- 2. IL- 2 is also secreted by the other activated T cells. IL-2 proliferates the T cells into blast cells. Blast cells release macrophage inhibition factor which stops the activity of macrophage around the antigen. Blast cells also release *macrophage aggregation* and *macrophage activation factors.* Macrophage aggregation factor helps in aggregation of macrophage on the surface of antigen and macrophage activation factor activates the macrophage for phagocytosis. Some macrophages contain human leukocyte antigen HLA on their surface along with antigens. This HLA and antigens are recognized by the cytotoxicity T cells and after recognition Tc cells kill the antigen by cytotoxicity. In this system suppressor T cells are also involved which regulate the function of Tc cells. Natural killer cells also kill the antigen by releasing perforin with the help of T_H cells.

POINTS TO REMEMBER

- Cell-mediated immune response performs different functions in the immune system.
- CMI is an immune response that does not involve antibodies but involves in the activation of macrophage form T cells.
- T lymphocytes recognize microbial antigens and produce cytokines which increase phagocytosis.
- $CD8^+$cytotoxic T lymphocytes are the mediators of cell-mediated immunity against viruses that survive and replicate within various cell types.
- The cell-mediated immune response occurs in two phases: The first phase involves the activation and proliferation of Tc cells and the second phase involves recognition of antigens.
- CTL cells store cytotoxic protein in the form of granules in their cytoplasm. The cytotoxic protein in CTL cells belongs to two categories-one belongs to perforins and other belongs to fragmentins.
- Natural killer cells are involved in immune defense against viruses and tumors. Natural killer cells are found in the circulation and in several tissues.
- The antibody molecule acts as a bridge between cytotoxic cell and target-cell and subsequently causes lysis of the target cell.

- This cytotoxicity is referred to as antibody dependent cell mediated cytotoxicity (ADCC).

REVIEW QUESTIONS

1. What is cell mediated immunity? Explain functions of CMI.
2. Describe various properties of effecter T cells.
3. Explain natural killer cell and its mechanism of action.
4. How are cytotoxic T cells activated?
5. Write short notes on:
   i. Antibody dependent cell-mediated cytotoxicity
   ii. Activation and differentiation of T Cells

Chapter - 9

Complement

The complement system is a group of about 20 heat labile plasma proteins which constitute nearly 10% of total serum proteins. These are important in activation of phagocytes, control of inflammation and cytolysis. They form one of the major immune defense systems of the body. The complement system plays important role both in innate and adaptive immune system. The main functions of complement are chemotaxis; opsonization and cell activation; immune clearance of immune complexes from circulation; lysis of target cells; triggering and amplification of inflammatory reactions and development of antibody responses. Evolutionary, the system is very ancient and is present even in invertebrate organisms like starfish and worms.

The complement proteins are mainly synthesized by liver cells. Besides liver hepatocytes, cells like monocytes, macrophages and epithelial cells of gastro intestinal and urinogenital tract also produce these proteins. Complement proteins act in a cascade manner. These proteins are present in inactive form and are activated and activate the next protein in a sequence like the blood clotting system. Complement proteins are designated by numerals C1 to C9. They were named in the order in which they were discovered. Peptide fragments formed during complement activation are indicated by small letters 'a' and 'b'. Usually 'a' denotes small fragment while 'b' designates the larger fragment. Complement complexes with enzymatic activity are designated by a bar over the number, like $\overline{\text{C4b2a}}$ (C3 convertase).

ACTIVATION PATHWAYS

There are three different pathways to activate the complement system. These are:

1. Classical pathway
2. Alternate pathway
3. Lectin pathway

The *classical pathway* is antibody dependent pathway and it is initiated when antibodies bind to antigens and antigen-antibody complex is formed.

The *alternate pathway* is antibody independent mechanism and does not require antigen-antibody complex to initiate the pathway.

The *lectin pathway* is also antibody independent and does not require antibody for its initiation.

All three pathways involve activation of C3, the most abundant complement protein. It starts a cascade of reactions in which one reaction triggers another which, in turn, triggers another and so on. All activation pathways converge in a common terminal pathway which leads to membrane dis ruption and lytic killing of the pathogen **(Figure 9.1).**

Classical Pathway

The classical pathway is initiated by the antibodies bound to the antigens. Formation of antigen-antibody complex activates C1 which is a complex molecule, consisting of a large unit C1q and two small molecules C1r and C1s. C1q consists of six subunits, each subunit contains three polypeptide chains**(Figure 9.2).** It binds to the Fc regions of the bound antibody through its globular head. The binding of C1q with antibody makes some conformational change in its shape and leads to autocatalytic activation of C1r. Activated C1r then activates C1s. Activated C1s has enzymatic activity and acts on C4 and C2. It cleaves C4 into two fragments C4a and C4b. C4a is a small fragment and is released. C4b is the large fragment. It binds to C2 and presents it for cleavage by C1s. C2 is cleaved into C2a and C2b. C2b is the smaller fragment and it diffuses away while C2a remains associated with C4b to form C4b2a. $\overline{C4b2a}$ (also known as *C3 convertase*) is the activation enzyme which cleaves C3, the most abundant complement protein into C3a and C3b. C3a is the small fragment and is released. C3b is the large fragment and binds to the activated surface of the microbe and promotes phagocytosis. Some of the C3b binds directly to the $\overline{C4b2a}$ and forms $\overline{C42a3b}$ (also called *C5 convertase*). This is the last enzymatic molecule of classical pathway. It binds with C5 and presents it for cleavage by C2a. A small fragment C5a is released while the large component C5b remains associated with $\overline{C4b2a3b}$ complex**(Figure 9.3).**

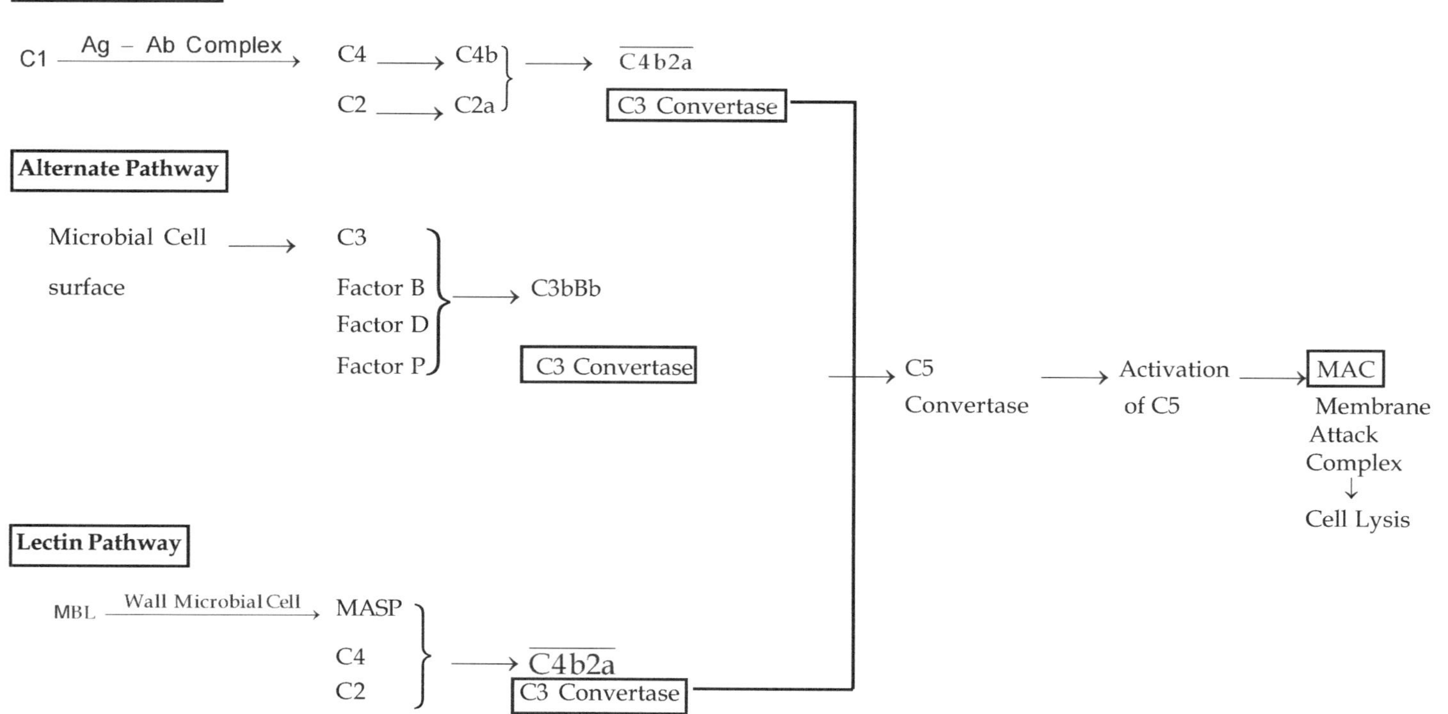

Fig. 9.1 : Pathways of complement Activation
(All the three pathways leads to formation of C3 convertase and C5 convertase.
This leads to the formation of MAC, membrane attack complex and lysis of the cell).

Globular head
Binding site for
exposed Cn2
domain of antibody
Clr
Cls
Stalk
Clq

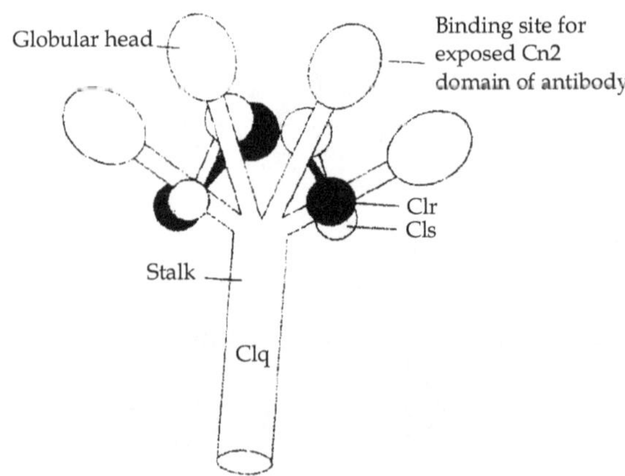

Fig. 9.2 : Structure of C1 complement
(C1 - a complex molecule consists of three components C1q, Clr and Cls)

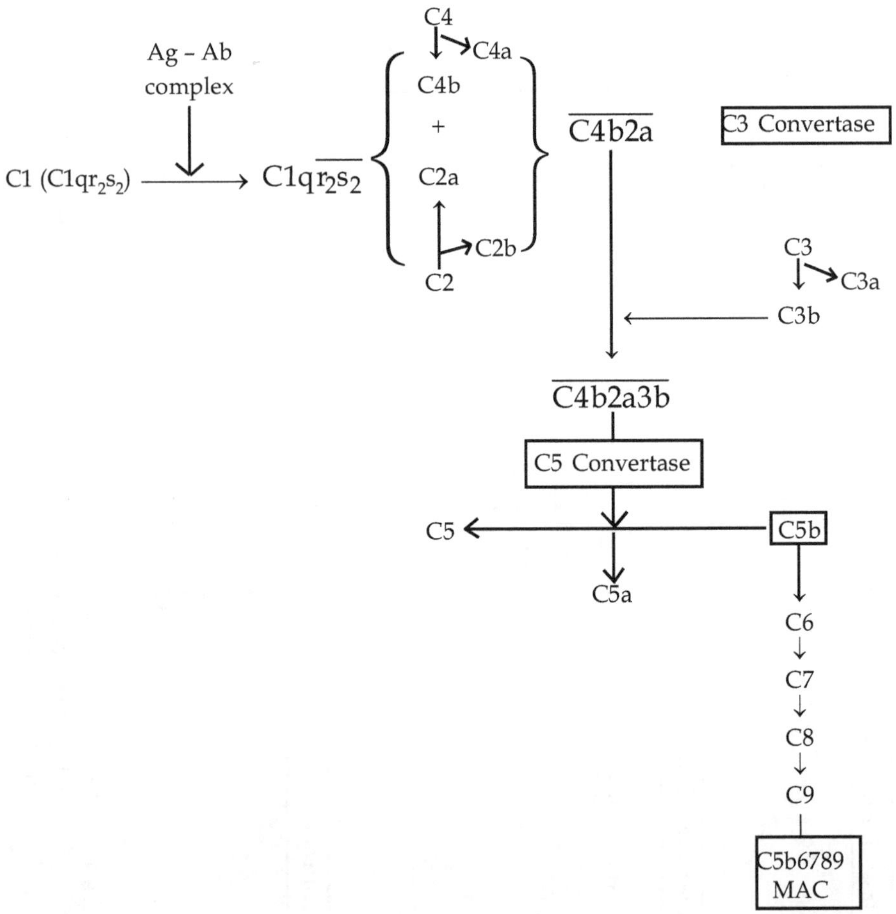

Ag – Ab
complex
C1 (C1qr2s2)
C1qr2s2
C4
C4a
C4b
+
C2a
C2b
C2
C4b2a
C3 Convertase
C3
C3a
C3b
C4b2a3b
C5 Convertase
C5
C5b
C5a
C6
C7
C8
C9
C5b6789
MAC

Fig. 9.3 : Classical Pathway of complement activation

The Terminal Pathway

After the cleavage of C5, the final or terminal stage is the formation of membrane attack complex (MAC). MAC is a transmembrane pore. This is the common terminal stage of all activation pathways. Attached with C5 convertase, C5b binds C6 and C7. Binding with C6 and C7 causes conformational changes in the molecule and causes its release form the C5 convertase. The membrane bound C5b67 binds C8 and finally multiple copies of C9 are associated in the molecule to form MAC. Formation of MAC creates a rigid pore in the membrane. The walls of it are formed from multiple copies of C9, arranged like barrel straws, around a central cavity **(Figure 9.4).** The pore allows free flow of solutes and electrolytes across the cell membrane and causes the cell to swell and rupture. The C5a is released and acts as a chemotactic factor, attracting phagocytes to the site of infection. It, along with C3a, binds to mast cells and basophils and activates them to release histamines and increase blood vessel permeability during inflammation.

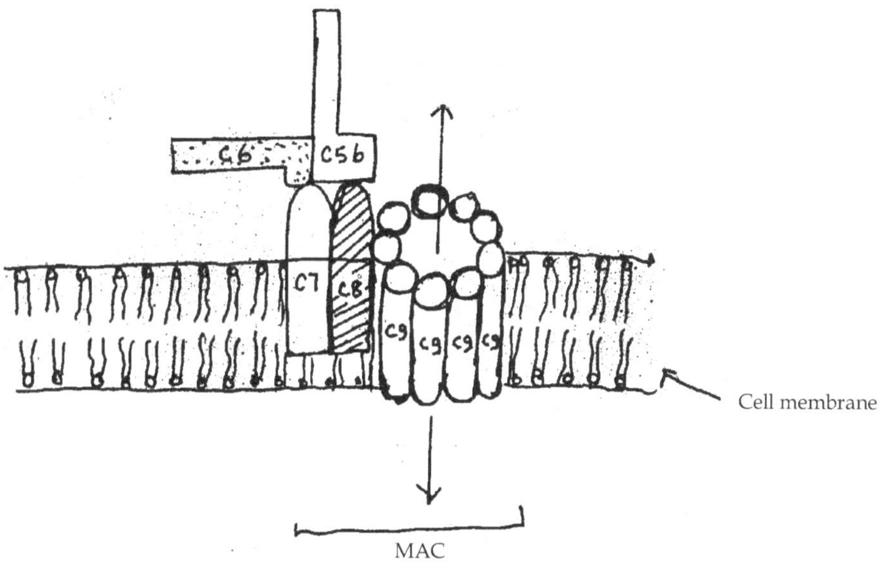

Fig. 9.4 : Formation of MAC (Membrane Attack Complex)
(Membrane bound C5b67 binds with C8 and multiple copies of C9 to form MAC)

Alternate pathway

The alternate pathway of complement activation is antibody independent and does not require antigen-antibody complex. The proteins of alternate pathway are called 'factors'- Factor B, Factor D and Factor P. C3 is slowly and spontaneously hydrolyzed in the plasma to form C3a and C3b. C3b binds to foreign surface (microbial surfaces) or even to the host's own cells. But membrane of most mammalian cells has high sialic acid content which causes rapid inactivation of bound C3b and thus further activation of complement is stopped. In case of microbes, the cell wall has low sialic acid content and thus, the bound C3b remains active for a longer time. C3b, bound on the microbial surface, binds to another serum protein Factor B. Once bound to C3b, Factor B can bind and activate Factor D. Factor D cuts factor B into Ba and Bb. Ba is released and C3Bb complex is formed. $\overline{\text{C3bBb}}$ complex is the C3 cleaving enzyme (C3 convertase) and cleaves C3 into C3a and C3b. C3b binds to C3bBb complex and forms $\overline{\text{C3bBb3b}}$ which is a C5 convertase **(Figure 9.5)**.

Lectin Pathway

Lectin pathway is also antibody independent. It is , in fact, a route for classical pathway activation that bypasses an antibody. The pathway begins by binding of mannose binding lectin (MBL). MBL is a lectin which binds to the mannose residues, present in the cell wall of various microorganisms. Binding of MBL to the carbohydrate residues on the cell surface of microbes induces several MBL associated serine proteases (MASP-1 and MASP-2) to bind to MBL. The active complex, formed by this association, causes cleavage and activation of C4 and C2 to continue activation of complement, exactly as in the classical pathway **(Figure 9.6)**.

REGULATION OF COMPLEMENT SYSTEM

Although complement system is intended to protect the body from foreign particles but there is also a threat of damage to the self (host) cells by the complement proteins. This is checked by regulating the uncontrolled activation of complement system. There are various substances/ways which regulate complement activation:

C1 Inhibitor

In the plasma, there is present a serine protease inhibitor (serpine) called *C1 inhibitor*. C1 inhibitor removes C1r and C1s, formed from C1. By removing C1r and C1s , it switches off the activation of classical pathway. C1 inhibitor can remove MASP enzymes from MBL complex, thus also regulating lectin pathway.

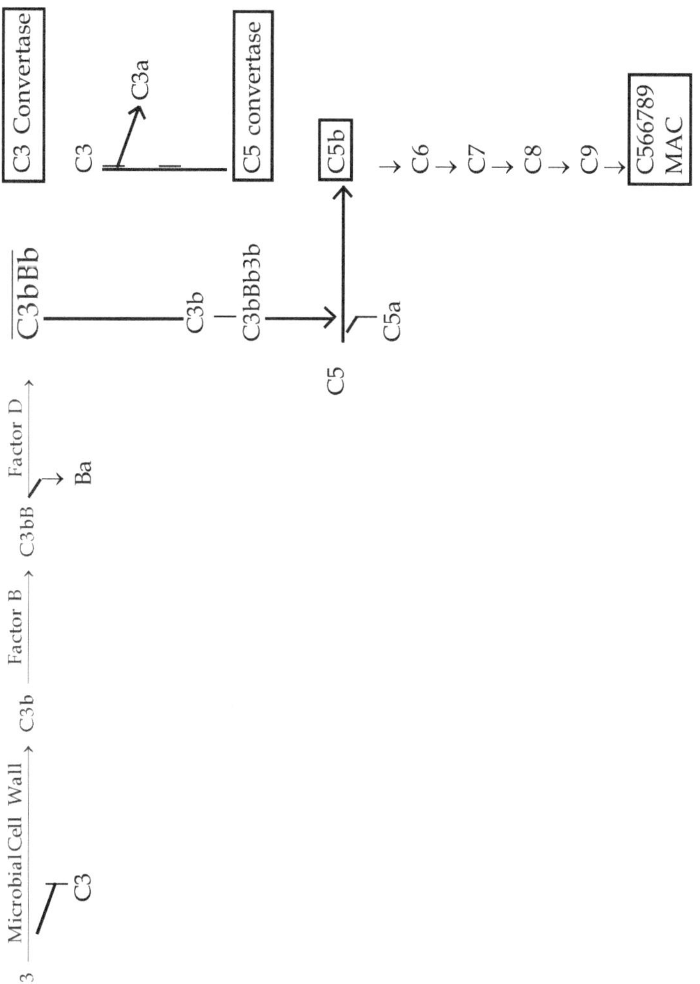

Fig. 9.5 : Alternative Pathway of complement activation

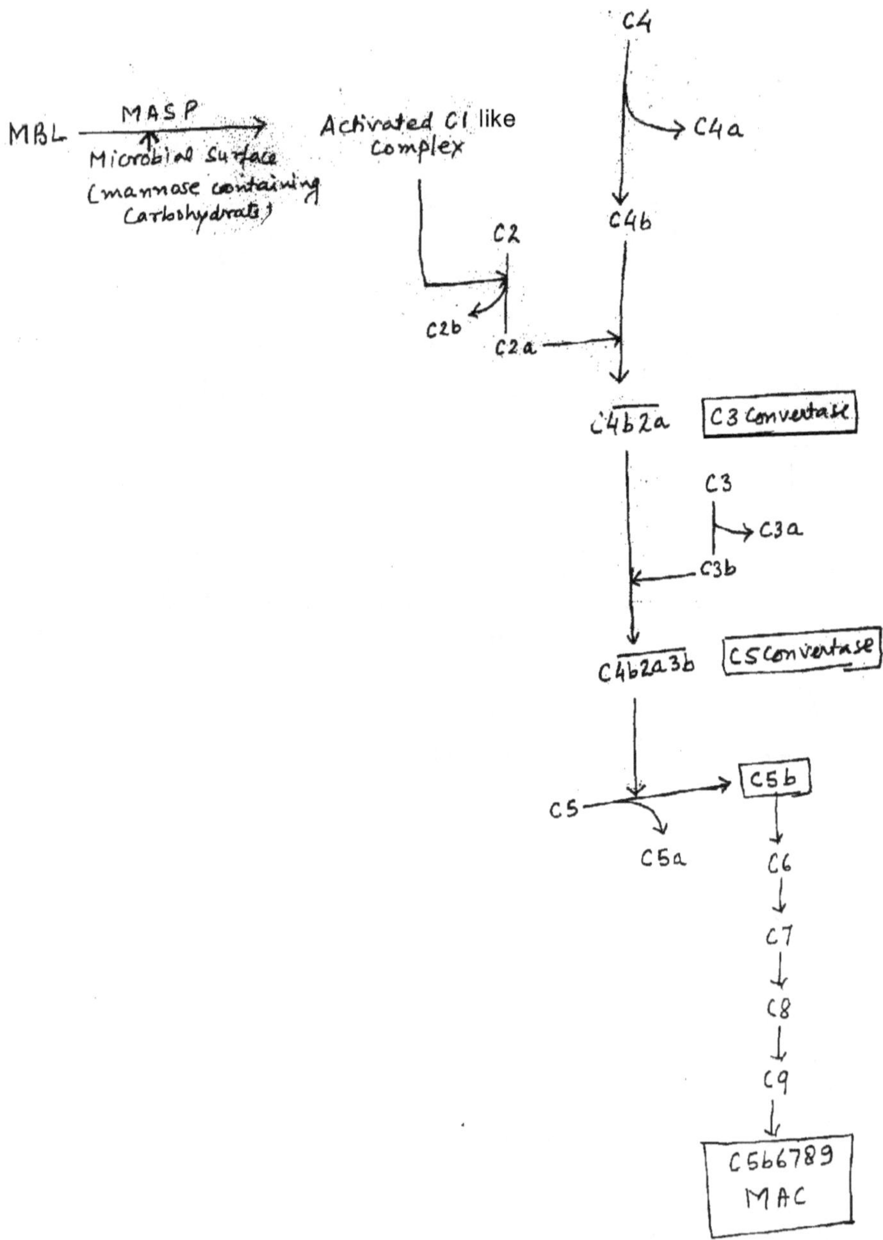

Fig. 9.6 : Lectin Pathway of complement activation

C3 and C5 Convertase

Another way of regulating complement activation is by controlling the activity of C3 and C5 convertase. The enzyme complexes are labile and can be destroyed within a few minutes of creation. Some fluid and membrane proteins can destroy these convertase and regulate the activation of complement. A plasma protein, C4 binding protein (C4bp) destroys C3 and C5 convertase ,in the classical pathway. Another proteins, Factor H and Factor H-like I destroy these enzyme of alternate pathway. Two other proteins, membrane cofactor protein (MCP) and decay accelerating factor (DAF) on cell membrane also destroys convertase enzymes of both classical and alternate pathways. Another factor, Factor I can cleave and inactivate C4b and C3b in presence of MCP. In alternate pathway protein, properdin acts as a positive regulator by stabilizing C3 convertase and increasing its life span.

Regulation of MAC Formation

C8 itself acts as a regulator by blocking the membrane binding site and formation of MAC. A membrane protein CD59 acts by preventing the binding of C8 and C9 during MAC formation and thereby preventing the pore formation. A soluble protein called S protein (vitronectin) binds with C5b7 complex and stops it inserting into cell membrane.

FUNCTIONS OF COMPLEMENT

The major functions of complement are:

- Chemotaxis
- Opsonization and cell activation
- Lysis of microbial cell by MAC
- Clearance of immune complexes
- Inflammatory mediators
- Priming of adaptive immune system **(Figure 9.7)**.

Chemotaxis

C5a is an important chemotactic agent for macrophages and neutrophils. It induces these cells to adhere to vascular endothelial cells and extravasate and migrate towards the site of complement activation. Binding of small fragments C3a and C5a, generated by cleavage complement proteins, to the mast cells and basophils causes activation of these cells to release histamines and other pharmacologically active mediators and inflammatory cytokines. They also increase the adhesive properties of these cells.

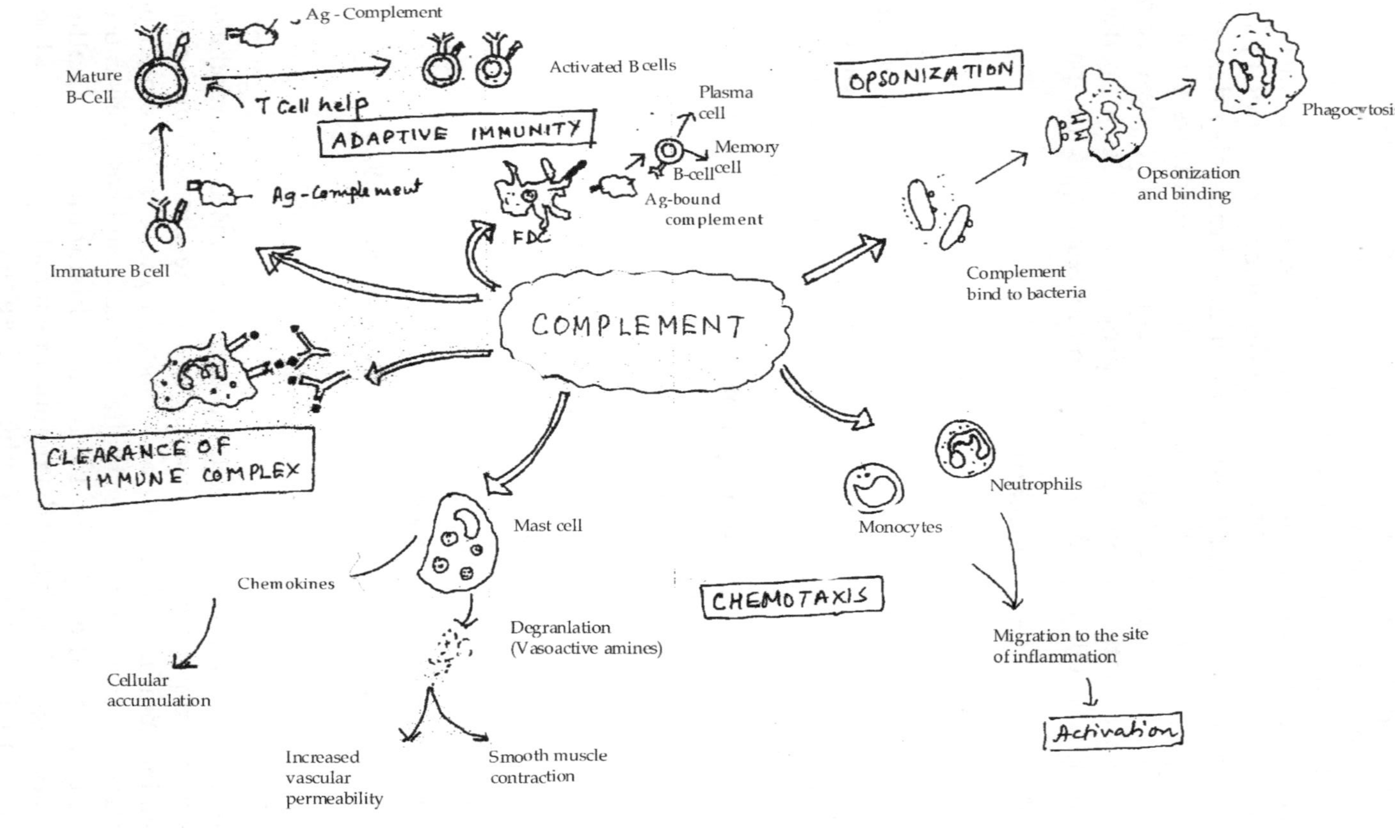

Fig. 9.7 : Functions of complement

Opsonization and Cell Activation

Complement plays important role in opsonization and phagocytosis. After the antibodies bind to the microbial surface, complement fragment coat their surface. C3 fragment function as opsonins and is recognized by phagocytes and other cells carrying complement receptors. These cells bind to target, causing cell activation and phagocytosis.

Lysis of Microbial Cells

The terminal pathway i. e. the formation of MAC, is the major defense mechanism which causes lysis of the pathogens, especially of Gram-negative bacteria. Assembly of MAC leads to the formation of pore in the membrane, causing cell lysis. MAC attack is particularly important in case of *Neisseria.*

Clearance of Immune Complex

Complement plays important role in clearance of immune complexes. Immune complexes, containing microbial antigens or antigens from death of host cells, are continuously formed both in health and diseases. These complexes may grow and get deposited in various tissues and cause inflammation. Complement fragments such as C3b bind to immune complexes and help in clearance of immune complexes. This prevents the deposition of these complexes in the tissues. Coating with C3b can also disaggregate large immune complexes by disrupting interactions between antigen and antibody. CR1 on erythrocytes can bind with C3b, bound on immune complexes and may take these complexes out of the plasma to liver or spleen for their removal.

Priming of Adaptive Immune System

Complement has significant role in priming adaptive immune system. These proteins help in B cell maturation, activation and proliferation. C3 fragments, bound to antigens, bind to complement receptors on B cells and increase their development. Complement opsonized particles may be 1000 folds as active as the unopsonized particles in triggering antibody formation.

Inflammatory Mediators

Fragments of complement proteins act as inflammatory mediators.

For example, C3a, C5a and C4a can activate mast cells and cause their degranulation. C5a also causes release of hydrolytic enzymes from neutrophils which cause inflammation.

COMPLEMENT DEFICIENCIES

Though complement deficiencies are rare but genetic deficiencies of many complement components have been reported in several cases. Deficiency of any component of complement has several adverse effects on the defense system, depending upon the pathway affected by those components.

Deficiency of C1, C4 and C2 causes a condition similar to systemic lupus erythematosus (SLE), an autoimmune disease. It causes collagen vascular disorders and marked increase in immune complex deposits in kidney, skin and blood vessels. Deficiency of C1 also causes the accumulation of apoptic cells in the tissues which release toxic cell components and cause inflammation.

Deficiency of MBL, the lectin which initiates the lectin pathway of complement activation, is associated with enhanced susceptibility to bacterial infection in infants. In adults, the deficiency of MBL is of little significance, except in people with weak immune system..

The individuals with deficiency of Factor B and Factor D may suffer from recurrent bacterial infection. Deficiency of C3 also causes severe recurrent bacterial infection and immune complex diseases. Deficiency of C3 may also occur due to deficiencies of Factor H and Factor I which are involved in alternate pathway. Deficiency of any of these factors leads to uncontrolled activation of alternate pathway amplification loop and complete consumption of C3. Deficiency of properdin and any of the terminal complements such as C5, C6, C7, C8 or C9 may lead to increased susceptibility to *Neisseria* infection. Deficiency of properdin is found only in males as it is inherited by X chromosome.

POINTS TO REMEMBER

- Complement proteins are a group of about 20 heat labile plasma proteins.
- The main functions of complement proteins are chemotaxis; opsonization and cell activation; immune clearance of immune complexes from circulation; lysis of target cells; triggering and amplification of inflammatory reactions and development of antibody responses.

- They are mainly synthesized in liver cells.
- Complement proteins are designated as C1 to C9.
- Complement proteins act in a cascade manner.
- There are three different pathways to activate the complement system: classical pathway, alternate pathway and lectin pathway.
- The classical pathway is antibody dependent pathway and it is initiated after formation of antigen-antibody complex.
- The alternate and lectin pathways are antibody independent and do not require antibody for its initiation.
- C3is the most abundant complement protein.
- All activation pathways converge in a common terminal pathway which leads to membrane disruption and lytic killing of the pathogen.
- Complement system is regulated by many substances such as C1 inhibitor, C3 and C5 convertase activity and by regulation of formation of MAC.
- Though very rare, deficiency of various components of complement may cause many adverse effects on the body.

REVIEW QUESTIONS

1. What is complement system? How is it important in protecting the body from foreign infection?
2. Describe the classical pathway of complement activation.
3. Which are the antibody independent pathways of complement activation?
4. Describe various functions of complement.
5. Why is it necessary to regulate complement system? Which are different ways to regulate complement system?
6. Describe different types of complement deficiencies.
7. Write short notes on :
   i. Membrane attack complex (MAC)
   ii. Lectin pathway

CHAPTER - 10

Major Histocompatibility Complex

Major Histocompatibility Complex (MHC) is a closely linked cluster of genes which have codes for cell surface proteins known as MHC molecules. These molecules have very important role in presentation of processed antigens to the T cells in initiation of an immune response as T cells recognize an antigen only when it is presented along with MHC molecule. These are highly polymorphic and are the major barrier to organ/tissue transplantation. All vertebrates have MHC. MHC differs among members of a species and are thus called alloantigens. MHC participates in the development of both humoral and cell-mediated immune responses.

ORGANIZATION OF MHC GENES

MHC genes are located on chromosome 6 in humans. They are organized into three distinct regions known as class I, class II and class III MHC gene loci. In humans they are also called **HLA(human leukocyte antigen) complex,** as they were first determined in leucocytes. In mice there MHC genes are located on chromosome 17 and are also called **H-2 complex**. These genes encode three classes of molecules known as class I, class II and class III MHC proteins /MHC molecules or MHC products. As these proteins were identified as antigen involved in transplant rejection, they are also called as MHC antigens. They all are nearly cell surface molecules which are involved in T cell recognition except, class III MHC molecules which have some other functions.

Class I MHC molecules are the glycoproteins which are encoded by class I MHC genes. These molecules are expressed on all nucleated cells and are involved in presentation of the peptide antigens to T_C cells.

Class II MHC molecules encoded by class II MHC genes are expressed on the surface of antigen presenting cells(APC) like macrophage, dendritic cells and B cells and function to present processed antigen peptides to CD4+ T_H cells.

Class III MHC molecules are encoded by class III MHC genes. They differ in structure and functions to class I and class II molecules. They include various complement proteins and inflammatory cytokines like TNFα and TNFβ**(Fig. 10.1).**

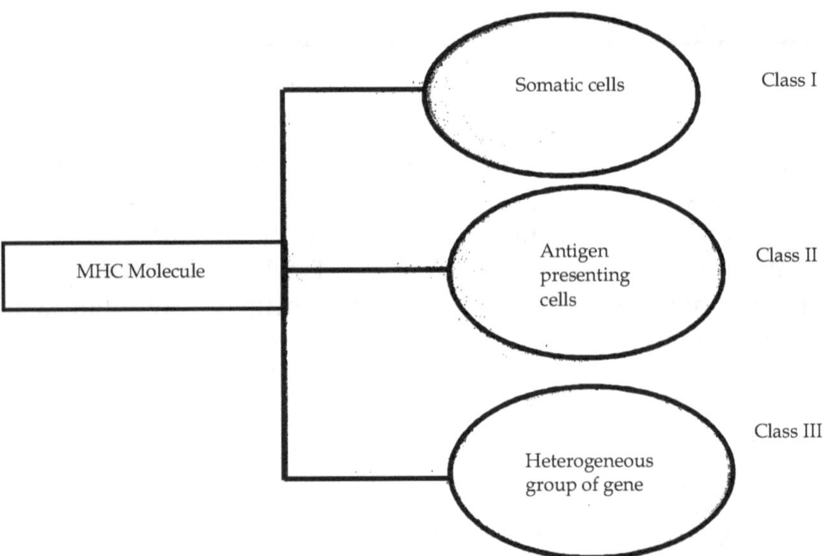

Fig. 10.1 : Location of MHC molecules on different cells

In humans class I MHC genes are designated as HLA-A, HLA-B and HLA-C **(Fig. 10.2).** They encode α chain of class I MHC molecule. The other smaller chain of class I MHC molecule, known as β2 microglobulin (β_2m) is encoded by non-MHC gene located on chromosome 15. In class I MHC molecule α chain pairs with β_2m non-covalently. This pairing is essential for expression of class I MHC on cell surface. In mice class I MHC genes are K , D and L on H-2 complex. As shown in figure 10.2, class I MHC region in mice is split into two parts.

Class II MHC genes are designated as HLA- DP, HLA-DQ and HLA -DR , in humans. They encode two glycoprotein chains α and β. In mice H-2 complex contains two class II MHC regions designated as

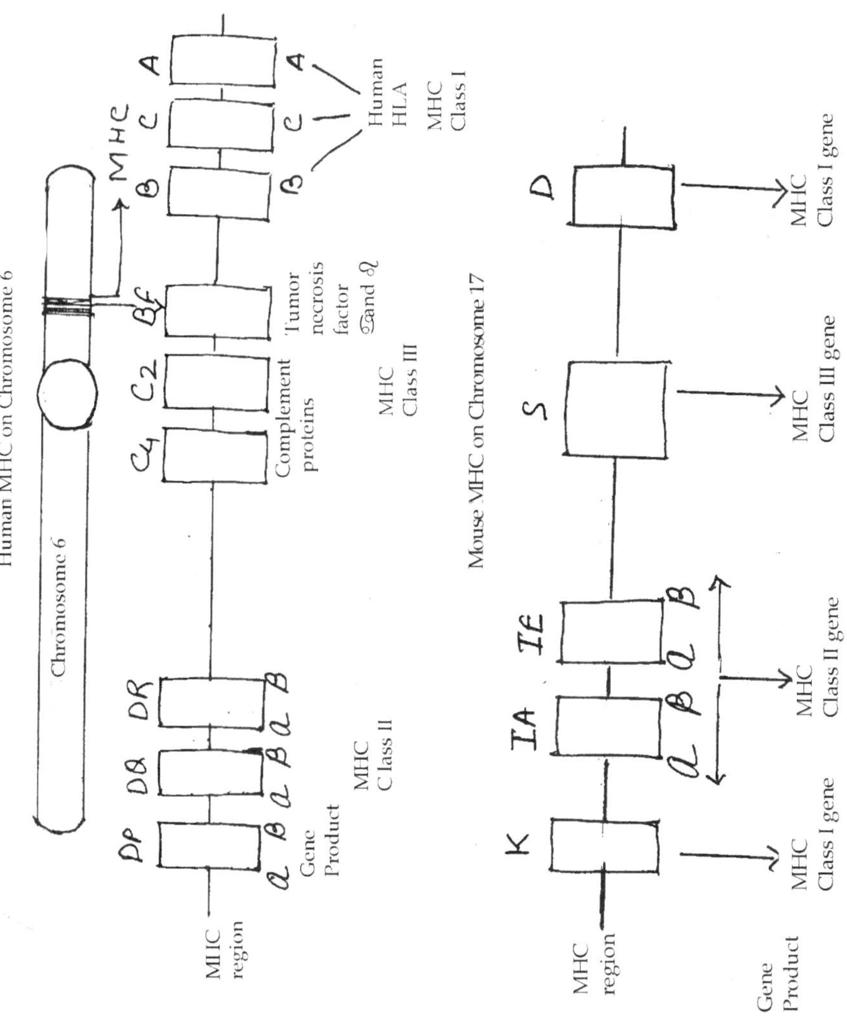

Fig. 10.2 : Organization of MHC gene in Human and Mice.

IA and IE. They encode IA-αβ and IE- αβ molecules. Class II MHC loci also have genes which encode some specialized roles in class I MHC antigen processing pathways.

Class III MHC genes encode products (complement proteins and inflammatory cytokines) which have no role in antigen presentation as class I and class II MHC products. They perform a variety of other functions in immune response.

MHC HAPLOTYPES

MHC are multiple variants of each gene and are highly polymorphic. The genes of MHC loci lie close together. The loci D.B.C and A are situated very close to each other on chromosome 6. For this reason, most individuals inherit the alleles encoded by their closely linked loci, as two set -one from each parent and each set of alleles is referred to as *haplotype*. Thus, an individual inherits one haplotype from mother and other from father. The offsprings are generally heterozygous and express both maternal and paternal MHC alleles and the MHC that are expressed in same cells, are called co dominantly expressed. If mice are inbred, each H-2 locus will be homozygous and all offsprings express identical haplotypes. If two mice, from inbred strains having different MHC haplotypes, are bred to one another, the F1 generation inherits haplotype from both parents. Inbred mice are genetically identical or syngenic and homozygous at every genetic locus. Two strains are congenic if they are genetically identical except at a single locus. For example, a congenic strain AB can have the genetic background of strain A but histocompatibility-2 (H-2) complex of strain B. During production of congenic mouse strains a crossover event sometimes occurs within the H-2 complex, yielding a recombinant strain that differs from parental strains. Detailed analysis of histocompatibility locus revealed that this locus contained several closely linked genes and each responsible for graft rejection.

DIVERSITY OF MHC MOLECULE

Diversity is exhibited by class I and class II MHC molecules. The diversity of MHC within a species stems from polymorphism. The presence of multiple alleles at a given genetic locus occurs within the species. Diversity of MHC molecules in an individual results not only from having different alleles of each gene but also from the presence of duplicated genes within similar functions not unlike the isotypes of immunoglobulins. MHC molecules have similar genes but are not identical in structure and function. Human class I molecule has 59 A alleles 111 B and 37 C alleles. So it is called polygenic. MHC have large number of different alleles at each locus and develop polymorphic genetic complex. These alleles have different DNA sequences at least 5 to 10%. The human class II genes are also polymorphic and there are different gene numbers in different individuals. MHC molecule is not randomly distributed along the entire length of polypeptide. It is clustered within α_1 and α_2 domains of class I and α_1 and β_1 domains of class II. Class III MHC molecules are functionally diverse and free proteins that play a role in immune response. Class III MHC molecules are strongly associated with some genetic disorders.

STRUCTURE OF MHC MOLECULES

MHC Class I and class II molecules are membrane bound glycoproteins and are almost same in structure and function. In both classes, the glycoproteinous membranes have specialized antigen presenting molecules and they develop stable complexes with antigenic peptides which are recognized by T cells. Class III molecules are functionally divers proteins.

Class I MHC molecule

Class I MHC molecules are glycoproteins consist of two polypeptide chains. The longer chain is called α chain. It is a membrane bound glycoprotein with molecular weight of 45kD. It is encoded by class I MHC polymorphic genes. The smaller chain, β_2 microglobulin (β_2m), remains attached to the áchain non-covalently. The β_2m is encoded by a highly conserved gene located on a different chromosome (chromosome 15, in humans). Structurally α chain of class I MHC molecule is organized into three external domains that is α_1, α_2 and α_3, each containing approximately 90 amino acids. The α_1 domain is located at the N-terminal end of the molecule and α_3 is located closest to the cell membrane. α_1 and α_2 α_1 and α_2 domains of a chain contain variable amino acids and form peptidebinding cleft. The β_2 microglobulin and α_3 domains are similar in size and organization and do not contain a transmembrane region and are non-covalently bound to the class I glycoprotein. The β_2 microglobulin interacts with amino acids of α_1 and α_2 class I MHC molecule, stabilized by disulfides bond. In the absence of β_2 microglobulin the class I MHC α is not expressed on the cell membrane. Stable cell surface expression of class I MHC molecule requires the presence of $\alpha\beta_2$ microglobulin chain and a bound antigenic peptide **(Fig. 10.3).**Class I MHC molecules are not capable of delivering intracellular signals.

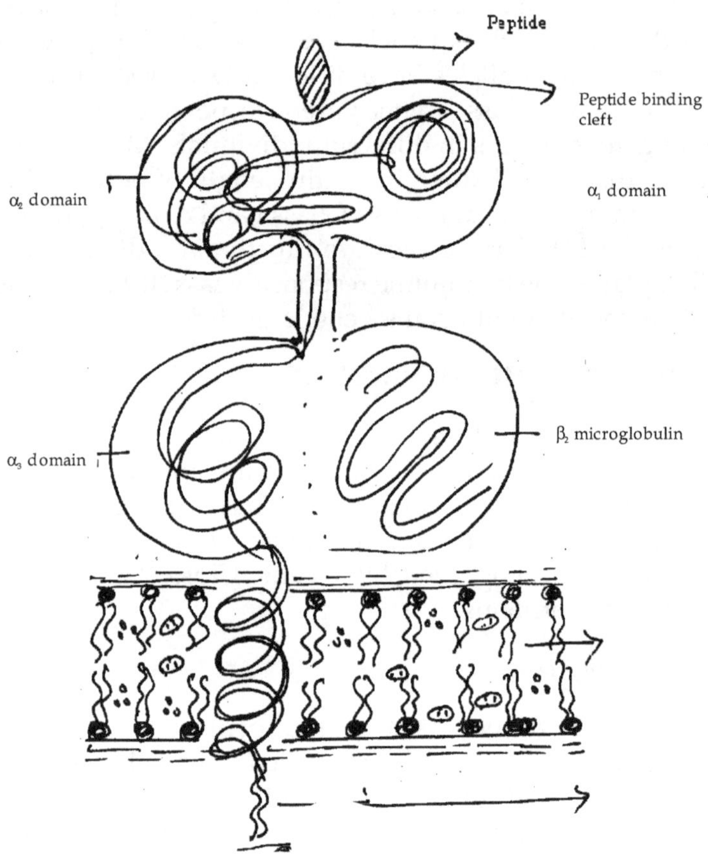

Fig. 10.3 : Diagramatic representation of class I MHC molecule.

Class II MHC molecules

Class II MHC molecules are also made up of two non-covalently associated polypeptide chains-α and β. Both these chains are coded by MHC genes. Both these chains have transmembrane regions and cytoplasmic tails. α chain is a membrane bound glycoprotein and contains two external domains α_1 and α_2. The β chain ,too, contains two external domain β_1 and β_2. The amino terminal of α_1 and β_1 domains of the class II MHC molecules interact and form the peptide binding cleft. α_1 and β_1 domains contain polymorphic amino acids and contribute to variation among different class II alleles in peptide binding. Class II MHC molecules also function in delivering intracellular signals. **(Fig. 10.4).**

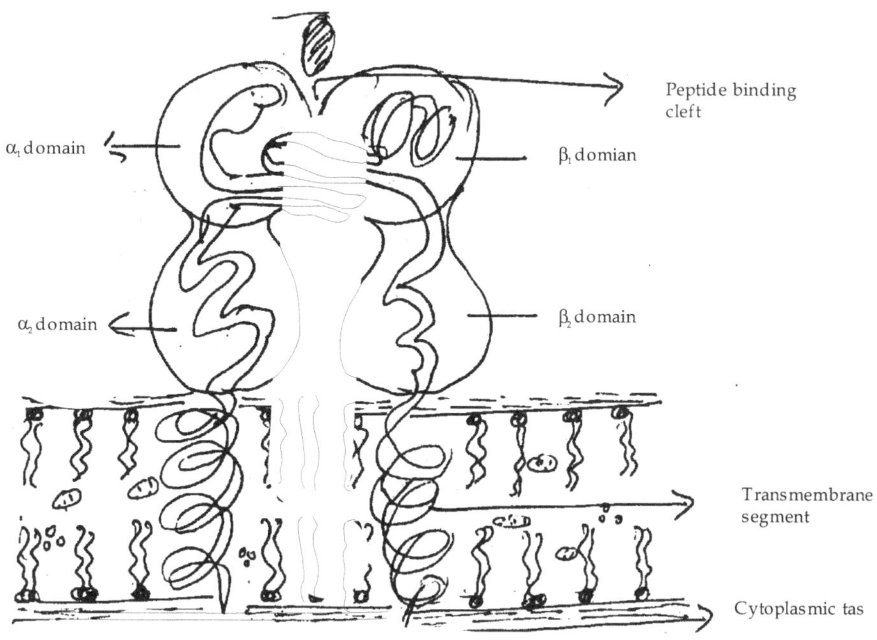

Fig. 10.4 : Diagrammatic representation of class II MHC molecule

PRESENTATION OF ANTIGEN TO T CELLS BY MHC MOLECULES

Both class I MHC molecule and class II MHC molecules have a similar function i.e. they help the T cells to recognize an antigen. Class I MHC molecules present antigen to CD8+ T cells and class II MHC molecules present an antigen to CD4+ T cells. In both cases, there is a groove on the top of the MHC molecules. The small peptides, obtained from processed antigen, bind to the grooves. These are then presented to T cells. T cell receptors recognize these peptides along with MHC molecules. A part of TCR region binds to antigenic peptide and some other part binds to MHC molecule. The variable region of α and β chains of the TCR fold to form antigen/MHC binding site. This binding of TCR to antigen/MHC complex involves same type of non-covalent bonding as occurs in antigen-antibody binding. The recognition of antigen/MHC molecules at the T cells is very specific. It is called MHC restricted recognition of antigen. T cells recognize an antigen only when it is associated with one MHC molecule i.e. it can recognize the same antigen being presented by same MHC molecules. If the same antigen is presented by different MHC molecule, it will be recognized by different T cell. During antigen processing, large protein molecules are broken down into small peptides (7-18 amino acids long). These small peptides get associated with MHC molecules intracellularly and are transported and expressed on cell surface. Whether an antigen will be presented by class I MHC or class II MHC molecules depends on the type of the antigen. For example, endogenous antigens that are produced within the cells are processed and presented by class I MHC molecules while exogenous antigens are presented along with class II MHC molecules.

Presentation by Class I MHC Molecules

During viral infection, a viral infected cell produces viral proteins within its cytoplasm. These proteins are processed by some proteoltyic enzymes and broken down into small peptides of 5 to 15 amino acids. Class I MHC are synthesized in rough endoplasmic reticulum (RER) and transported to cell surface. Antigen derived peptides are transported to RER and bind in the groove of class I MHC molecule. There are some special transporter molecules, called TAP transporters, which help in transporting peptides from cytoplasm to RER. The presence of a peptide in the groove of class I MHC molecule is necessary for the exit of the MHC molecule from ER through secretary pathway. The whole class I MHC molecule along with antigen peptide is then transported to cell surfacein a transport vesicle and is expressed on cell surface**(Fig. 10.5).** If antigen peptides do not bind to a class I MHC molecule they are unable to move to the cell surface.

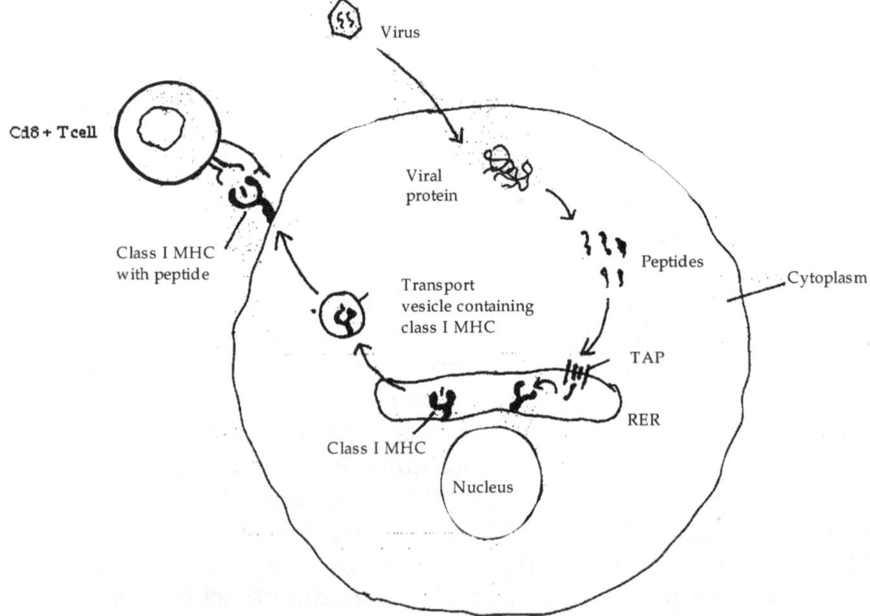

Fig. 10.5 : Processing and presentation of antigen to $CD8^{+}$ T_H cells by class I MHC molecules.

Presentation by Class II MHC Molecules

The exogenous antigens(antigens derived from extracellular pathogens) are taken up by the cells by endocytosis. Endosomes, the vesicles containing these antigens fuse with the lysosomes in the cytoplasm and forms endolysosomes.Inside the endolysosomes, these antigens are degraded into small fragments by proteoltyic enzymes of the lysosomes. Like class I MHC molecules, class II MHC molecules are also synthesized in RER. The peptides obtained from processing of antigens get associated with class II MHC molecules in a special vesicle called CPL(compartment for peptide loading) in the cytoplasm. During their synthesis in RER, in addition to their α and β chains, class II MHC molecules has an extra chain called invariant chain. This chain helps class II MHC molecule to move to CPL. CPL fuses with the endolysosomes. In the CPL, the peptide displaces the invariant chain and becomes bound in the groove of MHC molecule. The class II MHC molecules along with antigen peptide is transported to cell surface and are expressed on cell surface to be presented to the T cells **(Fig. 10.6).**

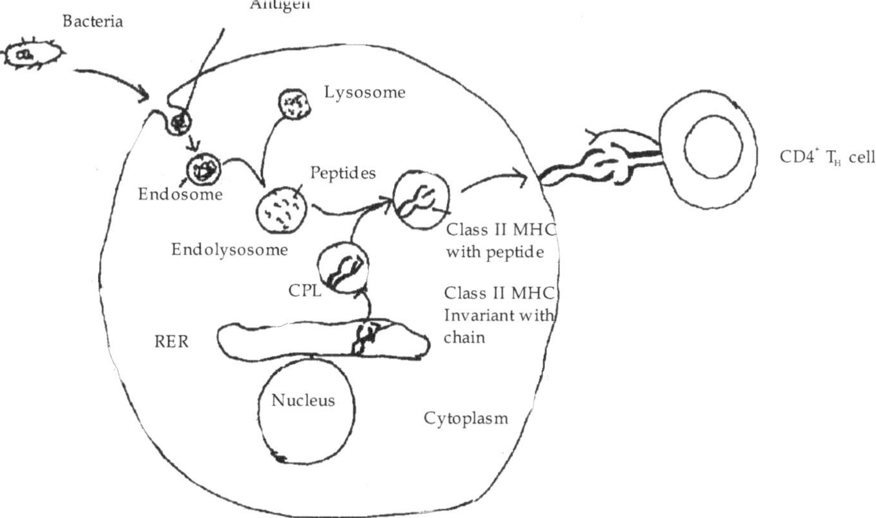

Fig. 10.6 : Processing and presentation of antigen to CD4+ T_H cells by class II MHC molecules.

EXPRESSION AND DISTRIBUTION OF MHC MOLECULES IN THE CELLS

The expression and distribution of MHC molecules is not the same in all cells. They differ in different cells. Class I MHC molecules are expressed almost on all nucleated cells though level of expression varies in different cell types. Differential expression of MHC molecules is likely because of a differential gene expression under regulated manner in different cell types. The highest level of expression is found in lymphocytes whereas the cells like liver cells, muscle cells and neural cells express low levels of these molecules. Expression of many class I MHC molecules on a cell allows the cell to display a large number of peptides in the peptide binding cleft of its MHC molecules and present them to T cells and initiating an enhanced immune response.Expression of class I MHC molecules in all nucleated cells in liver cells is essential in the considerable success of liver cell transplants by reducing the likelihood of graft-recognition by T cell of recipient.

In case of class II MHC molecules, their distribution is more restricted. Class II MHC molecules are expressed only by antigen presenting cells such as macrophage, dendritic cells and B cells.Expression of class II MHC cells, differ at different stage of development. Pre B cells do not express class II MHC but mature B cells express class II MHC. Class II MHC molecules may also be expressed on some non-immune cell types on stimulation by cytokines such as IFNγ.

MHC AND IMMUNE RESPONSIVENESS

A relationship exists between MHC molecules and immune response to an exogenous antigen. An immune responseis measured by the production of serum antibodies which is determined by its MHC. Class II MHC molecule play a main role in immune responsiveness. They present the processed antigen to the T_H cells which help the B cells to produce antibodies.Variability in immune response depends both on the MHC molecules and T cells. Two theories have been proposed to account for the variability in immune responsiveness, observed among different haplotypes. These are determinant selection model and holes in the repertoire model. According to *determinant selection model* structure of MHC molecule determines the strength of antigen association. The MHC polymorphism within a species generates a diversity of specificities. The haplotype of the class II molecule show the highest affinity for binding to a particular peptide.In this respect, different class II MHC molecules differ in their ability to bind to the processed antigen and to present it to T cells. In accordance with *holes in the repertoire model* T cells bear receptors that recognize foreign antigen which closely resembles to self antigen. The concept states that MHC molecules influence the development of the T cell repertoire through selection in the thymus.Functional T cells are capable to recognize the available MHC antigen complexes but they are absent because of clonal selection and active suppression in the thymus. In this respect, the T cell response to an antigen involves trimoleculer complex of the T cell receptor, an antigenic peptide and an MHC molecules.

REGULATION OF MHC EXPRESSION

Expression of MHC on different cell types is regulated by many factors at different levels. For example, some factors act at transcription level. There are transcription factors like CIITA and RFX which bind to the promoter regions of class II MHC genes and promote the synthesis of class II MHC molecules.

Many of the cytokines also play important role in regulation of expression of MHC genes. Cytokines like IFNα, IFNβ, and IFNγ and TNF up regulate the expression of class I MHC molecules. They increase the expression of class I MHC molecules. IFNγ stimulates the synthesis of some transcriptional factors which bind to the promoter regions of the MHC genes and stimulate the synthesis of MHC molecules. It also induces the expression of CIITA, transcriptional activator for class II MHC genes.

Certain viruses like human cytomegalovirus (CMV). HBV and adenovirus12(Ad12) also influence expression of MHC molecules. These viruses cause decreased expression of class I MHC on cell surface.

MHC AND DISEASE SUSCEPTIBILITY

There are so many diseases like autoimmune disease, viral disease, neurological disease and different types of allergies that are associated with MHC alleles. A much higher frequency of MHC has been observed in the people suffering from these diseases. This may cause differences in immune responsiveness due to variation in antigen presentation by different MHC alleles which determines the effectiveness of the immune response to pathogen. Various MHC alleles also code for binding site for specific viruses and bacteria. But the presence of an association between an MHC allele and a disease cannot be regarded as the cause of the disease as this association is a complex phenomenon. Multiple genetic and environmental factors have roles in the development of disease and autoimmune disease. This can be seen in identical twins who inherit MHC risk factors but both may not develop a disease.

POINTS TO REMEMBER

- A group of linked MHC genes is generally inherited as a unit from parents.
- MHC genes are located on chromosome 6 in humans and on chromosome 15 in mice.
- In humans they are also called HLA (human leukocyte antigen) complex and in mice these are known as H-2 complex.
- MHC genes encode three classes of molecules known as class I, class II and class III MHC molecules.
- The role of MHC molecule is to bind and present foreign antigens to T cells and thereby triggering a specific immune response.
- Class I MHC molecules consist of two glycoprotein chains- α and β_2m. Class II MHC molecules are composed of two non-covalently associated glycoproteins-α and β chains.
- Class III MHC molecules include a diverse group of proteins including complement.
- Class I molecules are expressed on most nucleated cells. Class II antigens are restricted to B cells, macrophages and dendrite cells.
- The outstanding feature of the MHC molecules is their extensive polymorphism.
- Most MHC allele differ from one another by multiple amino acid substitutions and these differences are focused on the peptide binding site and adjacent regions that make direct contact with the T cells receptor.
- Both class I MHC molecule and class II MHC molecules help the T cells to recognize an antigen.
- Class I MHC molecules present the processed endogenous

antigens to the T cells. Peptide derived from processed antigen binds to the class I MHC molecules are recognized by T cells in association with MHC molecules.

- Class II MHC molecules present the processed extracellular antigens to the T cells.
- The expression and distribution of MHC molecules is not the same in all cells. They differ in different cells.
- An association is observed between MHC molecules and immune response to an antigen.
- Two theories have been proposed to account for the variability in immune responsiveness, observed among different haplotypes: determinant selection model, and holes in the repertoire model.
- Expression of MHC on different cell types is regulated by many factors at different levels. These include transcription factors, cytokines and certain viruses.
- Many diseases are associated with MHC alleles.

REVIEW QUESTIONS

1. How do MHC molecules contribute to allograft rejection in mammals?
2. Describe general organization of MHC genes.
3. Describe the structure of class I MHC molecules.
4. How do class I MHC molecules present the antigen to the T cells? How does it differ from antigen presentation by class II MHC molecules?
5. Explain : MHC molecules' diversity.
6. Explain : Regulation of MHC expression
7. Write short notes on:
   i. MHC and immune responsiveness
   ii. MHC haplotypes

Chapter - 11

Cytokines

Cytokines are the chemical messengers that are involved in interactions between the cells of innate and adaptive immune responses. They are secreted by leukocytes and many other cells in the body. They are low molecular weight proteins or glycoproteins which are produced in response to various kinds of stimuli. Cytokines have important role in immune and inflammatory responses. Their functions are not limited to immune system only.They also stimulate the development of hematopoietic cells and also act as therapeutic agents.

Though, like hormones, cytokines act as chemical messengers but they differ from hormones in their mode of action. While hormones are carried via circulation and act away from their site of production (endocrine), cytokines act over both short and long range. They act mostly on the cells which are in close position to the producer cell (*paracrine*) or the same cell that has produced them (*autocrine*). However, a few cytokines act on distant cells and organs (*endocrine*) **(Fig.11.1).**

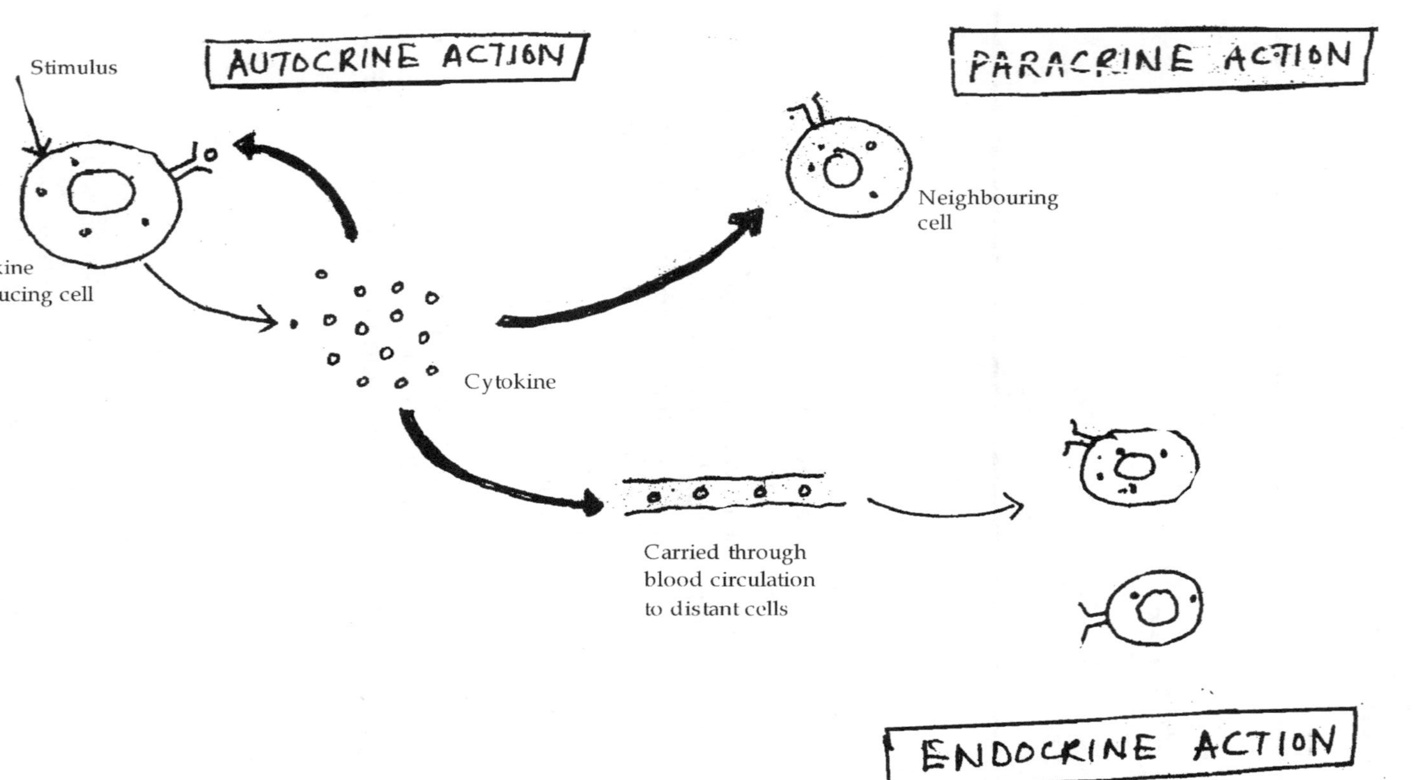

Fig. 11.1 : Different ways of actions of cytokines

GENERAL PROPERTIES

Cytokines are antigen nonspecific protein/glycoprotein molecules which are highly potent in their action. A cytokine may act on different cell types and can have different biological roles on different cells. This is called *pleiotropism*. Many cytokines, on the other hand, have similar biological functions (*redundancy*). Often the combined biological effect of two cytokines is greater than the summation of the effects of individual cytokine. Thus, they act *synergistically*. Cytokines also show *antagonistic* behavior i.e. one cytokine can inhibit the effects of another cytokine(**Fig. 11.2**).

Cytokines act by binding to their specific receptors on the cells. Binding leads to change in gene expression and modifies cell functions, initiating proliferation, migration or apoptosis (programmed cell death). Cytokines bind to their receptors with very high affinity and thus even a very minute quantity of cytokines can bring about the biological function. Their production is induced in response to a stimulus and is transient and highly regulated.

Types of Cytokines

Cytokines can be grouped into the following categories:

i. Interferons (IF)

ii. Interleukins (IL)

iii. Colony Stimulating Factors (CSF)

iv. Chemokines

v. Tumor Necrosis Factor (TNFα)

vi. Transforming Growth Factors (TGF-β)

Interferons

Interferons are antiviral proteins, produced very early in infection and are specifically important in preventing/delaying the spread of viral infections. There are two groups of interferons:

i. Type I which includes IFNα and IFNβ

ii. Type II or IFNγ

IFNα is a group of 20 closely related proteins which are produced by mononuclear phagocytes (macrophages, monocytes, lymphocytes and dendritic cells). IFNβ is a single protein, produced by many cells such as fibroblasts, some epithelial cells and dendritic cells. Type I

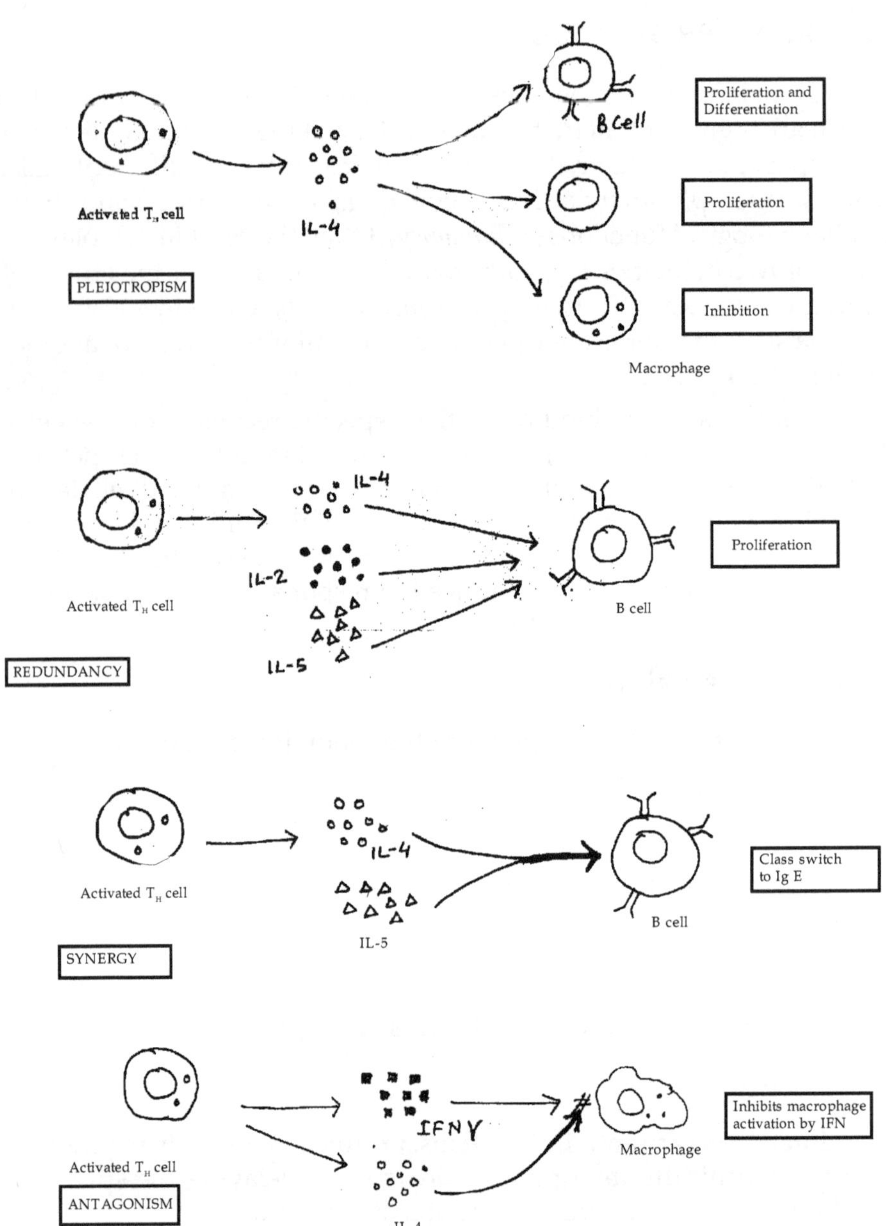

Fig. 11.2 : Properties of Cytokines

interferons (IFNα and IFNβ) are secreted by viral infected cells and other cells in the body, in response to dsRNA, produced by viruses during replication in the infected cells. They act both on viral infected cells, inhibiting viral replication in the cell and on uninfected adjacent cells, making them resistant to viral infection (antiviral state). They increase expression of class I MHC molecules on virus infected cells which help CD8+ cytotoxic T lymphocytes (CTL) to recognize and kill them. Type I interferons inhibit cell proliferation and thus may be used as anti-tumor agent.

Type II interferon, IFNγ is a single protein which is produced by NK cells and T_{H1} cells. It plays important role both in innate and cell mediated immunity. IFNγ activates macrophages to kill phagocytosed microbes. It induces differentiation of T_{H1} cells and inhibits proliferation of T_{H2} cells. IFNγ activates endothelial cells, induces T cell adhesion molecules and extravasation to the infection site. IFNγ stimulates antibody production and promotes class switching in B cells. It increases expression of both MHC class I and class II molecules.

Interleukins (IL)

Interleukins are a major group of cytokines, secreted by T cells and by mononuclear phagocytes. They are designated as IL-1, IL-2 etc. About 33 interleukins have been identified. They have wide variety of roles. IL-1 is the principal mediator of host inflammatory response to infection. IL-1 exists in two forms: IL-1α and IL-1β. Both forms bind to same cell surface receptors. At low concentration, IL-1 increases expression of cell surface adhesion molecules by endothelial cells. At high concentration, it induces fever and synthesizes acute phase plasma proteins in liver, effects similar to as caused by TNF. Interleukin-2 (IL-2), originally known as T cell growth factor, has a vital role in T cell proliferation. It is mainly produced by $CD4^+$ T cells. It acts on the same $CD4^+$ T cells which secrete them (*autocrine*) and nearby T cells (*paracrine*). IL-2 causes increase in the production of IFNγ and IL-4 from T cells. It stimulates growth and cytotoxic activity of NK cells and also stimulates B cell proliferation and antibody synthesis. It also promotes cell survival by inducing anti-apoptotic protein Bcl-2. IL-4 is the major stimulus for IgE production and T_{H2} cell development. It is produced by $CD4^+$ T cells, mast cells, basophils and also by neutrophils. It is a principal mediator of immediate hypersensitivity (Type I) hypersensitivity. IL-6, synthesized by mononuclear phagocytes, fibroblasts and other cell types in response to microbes and other cytokines, plays role both in innate and adaptive immunity. It stimulates synthesis of acute phase proteins by liver and also stimulates B cell differentiation and antibody formation. IL-6 also acts as growth factor for neoplastic plasma cells. IL-12 is produced by activated macrophages and dendritic cells. It is an important mediator of early innate immunity against intracellular microorganisms. It stimulates the secretion of IFNγ by NK cells and T cells. It also enhances lipolytic activities of NK cells and $CD8^+$ T cells (CTLs). Functions of some other interleukins are given in **table 11.1.**

Colony Stimulating Factors (CSF)

These are hematopoietic cytokines which are produced by activated T cells, macrophages, endothelial cells and bone marrow stem cells. They stimulate differentiation of bone marrow stem cells and increase production of inflammatory leukocytes. There are three main types of CSFs-GMCSF (Granulocyte Macrophage CSF), MCSF (Macrophage CSF) and GCSF (Granulocyte CSF). They have various functions. MCSF selectively stimulate the development of monocytes in bone marrow and macrophages in tissues. GCSF promotes the development of granulocytes. It increases production of neutrophils during infection. GMCSF not only promotes growth of granulocytes but also stimulates maturation of bone marrow cells and monocytes and dendritic cells.

Chemokines

Chemokines are a large group of cytokines that are involved in chemotaxis. They are produced by monocytes, macrophages, T cells, neutrophils, endothelial and epithelial cells and also by fibroblasts. Chemokines regulate migration of B cells and other leucocytes to different lymphoid tissues. They direct the migration of leucocytes, especially neutrophils, monocytes and lymphocytes to the site of infection. There are two subgroups of chemokines - a group with two adjacent amino terminal cysteine residues (CC chemokines) and another group with two cysteine, separated by an amino acid (CXC chemokines). CC chemokines induce migration of monocytes into the tissues where monocytes are transformed into macrophages. CXC chemokines act mainly on neutrophils and cause chemotaxis.

Tumor Necrosis Factor (TNFα)

Tumor Necrosis Factor alpha (TNF alpha), is an inflammatory cytokine produced by macrophages/monocytes during acute inflammation. Besides macrophages, TNFα is produced by $CD4^+$ T cells, NK cells and mast cells, when activated by bacterial lipopolysaccharides (LPS). It has a variety of roles and is potentially important in mediating inflammation and cytotoxic reaction. It is responsible for a diverse range of signaling events within cells, leading to necrosis or apoptosis. The protein is also important for resistance to infection and cancers. It promotes the recruitment of neutrophils and monocytes to the site of infection. It causes activation of endothelial cells and stimulates them to secrete chemokines. It also stimulates mononuclear phagocytes to release IL-1, IL-6, TNF and chemokines. It increases synthesis of acute phase proteins. It induces fever by increasing the synthesis of prostaglandins. At high level, TNF causes abnormal fall in blood pressure and also fall in blood glucose due to over utilization of glucose by the muscles.

Transforming Growth Factors (TGF-β)

TGF-β, secreted by antigen activated T cells, is mainly involved in inhibiting proliferation and activation of lymphocytes and other leucocytes. It is secreted in latent form in which it is complexed with two other polypeptides, latent TGF-beta binding protein (LTBP) and latency-associated peptide (LAP). Serum proteinases such as plasmin catalyze the release of active TGF-β from the complex. Inflammatory stimuli that activate macrophages enhance the release of active TGF-β by promoting the activation of plasmin. It also stops the activation of macrophages. TGF-β is a group of closely related molecules such as TGF-β1, TGF-β2 and TGF-β3.

Table No. 11.1 : Major cytokines and their effects

GROUP	CYTOKINE	SOURCE	EFFECTS
Interferons	**IFN-α**	Leukocytes, fibroblasts,	Induces antiviral state in most cells, increases expression of class I MHC molecules
	IFN-β	fibroblasts, epithelia	Induces antiviral state in most cells, increases expression of class I MHC molecules
	IFN-γ	T cells, NK cells, fibroblasts, epithelia	Activation of macrophage, increases expression of class I and class II MHC molecules
Interleukins	**Il-1**	B cells, fibroblasts, macrophage	Inflammation, induction of acute phase proteins
	Il-2	T cells	T cell proliferation, activation and proliferation of NK cells and B cells

Contd...

Il-3	T cells	Stem cell and myeloid progenitor cell growth
Il-4	T cells	IgE production and T_{H2} cell development
Il-5	T cells	Activation and generation of eosinophils
Il-6	T cells, B cells, fibroblasts, macrophage	Synthesis of acute phase proteins by liver , B cell differentiation and antibody formation
Il-7	Bone marrow stromal cells	Survival of T cells
Il-8	Monocytes, fibroblasts	Also known as neutrophil chemotactic factor, migration of neutrophils to the site of infection, phagocytosis
Il-9	T cells	Regulator of hematopoietic cells.
Il-10	T cells	Anti-inflammatory cytokine,enhances B cell survival, proliferation, and antibody production.
Il-11	bone marrow stromal cells, fibroblasts	Growth factor for megakaryocytes
Il-12	Monocytes	Mediator of early innate immunity, secretion of IFNγ by NK cells and T cells
Il-13	T cells	Stimulates growth and differentiation of B cells, inhibits T_{H1} cells
Il-14	T cells	Growth and proliferation of B cells
Il-15	Monocytes	Induces production of NK cells
Il-16	$CD8^+$ T cells, eosinophils	CD4+ chemoattractant
Il-17	$CD4^+$ T cells	Inflammation
Il-18	Macrophage	Induces production of IFNγ, increase activity of NK cell
IL-19	Monocytes	Monocyte cytokine release and monocytes apoptosis
Il-20	Keratinocytes	Proliferation and differentiation of keratinocytes
Il-21	T cells, mast cells	Co-stimulates activation and

Contd...

			proliferation of CD8+ T cells
Colony Stimulating factor	**M-CSF**	Macrophage	Development of monocytes and macrophages
	G-CSF	Macrophage	Development of granulocytes, increases production of neutrophils during infection
	GM-CSF	Macrophage , T cells	Production of macrophage and granulocytes, maturation and activation of dendritic cells
Tumor Necrosis Factor	**TNF- α**	Macrophage , T cells, lymphocytes	Inflammation, cytotoxic reaction
	TNF-β	Lymphocytes	Controls cellular proliferation, differentiation,
Transforming Growth factor	**TGF- β**	Macrophage	Inhibits proliferation and activataion of T cells
Chemokines	**BCA,BRAK, MCP,GCP**	Monocytes	Chemotaxis, migration of leukocytes to the site of action

CYTOKINE RECEPTORS

Cytokines act by binding to their specific receptors, present on different cells. The receptors consist of one or more transmembrane proteins**(Fig. 11.3)**. The extracellular part of these proteins bind to the cytokine molecule and the cytoplasmic portion initiates intracellular signaling pathways. Structurally, cytokine receptors can be classified into five families:

i. Immunoglobulin super family receptors
ii. Class I cytokine receptor family
iii. Class II cytokine receptor family
iv. TNF receptor family
v. Chemokine receptor family

Immunoglobulin Super Family Receptors

It includes the receptors for IL-1. There are two different receptors for IL-1: Type I IL-1R and Type II IL-1R. Type I IL-1R receptors are expressed on many types of cells. Type 2 IL-1R receptors are expressed on B cells. These receptors contain extracellular immunoglobulin domains and bind to diverse group of cytokines.

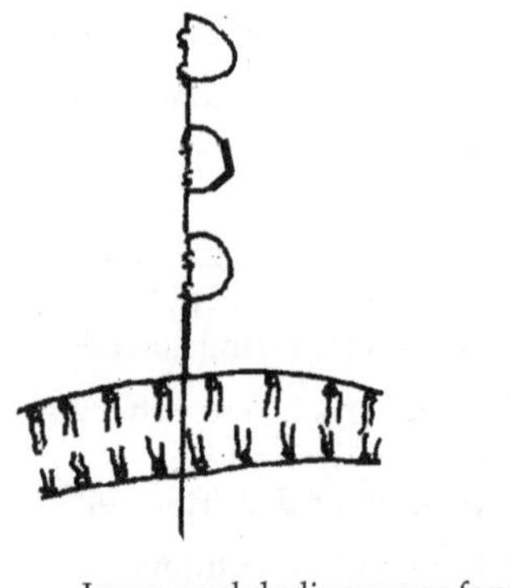

Immunoglobuline super family receptor

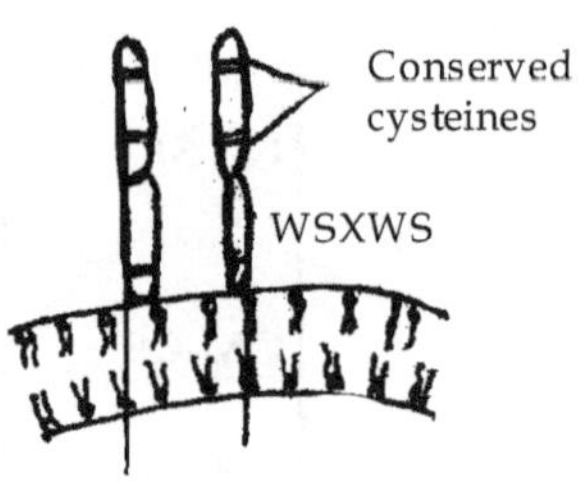

Class I Cytokine receptors

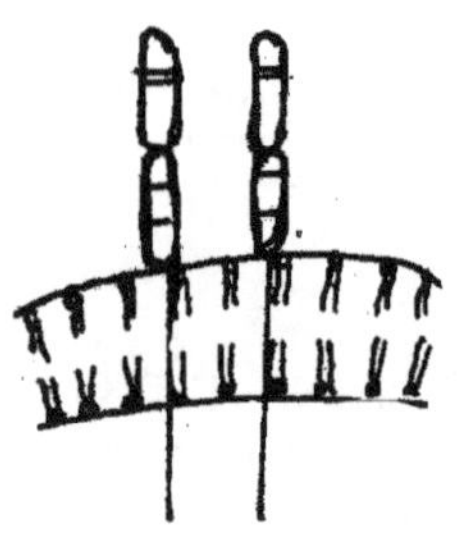

Class II Cytokine receptors

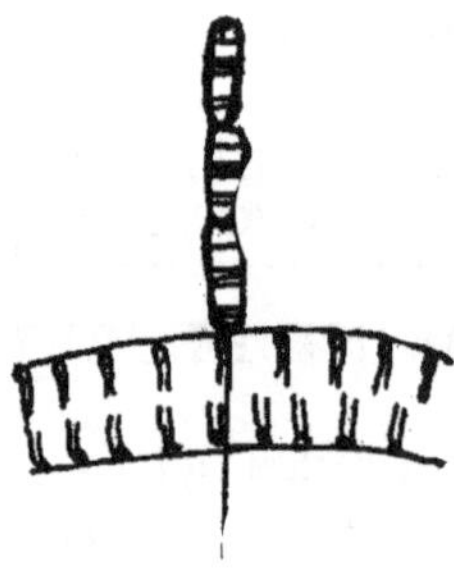

TNF receptors

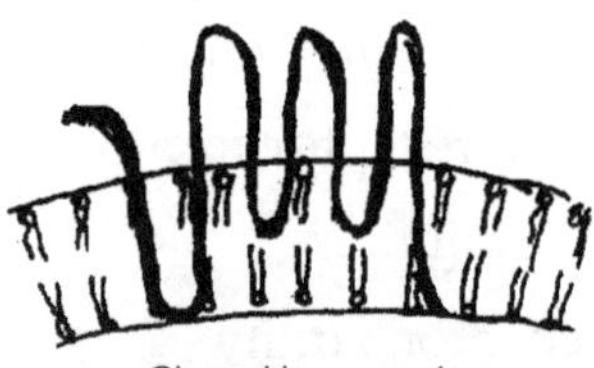

Chemokine receptors

Fig. 11.3 : Diagrams to show structures of cytokine receptors

Class I Cytokine Receptor Family

These are the receptors for hematopoietic cytokine, IL-2 and 13 other interleukins and some colony stimulating factors and growth hormone. These receptors are also called *hematopoeitin receptors*. The receptor for IL-2 consists of three chains α, β and γ. In addition to IL-2 receptors, IL-15 receptors are also trimeric in structure i.e. consist of three chains, whereas other receptors of class I family are dimeric in nature. All these receptors share a common γ chain. This chain functions in signaling. Il-2 receptors are expressed only at activated T cells i.e. the T cells which have been activated by antigens. These have conserved amino acid sequence motifs in the extracellular domains. It consists of four conserved cysteine residues and a conserved sequence of tryptophan-serine-tryptophan-serine (WSXWS), where X is not conserved i.e. any type of amino acid.

Class II Cytokine Receptor Family

These are the receptors for different types of interferons and interleukins. They bind to 27 different types of cytokines which include members of IL-10, Type I interferons and Type II interferons and members of interferon λ. They, too, consist of conserved cysteine residue motif but lack WZXWS motifs, as present in Type I cytokine receptors.

Tumor Necrosis Factor Receptor

Tumor necrosis factor receptor, also known as *death receptor,* is a trimeric cytokine receptor that binds tumor necrosis factors (TNF). It is activated by at least 18 different human homologues of TNF, referred to as the *TNF super family*. Signaling through these receptors is complex and involves interaction with cytoplasmic adaptor proteins. Some of these homologues lack transmembrane and cytoplasmic domains. They function as decoy receptors i.e. binding of ligand does not induce cell signaling.

Chemokine Receptors

Chemokine receptors are different from other cytokine receptors. Because of their specific structure, they are also known as *serpentine receptors*. They are grouped as CCR which recognize CC chemokines and CXCR which recognize CXC chemokines.

MECHANISM OF ACTION OF CYTOKINES

After binding to their receptors, cytokines initiate a series of events which lead to their specific effects such as antiviral effects, in case of IFN-γ. In case of class I and class II cytokine receptors, the receptors are composed of two separate subunits out of which one chain is required for cytokine binding and signal transduction while the other chain is mainly essential for signaling and has very little role in binding. A family of inactive proteins, called *Janus Kine (JAK)*, is associated with α chain of the receptor. It is activated in the presence of cytokine. These proteins have two functional sites. One binds to the cytokine receptors subunit while second is a catalytic site which exhibits tyrosine kinase activity. Binding of cytokine with its receptor brings about the association of two cytokine receptor subunits. A number of transcription factors, known as STATS (signal transducers and activators of transcription) bind to the phosphorylated tyrosine residues. Activated JAK phosphorylate STATS. It is followed by dissociation and dimerization of STATS. The phoshorylated dimmers of STATS are translocated into the nucleus where they induce expression of specific genes**(Fig. 11.4).** The expression of these genes mediates the specific effects of a particular cytokine. Defects in the cytokine signaling pathways may cause severe combined immunodeficiency (SCID). Mutation, in the γ chain of different receptors like IL-2, IL-4, IL-7, IL-9, IL-15 and IL-21, results in the failure of activation of JAKs.

CYTOKINE ANTAGONISTS

There are many protein molecules which inhibit the activity of cytokines. These are called *cytokine inhibitors.* They either bind to a cytokine receptor or bind directly to a cytokine and inhibit its activity. For example, IL-1 receptor antagonist (IL-1Ra) is a naturally occurring cytokine inhibitor. It is produced by mononuclear phagocytes. It binds to the IL-R on T cells. This prevents the binding of IL-1 to the receptor and blocks the activation of T cells. Some inhibitors bind to the cytokine ligand and reduce the concentration of cytokines, available to bind to the receptors.

OVER SECRETION OF CYTOKINES

Though cytokine play important roles, both in innate and adaptive immunity, but the excessive production of cytokines may be detrimental to the organisms. Over secretion of cytokines may lead to conditions like bacterial toxic shock and bacterial septic shock. Bacterial septic shock arises due to excessive production of TNFα ,IL-1 and other cytokines. Infection by Gram negative bacteria such as *Pseudomonas aeruginosa* and *E.coli* can lead to excessive production of these cytokines by activated macrophages. The condition is characterized by fall in blood pressure, fever and widespread intravascular coagulation. Bacterial toxic shock is a potentially fatal illness, caused by a bacterial toxin. Enterotoxins of some Gram positive bacteria such as

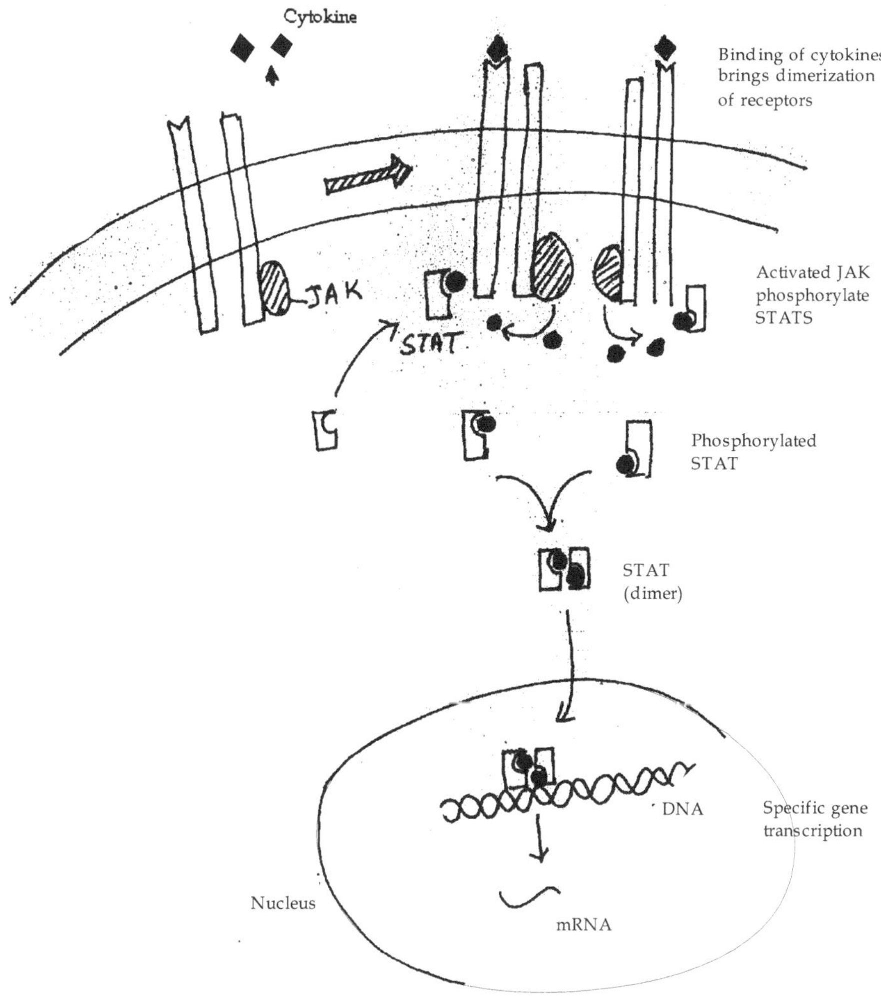

Fig. 11.4 : Mechanism of Action of Cytokines

Staphylococcus aureus and *Streptococcus pyogenes* can activate CD4+ T cells. This causes excessive production of TNF and IL-1. Symptoms of bacterial toxic shock syndrome vary, depending on causative organism. Infection with *Staphylococcus aureus* typically causes high fever, accompanied by low blood pressure, malaise and confusion which can rapidly progress to stupor, coma, and multiple organ failure. In case of *Streptococcus pyogenes*, infected individuals often experience severe pain at the site of the skin infection, followed by rapid progression of symptoms.

Besides having their roles in immune system, some cytokines have been associated with cancers. For example, IL-6 is associated with the development of many types of cancers. Abnormally high levels of IL-6 is secreted by different types of cancerous cells such as myeloma and plasmacytoma cells and cells of cervical and blood cancer.

THERAPEUTIC USE OF CYTOKINES

Some cytokines have therapeutic value and are used successfully in clinical therapy. Interferon α was the first interferon, used successfully. It has use in the treatment of hairy cell leukaemia (a rare blood cancer occurring mainly at older age), AIDS related Kaposi sarcoma, genital warts, hepatitis b and hepatitis c. In case of hepatitis c, interferons help to reduce the viral load. The effects may vary from patient to patient. Interferon β is used in the treatment of multiple sclerosis, an autoimmune disease. IFγ is used in the treatment of a rare inherited immune disorder, chronic granulomatous disease (GCD) which is characterized by recurrent infection by several bacteria. IFγ significantly reduces the incidence of infections. IFγ can also be used effectively in the treatment of osteopetrosis, a congenital disease in which overgrowth of bones occurs. In rheumatoid arthritis, anti IFNα drugs have proved to be very effective. They reduce pain, stiffness and joint swelling. These drugs also help in tissue repairing.

G-CSF and GM-CSF can be helpful to restore bone marrow functioning after bone marrow transplantation. IL-2 can be useful in the treatment of melanoma and renal cell carcinoma. Some studies have shown that IL-2 has potential in the treatment of AIDS. Studies have also revealed that it can help to restore CD4+ T cells in AIDS patients. Some soluble cytokines also help in the development and differentiation of RBCs and WBCs (hematopoiesis). Cytokines act as development signals and direct commitment of progenitor cells into their specific lineage. Many cytokines such as IL-3, IL-5, IL-6, IL-7, IL-11, G-CSF and GM-CSF act at different stages of hematopoiesis and stimulate production of different kinds of blood cells as shown in the **fig.11.5.**

Though interferon therapy is used in the treatment of many diseases, the use of interferons has some limitations. Short life of cytokines limit their use for clinical purpose. In addition, cytokines can cause several undesirable side effects. The side effects include flu-like symptoms, depression, irritation at the site of injection and weight loss. In some cases, cytokines may also cause nausea, respiratory distress or coma.

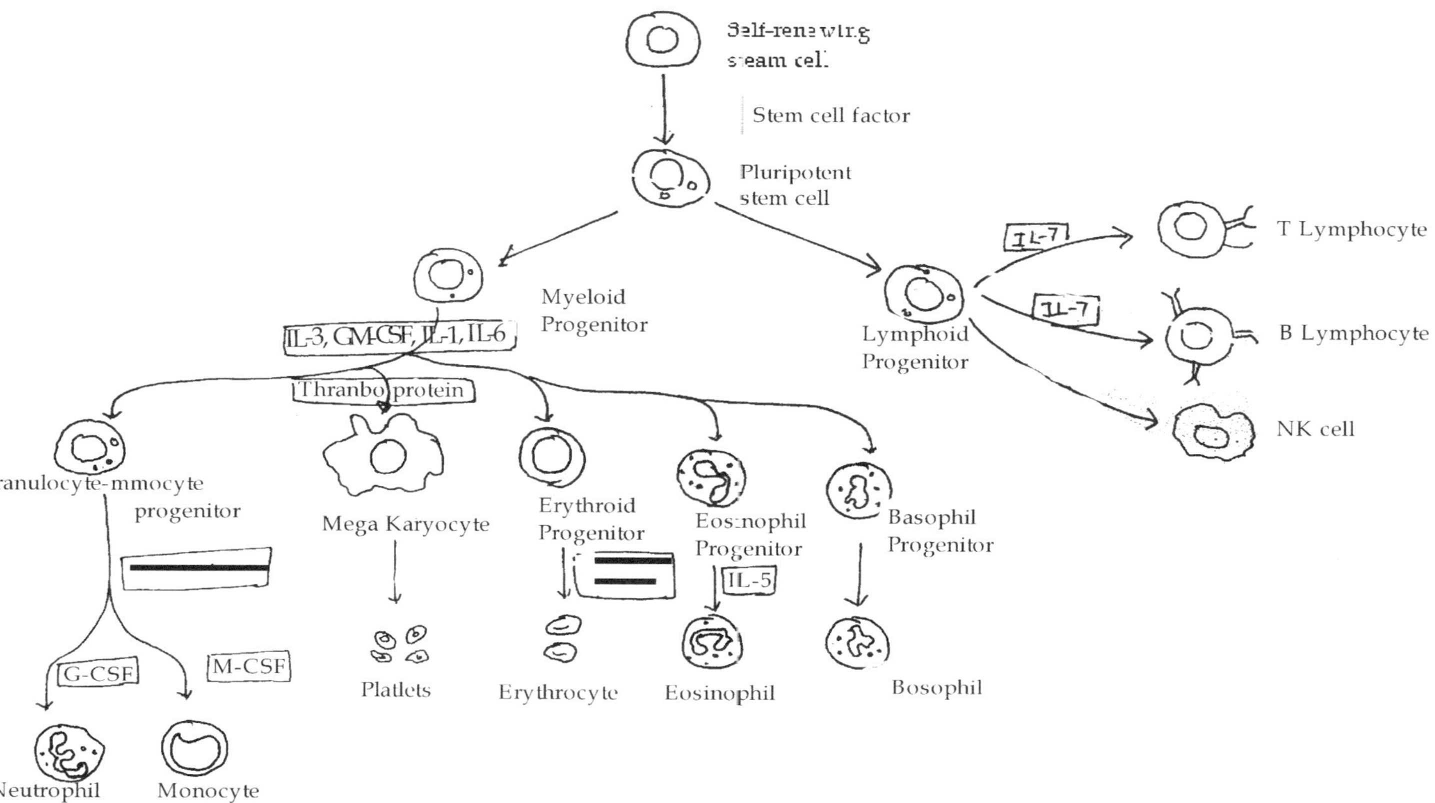

Fig. 11.5 : Role of cytokines in hematopoiesis

POINTS TO REMEMBER

- Cytokines are the chemical messengers that are secreted by leukocytes and many other cells in the body.
- Cytokines have important role in immune and inflammatory responses.
- They act in paracrine, autocrine and endocrine manner.
- Cytokines show pleiotropism and redundancy. They also act synergistically and also show antagonistic behavior.
- Cytokines can be grouped into the various categories: Interferons (IF), Interleukins (IL), Colony stimulating factors (CSF), Chemokines, Tumor Necrosis Factor (TNFα) and Transforming growth factors (TGF-β)
- There are two groups of interferons: Type I which includes IFNα and IFNβ and Type II or IFNγ.
- IFNα and IFNβ act as antiviral proteins and inhibit viral replication. IFNγ stimulates antibody production and promotes class switching in B cells.
- Interleukins are secreted by T cells and by mononuclear phagocytes.
- IL-1 is the principal mediator of host inflammatory response to infection and exists in two forms: IL-1α and IL-1β.
- IL-2 has a vital role in T cell proliferation.
- IL-4 is the major stimulus for IgE production and T_{H2} cell development.
- Colony stimulating factors are hematopoietic cytokines which are produced by activated T cells, macrophages, endothelial cells and bone marrow stem cells and stimulate differentiation of bone marrow stem cells and increase production of inflammatory leukocytes.
- Chemokines are the cytokines that are involved in chemotaxis.
- TNFα has a variety of roles and is potentially important in mediating inflammation and cytotoxic reaction.
- TGF-β is mainly involved in inhibiting proliferation and activation of lymphocytes and other leucocytes.
- Cytokine receptors can be classified into five families: i.

Immunoglobulin super family receptors, ii. Class I cytokine receptor family, iii. Class II cytokine receptor family, iv.TNF receptor family and v. Chemokine receptor family.

- Cytokines act by binding to their specific receptors and by initiating a series of events which lead to their specific effects.
- Cytokine antagonists are the molecules that inhibit the activity of cytokines.
- Over secretion of cytokines may be detrimental to the organisms and may lead to conditions like bacterial toxic shock and bacterial septic shock.
- Some cytokines have therapeutic value and are used in the treatment of some diseases.

REVIEW QUESTIONS

1. What are cytokines? Discuss their general properties.
2. Explain: Interferons and their role in immune system.
3. How do cytokines make their effects?
4. Describe various cytokine receptor families.
5. Write short notes on :
   i. Colony stimulating factors
   ii. Cytokine antagonists
   iii. Bacterial toxic shock

CHAPTER - 12

Hypersensitivity

The immune system has been designed to provide protection to the body from various diseases. But the same system which is aimed to provide protection may sometimes become harmful to the body and may even life-threatening. There are situations when immune response against harmless and inert antigens is exaggerated and actually causes diseases. Such a condition is called hypersensitivity or commonly as allergy. Though the term 'hypersensitivity' indicates an increased response but the immune response is not always heightened, instead it may be an inappropriate response to an antigen.

TYPES OF HYPERSENSITIVITY

Hypersensitivity reactions may be broadly categorized into two types: Immediate hypersensitivity and delayed type hypersensitivity. In case of **immediate hypersensitivity** the symptoms develop within minutes or hours after the exposure of antigen to a sensitized person. These reactions occur within the humoral branch of immune response. Such reactions develop by antibody or antigen-antibody complexes. In **delayed type hypersensitivity** the symptoms may take several days, after exposure, to develop. Such reactions occur within the cell-mediated immune responses. These reactions are mediated by T cells and not by antibodies.

Immediate type hypersensitivity can be further divided into three types:

1. **Type I or IgE-mediated hypersensitivity** which is commonly called as *allergy.*
2. **Type II or antibody-mediated hypersensitivity** is caused when cytotoxic antibodies IgG or IgM bind to antigens in the

normal or modified tissue components. Such reactions occur by activating complement or phagocytosis by neutrophils.

3. **Type III or immune-complex mediated hypersensitivity** is caused by deposition of antibody-antigen complex in the blood vessels of various tissues **(Fig. 12.1)**.

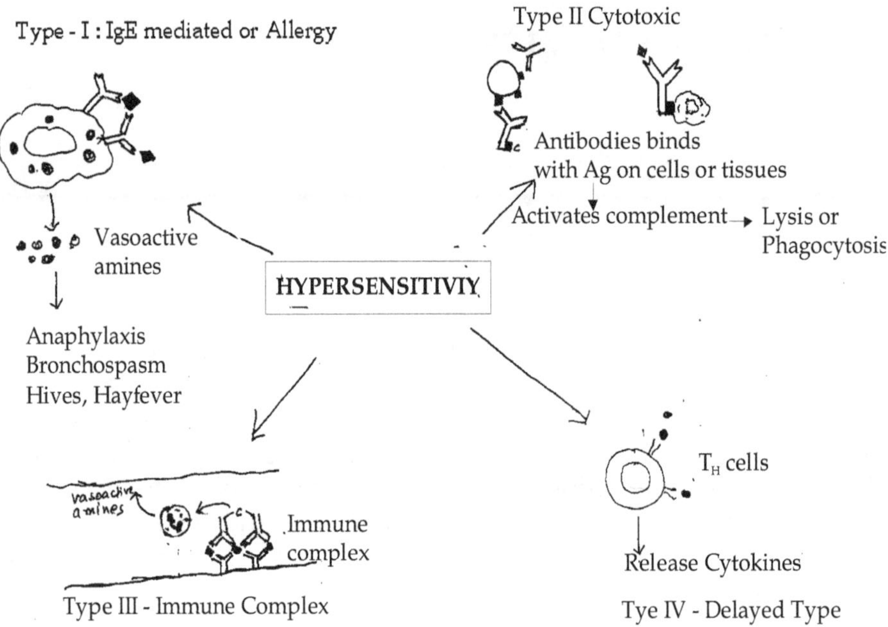

Fig. 12.1 : Different types of Hypersensitivity

Type I or IgE-Mediated Hypersensitivity

This is mediated by the production of IgE and is commonly known as allergy. The substances (antigens) that cause such responses are called allergens. There are several sources of allergens such as pollens, derived from plants, trees, and grasses; insects such as cockroach, wasp sting, bee sting, fire ants ; animal fur and excretory products of some animals (e.g. cats, dogs, rats etc.); drugs (e.g. vaccines, penicillin, sulfonamides, local anesthetics, salicylates, enzymes) and certain foods such as peanuts, sea foods, egg albumin and milk products. Out of these, pollens and allergens, derived from house dust, are the most common cause of asthma. Asthma is the most common allergic reaction. Besides asthma, other allergic reactions are rhinitis (hay fever), eczema, conjunctivitis and diarrhea. The type of allergic reaction occurring, depends on the type and nature of allergen. For example, air borne allergens such as pollens and dust are more likely to cause asthma and rhinitis while food allergens generally tend to cause gastric symptoms. Though food allergens, sometimes, may also cause skin reactions.

In addition to allergy, another term 'atopy' is also used to describe such reactions. But specifically 'allergy' is the term used for IgE mediated hypersensitivity to a particular allergen which manifests itself either clinically or as a positive skin prick test. The term 'atopy' is used to describe those individuals who have demonstrable evidence of IgE hypersensitivity against an allergen but may not have any clinical symptom.

Development of allergy

Allergy develops in three stages: sensitization, activation of mast cells and late phase reaction.

Sensitization

This is the first stage in the development of allergy. Sensitization refers to the production of IgE in response to exposure to an allergen, when allergen enters the body. It causes differentiation of CD4 cells into T_H2 cells. These T_H2 cells secrete IL-4 which stimulates B cells to differentiate into IgE. IgE binds to mast cells which are found in most tissues. Mast cells have specific receptors, called FcεRs, on their surface. IgE binds to mast cells to these receptors. These IgE remain bound to mast cells for months and years without causing any symptom.

Activation of mast cells

When the same allergen again enters the body, it binds to the IgE, bound to the FcεRs receptors on the mast cells. This causes cross linking of FcεRs. The cross linking of FcεRs results in a series of intracellular signaling events which lead to activation of mast cells. The activation of mast cells causes degranulation and release of some specific preformed mediators for mast cell granules. This also leads to synthesis and release of some new mediators. There are several preformed and newly formed mediators. These mediators cause many effects such as vasodilation, degradation of blood vessel endothelial membrane, platelet aggregation and degranulation, smooth muscle contraction and inflammation(**Table12.1**).

Table 12.1 : Various mediators associated with type I hypersensitivity

Mediator	Type	Effects
Histamines	Preformed	Increased vascular permeability, vasodilation, smooth muscle contraction
Eosinophil chemotactic factor (ECF-A)	Preformed	Attracts eosinophils from blood
Neutrophils chemotactic factor (NCF-A)	Preformed	Attracts neutrophils from blood
Enzymes (Protease, tryptase, chymase)	Preformed	Degradation of blood vessel basement membrane
Tumor necrosis factor-α	Preformed/newly synthesized	Inflammation
Prostaglandins	Newly synthesized	Vasodilation, platelet aggregation, pulmonary smooth muscle contraction
Leukotrienes	Newly synthesized	Increased vascular permeability, smooth muscle contraction
Cytokines	Newly synthesized	Increased IgE production, systemic anaphylaxis
Platelet activating factor	Newly synthesized	Aggregation and degranulation of platelets, pulmonary smooth muscle contraction
Bradykinin	Newly synthesized	Increased vascular permeability, smooth muscle contraction

Late phase reaction

Among various mediators, released by activated mast cells, TNFα activates the endothelium at the site and causes the expression of adhesion molecules. These molecules along with other chemotatic factors, released from the activated mast cells, promote the migration and accumulation of different types of leucocytes i.e. eosinophils, neutrophils, basophils and T cells to the site which produces inflammation **(Figure 12.2).**

SYMPTOMS OF ALLERGY

The clinical symptoms of the allergy can be either localized to the site of allergen or can be systemic.

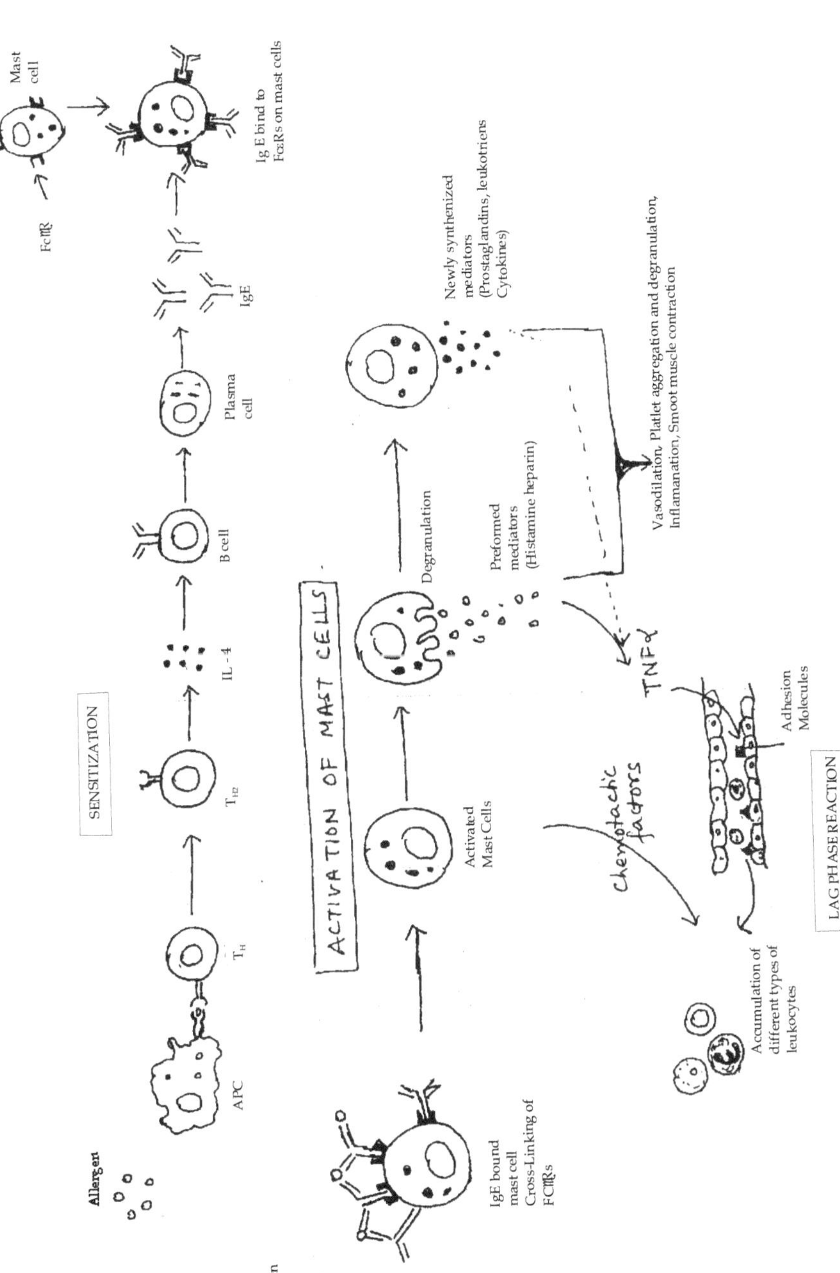

Fig. 12.2 : Development of allergic Response

Localized Reactions

The localized reactions are limited to a specific target tissue or organ and depend on where the mast cells are localized. These reactions are asthma, rhinitis, food allergies and skin allergies.

Asthma

In asthma , there is narrowing of the large and small airways, known as bronchospam which results in the shortness of breath. It is caused either by airborne or blood borne allergens such as pollen, dust, fumes, insect bites or viral antigens. It can also be induced by exercise or cold. Asthmatic attacks can be due to release of histamine, LTC4 and PGD2 from mast cells which cause smooth muscle contraction and result into bronchocontraction. Histamine and LTC4 also stimulate the secretion of mucus into the airways, further narrowing the passage. These responses are known as *early responses*. A few hours later, there occurs expression of adhesion molecules on endothelial cells and recruitment of inflammatory cells such as eosinophils and neutrophils causing further blockage of airways. These responses are mediated through some additional molecules such as IL4, IL5 , I-16, TNFα, eosinophils chemotactic factor (ECF) and platelet activating factor (PAF).

Rhinitis

Rhinitis or hay fever is caused by airborne allergens, most often, pollen. These cause activation of mast cells in the nose and eye conjunctivae. This causes release of mediators from the mast cells which cause vasodilation and increased vascular permeability. Hay fever is characterized by a blocked and running nose, coughing, sneezing and watery exudation of the conjunctive.

Food allergies

Food allergies can induce smooth muscle contraction and vasodilation, in the gut,which results in vomiting/or diarrhea. Degranulation of mast cells along the gut can increase the permeability of mucus membrane thus allowing the entry of food allergens into blood streams where they can activate mast cells at other sites, causing asthma or skin allergies.

Skin allergies

Hives or urticaria (itchy red swelling on skin) and allergic eczema (itchy skin rash) are caused by activation of mast cells in the skin by certain allergens. This results in vasodilation and increased permeability, causing various atopic dermatitis.

SYSTEMIC REACTIONS OR ANAPHYLAXIS

The allergic responses which spread throughout the body are the systemic reactions. A systemic allergic reaction is called anaphylaxis. It may be a shock like and can be life threatening. A number of antigens such as drugs like penicillin, insulin, antitoxins, venom from wasp, bee, stings and some sea foods and nuts have been shown to trigger such reactions in susceptible individuals. The mildest manifestation is hives and more severe forms result in swelling of lips, tongue and larynx, nausea, vomiting and asthma. If not treated, such reactions can be fatal. Death can result from respiratory or circulatory failure. Epinephrine is used to treat systemic anaphylaxis.

FACTORS CAUSING TYPE-I HYPERSENSITIVITY

Both genetic and environmental factors are responsible for causing Type -I hypersensitivity. At the genetic level, several genes are involved in causing allergy. Several genes have been proposed to affect the allergy. These include both MHC and non-MHC genes. MHC genes influence both IgE level and IgE response to individual allergens. Non-MHC genes also affect IgE production and some other aspects of allergy. For example, IL-4 is reported to affect IgE level. Similarly, FcεR receptors consist of four chains an α , a β and two identical γ chains. An association has been found between regions of chromosomes 11 containing the gene for β chain and allergy.

Environmental factors also contribute to the development of allergy. Many environmental factors have been implicated in enhancing IgE production. These factors affect both the incidence and manifestation of allergy. Pollutants such as diesel exhaust particles, ozone, nitrogen oxide, tobacco smoke and mercuric chloride have been associated with allergy. Increased vaccination has changed the spectrum of infection and has also contributed in increasing the allergy.

TREATMENT

There are three ways of treatment of allergy**(Fig. 12.3).** These are:

i. Avoidance of allergens

ii. Use of medicines

iii. Modifying immunological response

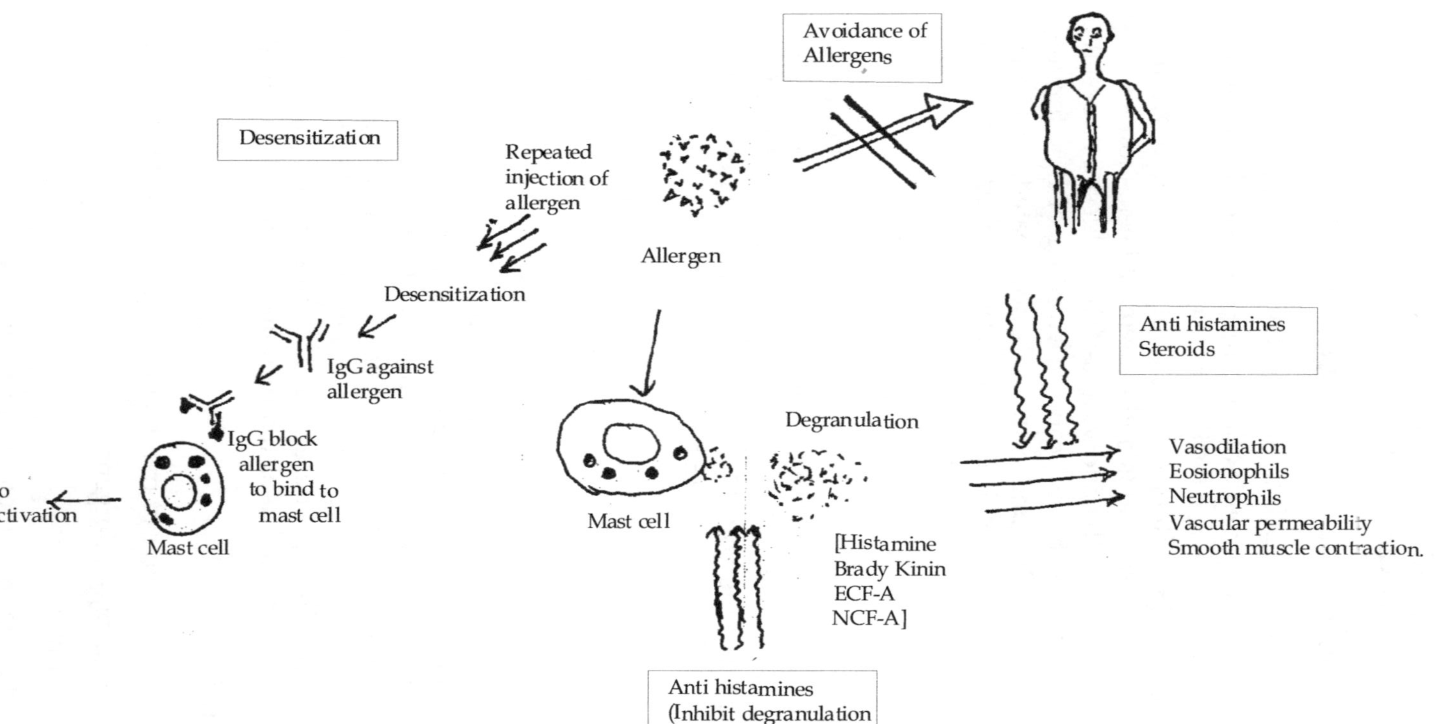

Fig. 12.3 : Different ways of treatment of allagy

Avoidance of Allergens

Some allergies can be prevented simply by avoiding the exposure to the allergens such as dust, contact with pets or foods causing allergy. When allergens causing allergy are avoided for certain period, the symptoms of allergy disappear.

Use of Medicines

Many medicines are used in the treatment of allergy. For example, antihistamines block the actions of histamine and are inhibitors of mast cell degranulation. Steroids inhibit the production of cytokines or other inflammatory mediators by various types of cells. They also inhibit the migration of inflammatory cells to the site of reaction. Adrenaline is used to treat anaphylactic shock.

Modifying Immunological Response

Another approach in the treatment of allergy is by preventing the production or effects of IgE. Anti IgE antibodies inhibit the production and activity of IgE. They also prevent IgE from binding to the mast cells and activating them. Desensitization is another way to prevent allergy. When a small amount of allergen is repeatedly injected, it induces the production of IgG rather than IgE. Subsequently, when the patient encounters the same allergen, IgG acts as a blocking antibody and competes with IgE to bind with allergen and preventing the activation of mast cells.

DETECTION OF TYPE-I HYPERSENSITIVITY

Several methods are used to detect type I hypersensitivity reactions. Commonly, allergy is detected by skin testing or by determining the serum level of total IgE antibodies. The most common and widely used method is skin testing. In this method, small amount of allergens is injected at a specific skin site i.e. on the forearm or back of the individual. A number of allergens can be applied at one time. If the person is allergic to the allergen, a wheal and flare appears within 30 minutes. This is because of local mast cells degranulation and release of histamine and other mediators. This is relatively an inexpensive method in which a large number of allergens can be tested at one time. But in certain cases the skin test can sensitize the allergic individual to new allergens or can induce systemic anaphylactic shock.

Total serum IgE can be determined by radioimmunosorbent test which is based on radioimmunoassay. It is a very sensitive method that can detect nanomolar level of total IgE.

ANTIBODY-MEDIATED CYTOTOXIC (TYPE II) HYPERSENSITIVITY

In type II hypersensitivity, antibodies (IgG or IgM) react with antigens, present on the cell or tissues. This activates complement which opsonizes the target cell and enhances phagocytosis. In certain cases, autoantibodies bind to the antigens, present on the self-cells, causing autoimmune diseases. There are three major examples of type II hypersensitivity:

i. Blood transfusion reactions
ii. Hemolytic disease of newborn
iii. Drug induced hemolytic anemia

Blood Transfusion Reactions

In ABO blood groups system, there are four types of blood groups: A, B, AB and O. An individual with blood group A has antigen A and antibodies against B. Individuals with blood group B have antigen B and antibodies against A. Individuals with blood group O have no antigen and make antibodies against both A and B. AB individuals have both A and B antigens but do not have A or B antibodies in their blood. Such individuals can receive any type of blood and are considered as universal recipient **(Fig. 12.4).** During blood transfusion, it is necessary to match the blood groups. When an incompatible blood group is transfused (say blood group A is transfused to B), the antibodies bind to transfused blood causing clumping of the foreign cells. This is followed by phagocytosis and hemolysis inside the blood vessels. Symptoms include fever, chill, vomiting, nausea, lowering of blood pressure and chest and lower back pains.

Hemolytic Disease of Newborn

Red blood cells of some individuals carry an antigen which is also present on RBCs of Rhesus monkey. These are called Rh antigens and individuals carrying these antigens are called Rh+. Individuals lacking Rh antigens are called Rh-. If an Rh- mother carries an Rh+ child, some of the Rh- fetal cells reach the maternal circulation at the time of child birth and sensitize mother's immune system. The mother produces IgG antibodies to Rh factor. But complications do not occur

Blood Group	Genotype	Antigen on erythrocytes	Serum Antibodies	Reaction with Blood Groups			
				A	B	AB	O
A	AA or AO	A	Anti - B	No Clumping	Clumping	Clumping	No Clumping
B	BB or BO	B	Anti - A	Clumping	No Clumping	Clumping	No Clumping
AB	AB	A and B	None	No Clumping	No Clumping	No Clumping	No Clumping
O	OO	None	Anti-A Anti-B	Clumping	Clumping	Clumping	No Clumping

Fig. 12.4 : ABO Blood Groups and possibilities of blood transfusion

during the first pregnancy. If the mother conceives another Rh+ child, some of the anti Rh IgG antibodies cross the placenta and bind to fetus' red blood cells. This results in the opsonization of red blood cells and their destruction. This is called ***hemolytic disease of newborns*** **(Fig. 12.5)**. Destruction of RBCs results in the accumulation of toxic compound, bilirubin which, in severe case, could be fatal. Hemolytic disease of newborn could be prevented by giving an injection of anti Rh antibodies (Rhogam), shortly after the birth of Rh+ child. These antibodies bind to the fetal blood cells which could have entered the mother's blood and destroy them before they could sensitize mother.

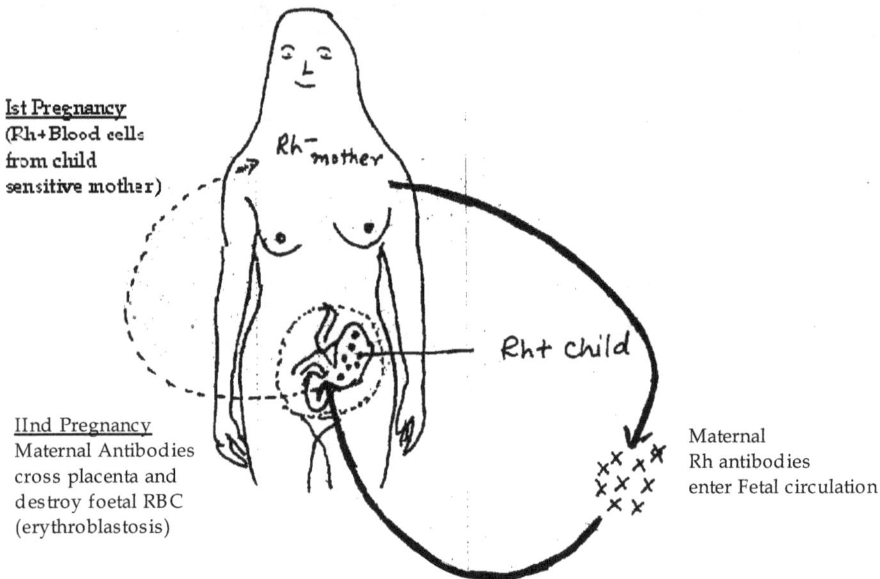

Fig. 12.5 : Erythroblastosis foetalis (Hemolytic diseases of newborn

Drug Induced Hemolytic Anemia

Some drugs (e.g. penicillin, quinidine, streptomycin) can bind to proteins in red blood cells or platelets. These drugs cannot stimulate the immune response themselves but by binding to the proteins on RBCs, these drugs form drug-protein complex. In some people these drug-protein complex (like hapten-carrier complex) stimulate antibody production against the drug. These antibodies bind to the drug and induce complement mediated lysis or opsonization, resulting in anemia. Usually, when the drug is withdrawn, the symptoms disappear.

IMMUNE COMPLEX (TYPE III) HYPERSENSITIVITY

Under normal conditions, when an antibody reacts with an antigen it forms immune complex which helps in the clearance of antigen. But in certain situations, when the amount of immune complex formed is large, this causes immune complex hypersensitivity or type III hypersensitivity.

These hypersensitivity reactions occur due to complement activation, leading to inflammation. During these reactions when the activated complement binds to immune complex, it leads to degranulation of mast cells releasing in histamines. This histamine causes vasodilation and vascular permeability. Complement C3a and C5a also activate neutrophils which release lytic enzymes causing tissue damage. The tissue damage, occurring due such reactions, is caused by the lytic enzymes released by neutrophils **(Fig. 12.6)**.

Type III hypersensitivity reactions may be local or systemic, depending on the deposition and distribution of immune complex.

Local Immune Complex Reactions

When the immune complexes are deposited in tissues, near to the site of entry of antigen, a localized reaction occurs. Such reactions result in edema (accumulation of fluid) and erythema at the site of reaction. If the reaction is severe, it can lead to swelling and redness to tissue necrosis. The examples of localized reactions are farmer's lung, caused by inhalation of mold spore and pigeon farcier's disease, caused by inhalation of serum proteins, derived from pigeon feces.

Systemic Immune Complex Reactions

When immune complexes are formed in the blood, it can lead to systemic immune complex reaction. In these cases, reactions can occur wherever immune complexes are deposited. More frequently these complexes are deposited in blood vessels, kidney, synovial membrane of joints, skin and brain. The deposition of immune complexes results in the activation and migration of neutrophils which causes tissue damage. The symptoms of such reactions are fever, rashes, arthritis and kidney malfunctions. Example of systemic immune complex reaction is serum sickness. When animal serum (e.g. horse serum) containing antitoxin antibodies to diphtheria or tetanus was given to a patient, the patient formed antibodies specific to serum proteins as the serum came from another species. These antibodies formed

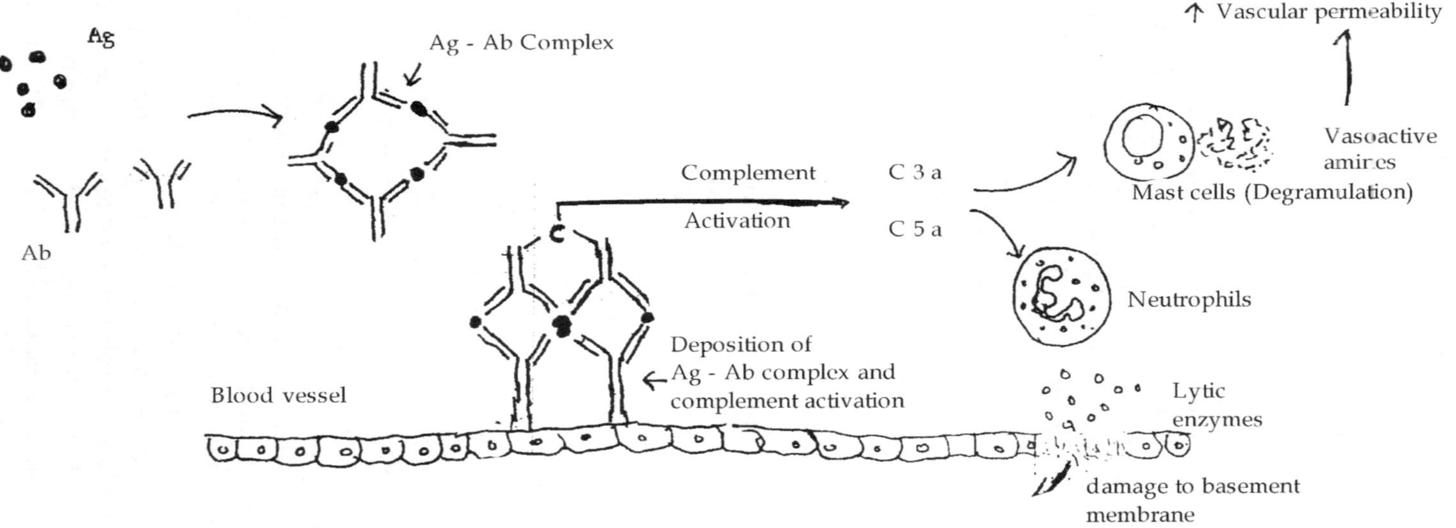

Fig. 12.6 : Type III (Immune Complex) Hypersensitivity

(Immune complex activates complement C3a and C5a activate mast cells and cause degranulation. C3a and C5a activate neuralphils to release lytic enzymes which cause damage to basement membrance) circulating immune complexes and these were deposited in various tissues that resulted in symptoms such as fever, rashes, edema, erythema and arthritis.

DELAYED TYPE (TYPE IV) HYPERSENSITIVITY REACTIONS

Delayed type or Type IV hypersensitivity is called so as it takes generally 24 to 72 hours to develop. It is a localized inflammatory reaction where there is no formation of antibodies. It involves role of T cells, macrophages and dendritic cells. T_H cells are the most crucial cells in such reactions.

When T cells, specifically T_H cells, encounter certain types of antigens [intracellular pathogens such as mycobacteria, viruses, fungi; chemicals like nickel, chromium; allograft (foreign organs or tissue transplants)], they secrete cytokines which induce inflammatory reactions **(Fig. 12.7).** In some cases, a delayed type hypersensitivity response causes extensive tissue damage while in many cases tissue damage is limited.

Development of Delayed Type Hypersensitivity

There are two phases in the development of Delayed type hypersensitivity. First, there is *sensitization phase* followed by *effector phase.*

Sensitization phase

On primary contact with the antigen, the antigen processing cells (APCs) process the antigen and present them to the T_H cells. The T_H cells are activated and differentiated to T_{H1} cells.

Effector phase

On subsequent exposure to the same antigen, sensitized T_H cells are activated. These cells release a variety of cytokines which recruit and activate macrophages and other nonspecific inflammatory cells. The monocytes are drawn from the blood vessels to the surrounding tissues and are differentiated into macrophages.

Macrophages, on activation by IFN-γ (interferon gamma) and TNF-β (tumor necrosis factor beta), show enhanced capacity to phagocytose and destroy microbes. The DTH response takes at least 24 hours after contact with the antigen, to appear. This is the time,

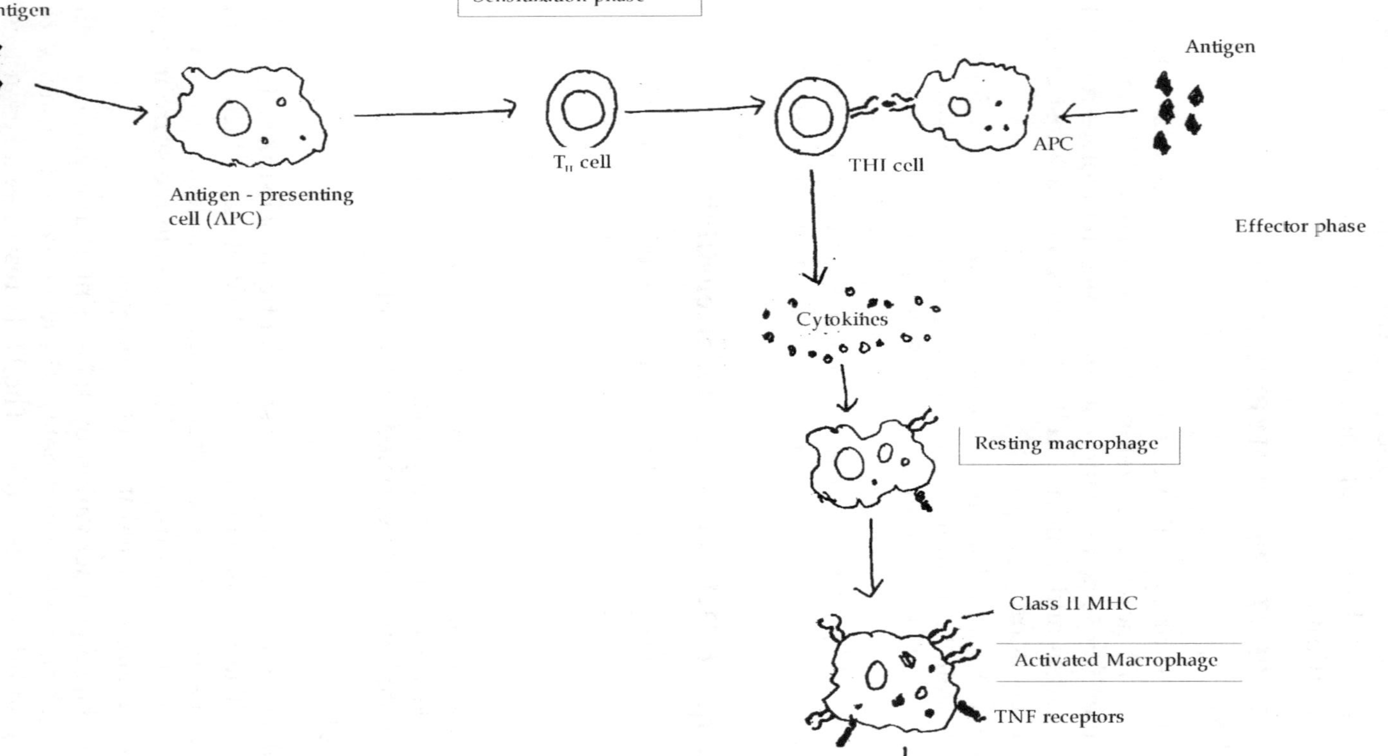

Fig. 12.7 : Delayed type (IV) Hypersensitivity

required for the cytokines to induce recruitment of macrophages and their activation.

Contact dermatitis, caused by various chemical agents is a type of Delayed type hypersensitivity. Certain metals such as nickel, gold, chromium, used in jewelry and chemicals used in soaps, cosmetics, hair dye and rubber products give rise to contact hypersensitivity reactions. The reactions result is skin swelling, redness and itching. Poison ivy is a major cause of contact dermatitis, especially in the United States. The actual chemicals that induce contact hypersensitivity reactions are too small in size to induce immune response. These are haptens which bind with proteins of the host and induce immune response. These hapten-proteins are taken up by antigen processing cells such as Langerhans' cells and are presented to CD4+ T cells in the regional lymph nodes. These cells differentiate to T_{H1}cells. On subsequent encounter with the agent, the sensitized T_H cells are activated and initiate a rapid immune response.

POINTS TO REMEMBER

- Hypersensitivity is a condition in which a normally harmless and inert antigen is exaggerated and causes diseases. It is also called allergy.
- Hypersensitivity can be broadly classified as immediate hypersensitivity and delayed type hypersensitivity.
- Immediate hypersensitivity develops within minutes or hours after the exposure of antigen to a sensitized person.
- Delayed type hypersensitivity may take several days to develop,after exposure.
- Immediate type hypersensitivity is of three types: Type I or IgE-mediated hypersensitivity,Type II or antibody-mediated hypersensitivity and Type III or immune-complex mediated hypersensitivity.
- Type I or IgE-mediated hypersensitivity,commonly known as allergy,is mediated by the production of IgE. Asthma is the most common allergic reaction.
- Allergy develops in three stages: sensitization, activation of mast cells and late phase reaction.
- The clinical symptoms of the allergy can be either localized to the site of allergen or can be systemic.

- The localized reactions are limited to a specific target tissue or organ and include asthma, rhinitis, food allergies and skin allergies.
- Systemic allergic reactions spread throughout the body. A systemic allergic reaction is called anaphylaxis.
- Allergy can be treated by avoidance of allergens, use of medicines, and by modifying immunological response.
- Commonly used methods to detect allergy are : skin testing or by determining the serum level of total IgE antibodies.
- The major examples of Type II or antibody-mediated hypersensitivity are : blood transfusion reactions, hemolytic disease of newborn and drug induced hemolytic anemia.
- Type III or immune-complex mediated hypersensitivity is caused by formation of immune complexes. Such reactions can be local or systemic.
- Delayed type or Type IV hypersensitivity takes generally 24 to 72 hours to develop and involves role of T cells, macrophages and dendritic cells. Contact dermatitis, caused by various chemical agents, is a type of Delayed type hypersensitivity.

REVIEW QUESTIONS

1. What is hypersensitivity? Differentiate between immediate and delayed type hypersensitivity.
2. What is allergy? Describe various factors which can induce allergy?
3. Describe various stages in the development of allergy.
4. Describe different strategies in the treatment of allergy.
5. How does type II hypersensitivity differ from type III hypersensitivity?
6. What is delayed type hypersensitivity? Describe the mechanism of DTH.

CHAPTER - 13

Immune Response to Infections

Infection refers to the invasion of harmful microorganisms into the body and causing disease. The most important function of the immune system is to protect the body of the host against various viruses, bacteria, fungi and parasites. The grounds of immune surveillance stand on the basis of innate and specific immunity that build up against the evasive mechanisms of the disease causing organisms. Infectious disease is result of a complex interaction between the host and the microbes. Evasions of protective immunity by the parasites critically determine the survival and pathogenicity of the microbes. The invasion of the body by harmful microorganisms is called infection. Infections are dangerous in any part of the body. Microorganisms spread vary easily into the blood stream and cause blood poisoning which is called *septicemia*.

INFECTIONS TOWARDS VIRUSE INFECTION

Viruses are obligate intracellular organisms which reside inside the host cells and use host cell protein and replicate within the cell. The early stage of a viral infection is a race between the virus and the host's defense system. The initial defense against virus invasion is the integrity of the body surface. Viruses which cause infections are of two types: *cytopathic viruses* and *non-cytopathic viruses*. In case of cytopathic viruses, viruses enter into a host cell and uncoat. So the nucleic acid is released and transcription starts which is followed by translation. The viral nucleic acid is replicated and released through cytolysis and infects the neighboring cells. These viruses damage the cell and cause cell death. In non-cytopathic viruses, viruses hide in inside the host cell and synthesize foreign protein. An infected man, who has suffered from viral infection, is also susceptible to bacteria and other infections. It is due to depressed immunity. Sometimes viral infection causes permanent depression of immunity and this condition causes AIDS **(Fig. 13. 1)**.

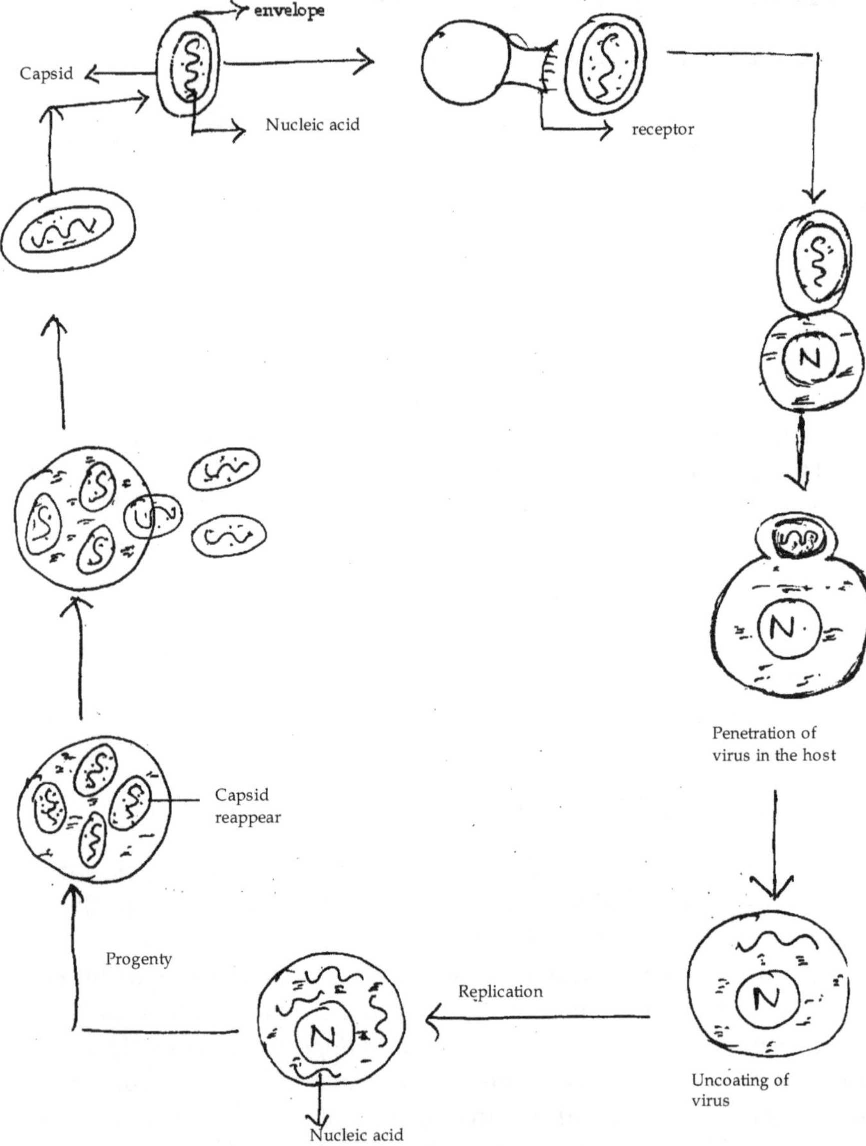

Fig. 13.1 : Infection of virus.

INNATE IMMUNE RESPONSE AGAINST VIRUSES

A number of immune effectors mechanisms with nonspecific defense mechanisms work together to eliminate an infecting virus. In the meantime virus activates some other mechanisms to prolong its own survival. The infection of virus depends upon how effectively host's defensive mechanisms resist the offensive practice of virus.

There are two mechanisms of non-specific or innate immunity against viral infection. These are primarily through the induction of type1 interferon's (IFN) and secondly through natural killer cells (NK).

Types of Interferons

(a) Leukocytic interferon IFN-α : Product of a family of 21 genes on chromosome 9

(b) Fibroblast interferon IFN-β: Product of a single gene on chromosome 9

(c) Immune interferon IFN-λ : Product of a single gene on chromosome 12

Virus infection of a cell leads to the production of IFN α and IFNβ. IFN α and IFN β activate antiviral mechanism in adjacent cells. Interferons activate a number of genes with direct antiviral activity. IFN α and IFN β induce transcription of several genes. One of these genes encodes an enzyme known as *2-5 oligoadenylate synthatase,* which activates a *ribonuclease* that degrades viral RNA. IFN $\alpha\beta$ and IFN β can induce an antiviral response by binding to the IFN α / IFN β receptor which inhibits the viral replication **(Fig. 13.2).** Natural Killer Cells (NK Cells) also take part in the lysis of viral infected host cells. Active NK cells are detected within 2 days of a virus infection. They have been identified as major effector cells against hyperviruses and murine cytomegalovirus (MCMV). NK Cells are first immune effectors which are detected at very early stage of viral infection. The binding of IFN α and IFN β to NK cells induces lytic activity, killing virally infected cells. NK cells are also one of the main mediators of antibody dependent cellular cytotoxicity (ADCC). Macrophages are present in the tissues of the body and act as a defense mechanism against viral infection. They destroy the virus in different ways by :

- Phagocytosis of virus and virus infected cells.
- Killing of virus infected cells.
- Production of antiviral molecules.

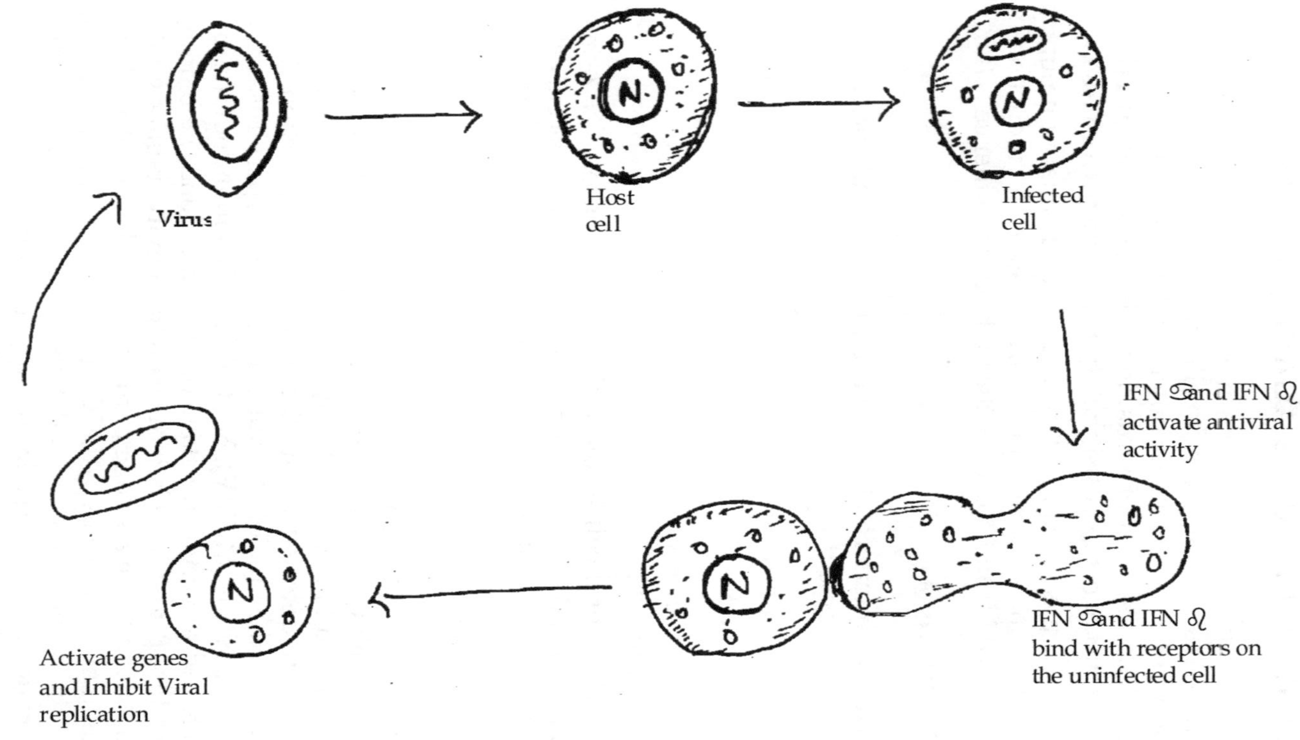

Fig. 13.2 : Induction of antiviral activity by IFN α and IFN β

SPECIFIC IMMUNE RESPONSE AGAINST VIRUSES

Immune defense mechanism against viral infection occurs as a joint function of humoral and cellular immunity. Antibodies specific to viral surface antigens are often crucial in containing the spread of a virus during acute infection and in protecting against reinfection. Antibodies provide a major hindrance to the spread of viruses between cells and tissue. Viruses have surface receptor molecules that enable them to initiate infection by binding to specific host cell. Antibodies provide a major barrier to virus spread between cells and tissue, restricting their spread in the blood stream. Secretion of IgA in mucosal surface plays an important role in host defense against virus by blocking viral attachment to mucosal epithelial cell. Secretary forms of IgA may be active in neutralizing viruses which try to gain mucosal surface. In some cases, antibodies may block viral penetration by binding to epitopes that are necessary to mediate fusion of viral cell with plasma membrane.

Antibody and complement dependent mechanisms also promote phagocytosis. Killed or attenuated viruses, when used as prophylactic vaccines against viral infection, stimulate specific antibody responses. Antibody or complement can also agglutinate viral particles. Complement can also damage virion envelope through virolysis. Some viruses can directly activate the complement, so complement is not considered as a major part in the virolysis because individuals with complement deficiencies are not predisposed to severe viral infection. Table 13.1 shows effects of antiviral antibodies on virus and viral infected cells.

Table 13.1 : Effects of antiviral antibody

Virus Cell	Factor	Effects
Free Virus	Antibody	Blocks binding to cell. Stops entry into cell. Stops viral uncoating
	Antibody+Complement	Blockage of viral receptor. Damage of virus envelops.
Virus infected cell.	Antibody + Complement	Phagocytosis of infected cell. Opsonization and virolysis.
	Antibody bound to virus	Lysis of viral infected cell. ADCC by NK cells and macrophages.

T CELL-MEDIATED VIRAL IMMUNITY

Cell mediated immune mechanisms are most important in host defense. T cells exhibit a variety of functions in antiviral immunity. Most of the antibody response is T-dependent. CD_8+ T_C cells and CD_4+ T_H cells are the main components of cell -mediated antiviral activity. CD_4+ T cells also help in the induction of CD_8+ cells and formation and activation of macrophages at virus infection. CD_8+ T cells are effective in the prevention of reinfection of viruses such as influenza virus and respiratory syncytial virus. In most viral infections cytolytic T cells (CTLs) activities arise within 3 or 4 days after infection. Within 7 to 10 days of primary infection most virions have been eliminated. The CTLs system operating against viruses is highly efficient and selective. CTLs cells focus at the site of virus replication and destroy virus infected cells. CD_4+ T cells are one of the primary effector cells in immune system towards Herpes simplex virus-1 (HSV-1). At this time macrophages are generated, resulting in the delayed hypersensitivity and faster elimination of virus. Sometimes, CTLs are involved in tissue injury while defending non-cytopathic viruses. Macrophage produces nitric oxide which acts as a viricidal agent. CD_4+ T cells produce cytokines. Key cytokines in the T cell- mediated response are immune interferon (IFN λ) which is necessary for activation of monocytes and tumor necrosis factor α (TNFα). TNFα which hasan antiviral activity **(Fig. 13.3)**.

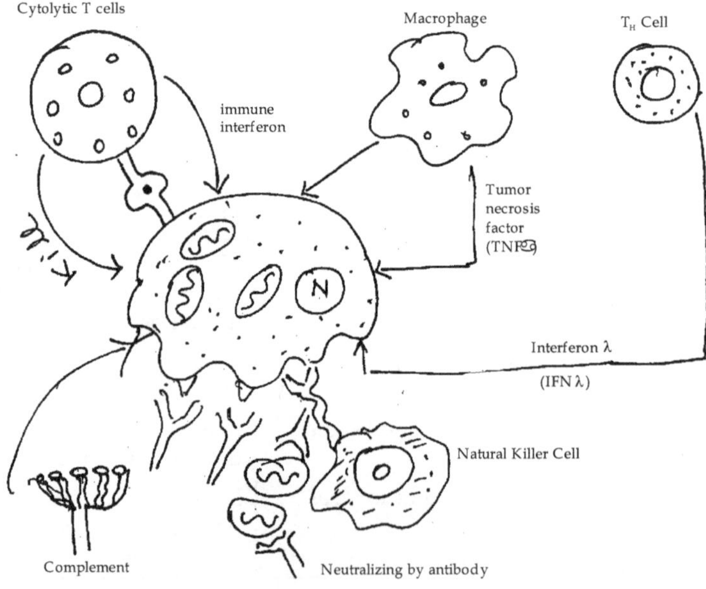

Fig. 13.3 : T cells mediated viral immunity

STRATEGIES OF VIRUS TO EVADE DEFENSE MECHANISMS

A number of viruses have been found to encode proteins that interfere at various levels with specific and nonspecific defense mechanisms. Many viruses have adopted mechanism to evade complement dependent destruction. IFN α and IFN β are major innate defense against viral infection but some viruses have developed some mechanisms to evade the action of IFN α and IFN β. Herpes simplex viruses (HSV) have an immediate early protein, synthesized after viral infection, which inhibits the human transporter molecule, needed for antigen processing. Many viruses have the capacity to change their antigens. Influenza virus can change its surface antigens so the immune system cannot protect from virus at initial stage. Influenza virus undergoes small changes per year. This is called antigenic drift. Sometimes two influenza viruses, from different species, co- infect an animal. When this happens, large antigenic shifts occur and host does not have a protection against this type of virus. Antigenic variations are found in case of human immunodeficiency virus(HIV) and influenza virus. Virus proteins have the capacity to modify MHC class 1 molecules, expressed on the cell surface. Virus has the capacity to stop the transfer of class I MHC from the site of synthesis to cell surface. Some viruses, like HIV, directly affect lymphocytes. Viruses are capable of changing intracellular movement of molecules, without any cytoplasmic change. Viral infection evokes autoimmune diseases. If there is any viral infection, infected tissues get damaged, provoking inflammatory reaction and so many hidden antigens are exposed on cell surface.

IMMUNITY TOWARDS BACTERIAL INFECTIONS

Bacterial infections have an enormous impact on human society and despite the discovery of antibiotics, continue to be a major threat to public health. Once a bacterium has entered the body, there are a number of ways by which it can cause disease. It may produce damage to local tissues through the action of bacterial toxins. Bacteria enter the body either through a number of natural routes like respiratory tract, gastrointestinal tract, and genitourinary tract or through mucosa membrane of skin **(Fig. 13.4)**.

Immune Response to Extracellular Bacteria

Extracellular bacteria have the capacity to multiply outside the host cells. Infection by extracellular bacteria induce production of

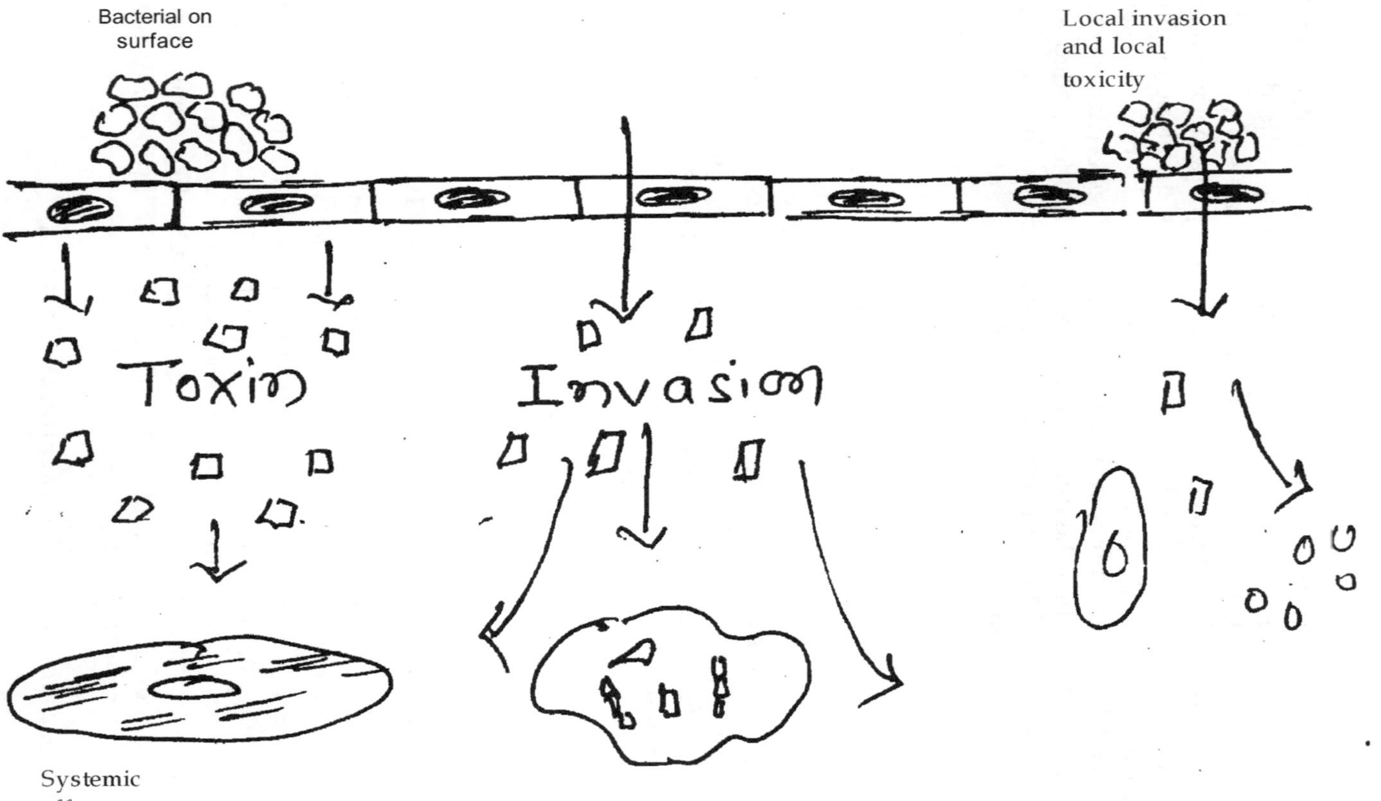

Fig. 13.4 : Bacterial infection

humoral antibodies which are secreted by plasma cells in lymph nodes and submucosa of respiratory and gastrointestinal tracts. The humoral immune response is the main response against extracellular bacteria. The antibodies act in several ways to protect the host from microorganisms, including removal of bacteria and inactivation of toxins. Extracellular bacteria produce inflammation and cause tissue damage and destruction at the site of damage. Bacteria produce toxins which may be exotoxins or endotoxins. Endotoxins of gram-negative bacteria are a lipopolysaccharide (LPS), which act as a potent stimulator of cytokine. In some gram-negative bacteria, complement activation can lead directly to lysis of the organisms. Excess release of cytokines is stimulated by LPS in gram-negative bacteria which lead to diffused intravascular coagulation with resultant clotting problems, hemorrhagic and necrosis in the intestine. Some bacteria produce neurotoxins that bind to neuromuscular junction and cause tetany. Immunopathogenicity of bacteria can occur through a single toxin as they have ability to attach to the epithelial surface, without being invasive and produce systemic effect. Gram-positive bacteria do not possess LPS but induce intense inflammatory response and severe infection like colitis and diarrhea.

The humoral immune response is the main protective response against extracellular bacteria. Different immunogens of cell walls and capsules of the microbes are thymus independent TI antigens, which can directly stimulate B lymphocytes leading to strong and specific IgM response. The antibody acts in several ways to protect the host from the invading organisms. Antibodies remove the bacteria and neutralize toxin by blocking the attachment of the binding protein to the target host cell. Antibodies bind to antigens on the surface of a bacterium with the component of complement, which increases phagocytosis. In some bacteria, complement activation directly lysis the bacteria. Regarding T cells, CD_4^+ T cells respond to protein antigen that is presented along with MHC class II molecule. The CD_4^+ T cells are activated through the secretion of cytokine. If bacteria are present in a host cell, cytokines induce inflammation. There are some bacterial toxins which stimulate CD_4^+ T cells. These toxins are called *super antigens*. In rheumatic fever and B hemolytic infection, bacteria lead to secretion of antibodies against bacterial cell wall. Bacterial endotoxins and superantigens develop auto-immune disorders.

Innate immunity is not very effective against bacterial infection. Bacteria activated T cells provide an early defense mechanism. Bacteria are phagocytosed by neutrophils and lysosomal digestion occurs in macrophages. Complement system is activated in the absence of antibody and plays its role in the eradication of the microbes **(Fig. 13.5).**

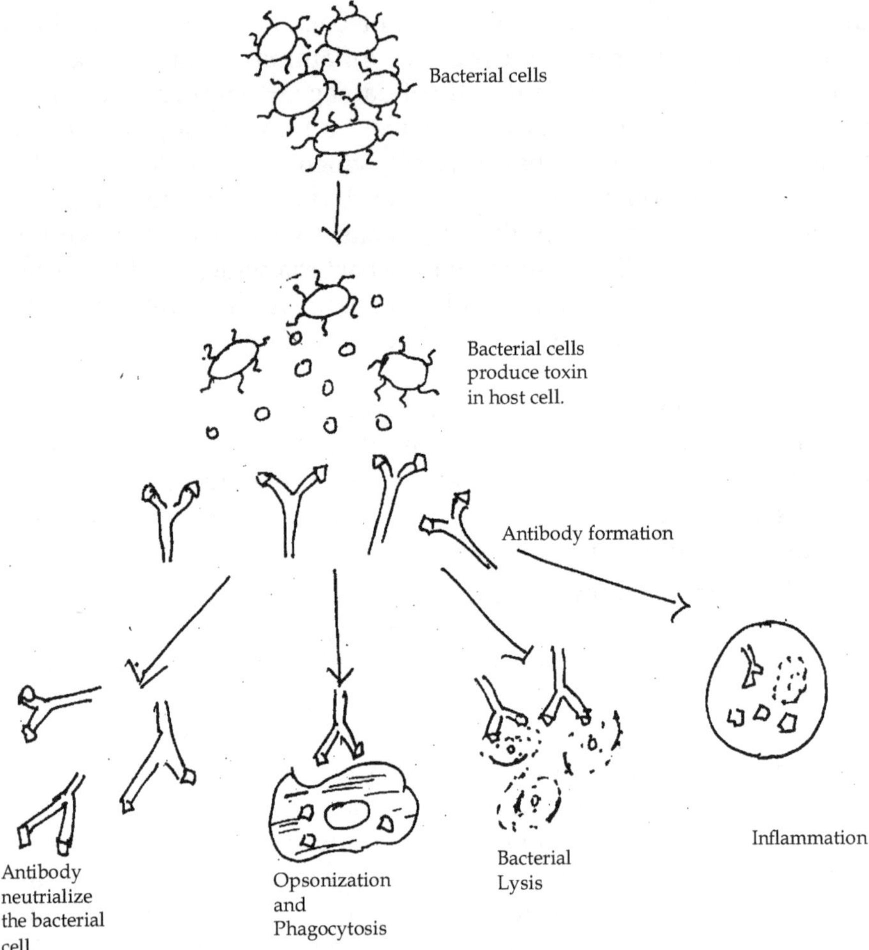

Fig. 13.5 : Immune response against bacteria.

Immune Response to Intracellular Bacteria

Intracellular bacteria have the power of multiplication in a host cell. Intracellular bacterial infections induce a cell-mediated immune response. This infection induces delayed hypersensitivity. After infection, phagocytosis and endocytosis start and kill the bacterial cells. NK cells stimulate macrophage. Some bacteria e.g. *Streptococcus pneumoniae* have some surface structures which inhibit macrophages. Protein of intracellular bacteria stimulates CD_4^+ T and CD_8^+ T cells. T_H cells secrets interferon λ (INF λ) to activate the production of antibody and this antibody activates complement for opsonization. If intracellular pathogens are not quickly eliminated, activity of the macrophages and T cells against microbes, causes tissue injury resulting in the formation of granuloma. It is generally associated with chronic bacterial infection, like tuberculosis and syphilis. Neutrophils and dendritic cells can also found in granuloma. The presence of activated macrophages in the host cell ensures to control action against bacterial growth. It has also been found that specific condition of the individuals is also responsible for immune response against bacteria. For example, AIDS and diabetes are important risk factors for loss of control of tuberculosis. Tuberculosis is the major cause of death in the world. One third of the world population is infected with the *Mycobacterium tuberculosis*. After the infection of *Mycobacterium tuberculosis,* the most common disease is pulmonary tuberculosis. CD_4^+ T cells are activated within 2 or 6 weeks after infection of *Mycobacterium tuberculosis*. These bacteria induce the infiltration of large number of activated macrophages inside the cells, forming a granulomatous lesion, called a *tubercle*. A tubercle contains lymphocyte, macrophage and multinucleated giant cells. Macrophages release lytic enzymes in a tubercle. These enzymes dissolve the bacteria as well as nearby healthy cells. The CD_4^+ T cells control the infection and later protects against reinfection **(Fig. 13.6).**

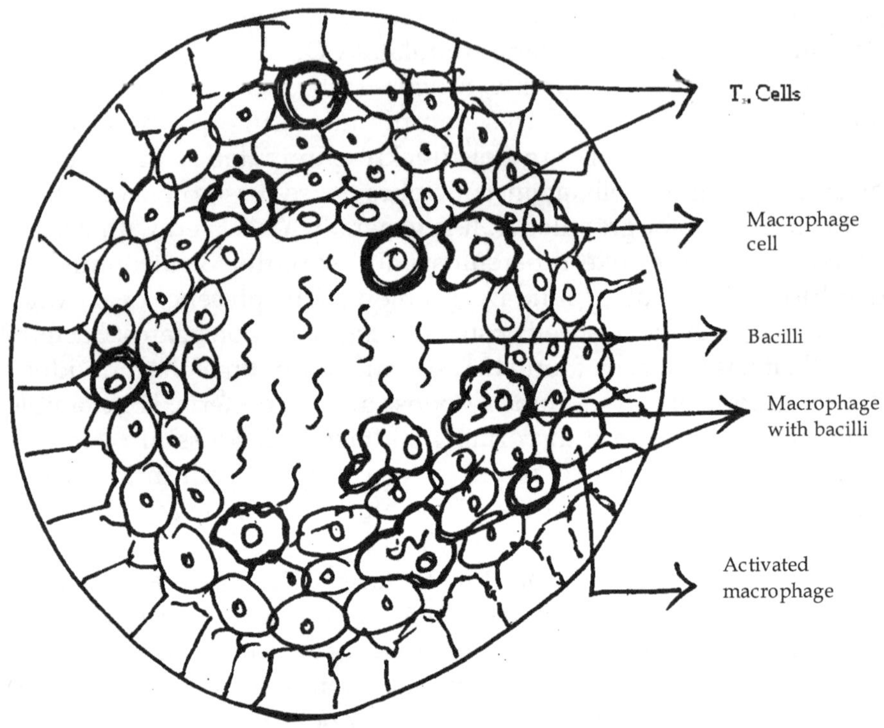

Fig. 13.6 : Bacterial Infection in tubercle.

Strategies of Bacteria to Evade Defense Mechanisms

Some bacteria have specific surface structures that enhance their ability to attach to the host cell. Gram-negative bacteria have hair like projections which enable them to attach to cell membrane. Some bacteria play anti-phagocytic mechanism and inactivate complement system. Some bacteria may hide in metabolically damaged host cells. Secretary IgA antibodies, specific for such bacterial structure, can block the bacterial attachment. However, some bacteria like *Haemophilus influenzae* and *Neisseria meningitidis* evade the IgA response. *Listeria monocytogenes* release enzymes that dissolve the membrane of phagosomes. *Straptococus pneumoniae* secrets an enzyme *coagulase* which blocks the function of macrophages. Some bacteria e.g. *Neisseria gonorrhea* have variation in surface antigens which help them to escape antibody attack. Some staphylococci are capable to assemble a protective coat from host proteins. These bacteria secrete coagulase enzyme that precipitates a fibrin coat around them and protects them from phagocytic cells. *Listeria monocytogenes* produces hemolysin which blocks microbicidal function of the macrophage. *Pseudomonas* secrete an enzyme (*elastase*) that inactivates the inflammatory reaction.

IMMUNITY TOWARDS PROTOZOANS INFECTIONS

Protozoan parasites are unicellular intracellular or extracellular pathogen that causes many diseases in human such as malaria, leishmanisis, and sleeping sickness. Diseases depend upon type of parasites, location of parasites in the host body and host response. Diseases usually develop either from immunopathological side effects from unchecked parasite growth due to immunosuppression or due to parasites resistance.

Immune Response to Protozoans Infection

A protozoan parasite enters into circulation or tissue and is able to survive and replicate. The intermediate stages of the parasites are present in invertebrate host and stimulate alternative complement system. Invertebrate hosts are resistant against complement-mediated lysis. Macrophages have the ability to engulf protozoa. Many protozoan have induced remarkable enhancement in NK cell activity in some animals. Certain African population lacks erythrocytic antigens that are Duffy antigens, so they are resistant totally to malaria parasite. Malaria is caused by various species of *Plasmodium*. The pathogens are transmitted through anopheles mosquito. In many parts of the world malaria is endemic to the regions where immune response to the pathogens is very poor. In these regions children with less than 14 years of age show low immune response and 50% rate of mortality is observed. Low level of immune response is due to many factors. First is the changing of surface of antigens at different stages of parasite's lifecycle. Intracellular phases of lifecycle in liver and erythrocyte reduce the immune activity. If antibody is developed, the parasite has evolved a process of sloughing off the surface antigens, so antibodies cannot bind to their body. Flagellated protozoa *Trypanosoma* which causes sleeping sickness in Africa spreads from blood to the nervous system and causes meningoencephalitis. On the surface of *Trypanosoma,* glycoprotein antigenis present which stimulates humoral antibody response. Antibodies eliminate the parasite from blood through the process of phagocytosis, opsonization and lysis but few parasites escape from the antibodies and they again proliferate in blood and cause new infections. *Entamoeba histolytica* is responsible for colitis and liver abscess. It kills the gut epithelial and immune cells. This killing is contact dependent and is mediated by the trophozoite galactose and galactose N acetylgalactose lectin, a protein that binds host cell glycoconjugate. A family of pore forming proteins also participates in cytotoxic activity.

Many protozoans have been shown to induce NK cell activity. Major functions of NK cells include lysis of target cell and the production of cytokines, involved in the stimulation of immune cells. The destruction of protozoa by NK cell is through lysis of extracellular organisms. NK cells also destroy host cells which are infected by protozoan like *Plasmodium* or *Trypanosoma*. Immune response against *Trypanosoma cruzi* depends on NK cells along with CD_4^+ and CD_8^+ T cells and also on antibody production. NK cells are also important for the production of Interferon λ (INFλ) and tumor necrosis factor λ (INFλ) cytokines. Cytokines activate macrophages and kill both extracellular and intracellular protozoan. NK cells also produce mediators which influence host resistance to infection.

T cells also control infection. CD_4^+ and CD_8^+ T cells protect against different phases of *Plasmodium* infection. CD_4^+ T cells mediate immunity against blood stage of *Plasmodium yoelii*. CD_8^+T cells protect against liver stages of *Plasmodium bergbei*. CD_8^+ T cells inhibit the multiplication of parasite by secreting interferon λ. Helper T cells have a significant role in removing an infection. T_{H1} cells mediate killing of intracellular pathogens and, T_{H2} cells eliminates extracellular ones. Regulatory T cells are able to modulate the extremes of T_{H1} and T_{H2} response. Cytokines act not only on effector cells but also act as growth factor of cell number. In case of malaria, the enlargement of spleen is related to increased number of cells. Chemokines molecules play an important role in immune mechanisms to produce chemotoxicity and act in leukocyte activation, inflammation and anti-parasitic immunity.

In some cases, parasites evoke hypergammaglobulinemia due to substance acting as B cell mitogens. There is rise in IgM in malaria and trypanosomiasis and IgG in malaria and visceral leishmaniasis. Further infection, induces effective anti-parasitic immunity through development of individual resistance and production of high affinity antibodies which inhibit the production and infection of protozoa. Antibodies act directly on protozoa and damage them either by itself or by activating the complement system. Antibodies neutralize parasite directly by blocking its attachment to a new host cell. Antibody can increase phagocytic activity of macrophages and it is also involved in antibody dependent cell-mediated cytotoxicity (ADCC) **(Figure 13.7).**

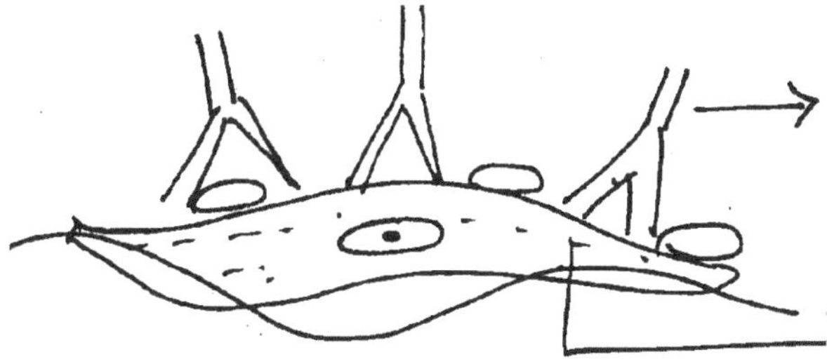

A. Antibody activates classical complement pathway causing direct damage or lysis of parasite. e.g. *Trypanosoma* .

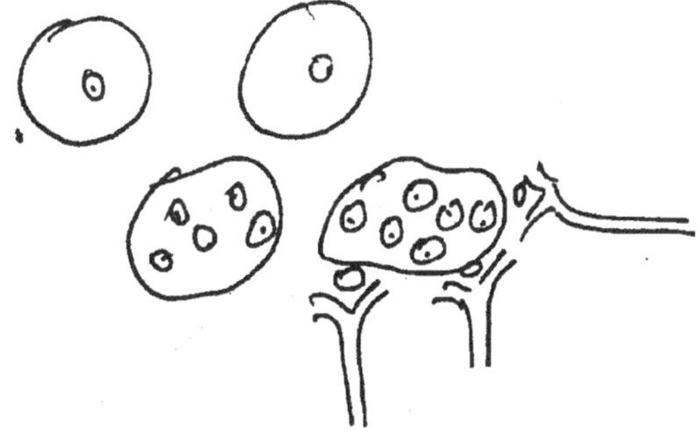

B. Antibody neutrializes the attechment sites and prevents escape from lysosome.

e.g. Plasmodium sporozoite and merozoite

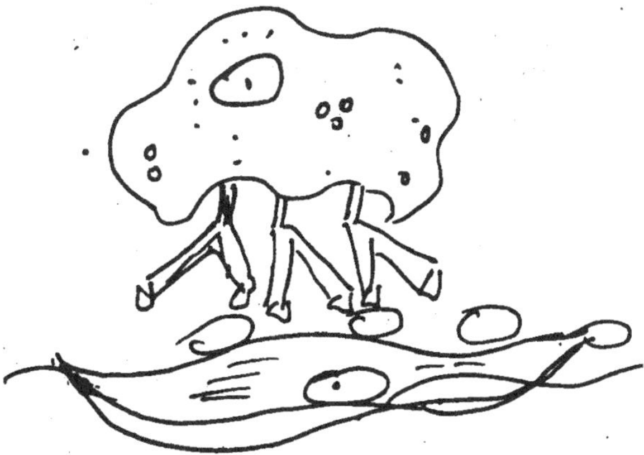

C. Antibody enhances the process of phagocytosis and opsonization e.g. *Trypanosoma* and *Plasmodium.*

Fig. 13.7 : Antibody dependent cell mediated cytotoxicity.

Strategies of Protozoa to Evade Defense Mechanism

Different protozoa have developed remarkable effective ways of resisting specific immunity. Some even exploit cells and molecules of the immune system to their own advantage. *Leishmania* spp., use complement receptor to affect their entry into macrophages and avoid triggering the oxidative burst and thus destruction by its toxic products. Malaria parasite survives in its own vesicles and avoids phagocytosis. *Entamoeba* makes a cyst, which is resistant to immune system. *Trypanosoma cruzi* bears a surface glycoprotein that has activity, resembling the decay accelerating factor (DAF) that limits the complement reactions. *African Trypanosoma* maintains a surface coat to protect it from the host defense mechanism. Some species of malaria parasite have also developed antigenic variations. In leishmaniasis, T cells from host infected by *L. donavani*, when cultured with specific antigen, do not secrete interferon λ. T_{H1} cell response is also deficient in some protozoan infections. In malaria and African trypanosomiasis, the increased number and maximum activity of macrophages and lymphocytes in the liver and spleen, lead to enlargement of these organs. The formation of immune complexes is common. They may be deposited in the kidney as in the nephritic syndrome of quartan malaria and cause so many pathological changes. Excessive production of cytokines may contribute to some of the manifestations of diseases. A single parasite protein may produce multiple pathological effects.

IMMUNITY TOWARDS HELMINTHES INFECTION

Helminthes are large multicellular organisms. They do not multiply in large numbers in humans and they are not intracellular pathogens. They are more accessible to the immune system. Helminthic parasites enter into circulation and are able to survive and replicate.They are well adapted to resist natural host defense mechanisms.

Mostly, people are infected with *Ascaris*, which is a roundworm that infects the small intestine. Parasitic worms that infect humans include trematodes of flukes, cestodes and nematodes or round worms. Tapeworms and hook worms inhabit at the gut. *Schistosoma haematobium* is a blood parasite of intestinal wall, liver and bladder. Several nematodes and platyhelminthes are parasites of domestic animals like pig and sheep.They enter the human body through improperly cooked meat, e.g. *Taenia*. Many parasitic worms pass through complicated life cycles through various parts of the host body. Hook worms and schistosome larvae invade their hosts directly by penetrating the skin. Tapeworms, pinworms and roundworms are ingested and filarial worms depend upon an intermediate insect host or vector to transmit them from person to person.

Immune Response to Helminthes Parasite

There are three species of *Schistosome,* which infect humans in different regions of the world. They are *Schistosome manoni, S. japonicum* and *S. haematobium.* When *S. mansoni* infection occurs, IgE titer in the blood goes up and shows localized increase in mast cells degranulation and increased number of eosinophils. Degranulation of mast cells release pharmacological mediators that increase infiltration of cells like macrophages and eosinophils. The eosinophils have IgE and IgG and bind to antibody coated parasites. Once bound to parasite, eosinophils release basic protein and sulfotransferase. Basic proteins released by eosinophils are toxic to helminthes.This is antibody dependent-cell mediated cytotoxicity ADCC.

Helminthes are resistant to cytocidal action of neutrophils and macrophages. In some helminthes' infections, a process occurs where by an initial infection is not eliminated but the body develops resistance to invasion of new worms. In acute and chronic *Schistosoma mansoni* infection, formation of granulomas around eggs in the liver is the main pathogenic condition. T_{H1} and T_{H2} also play important role in production of IgE for removing infection T_{H2} cytokines are very important for elimination of intestinal worms. Many cells are involved in generating immune response against parasite. These cells are phagocytic cells and NK cells. Antigen presenting cells (APCs) recognize the parasite. Innate immune recognition relies on a growing numbers of receptors i.e. pattern recognition receptors (PRRs). PRRs recognize pathogen associated molecular patterns (PAMPs) **(Figure13.8)**.

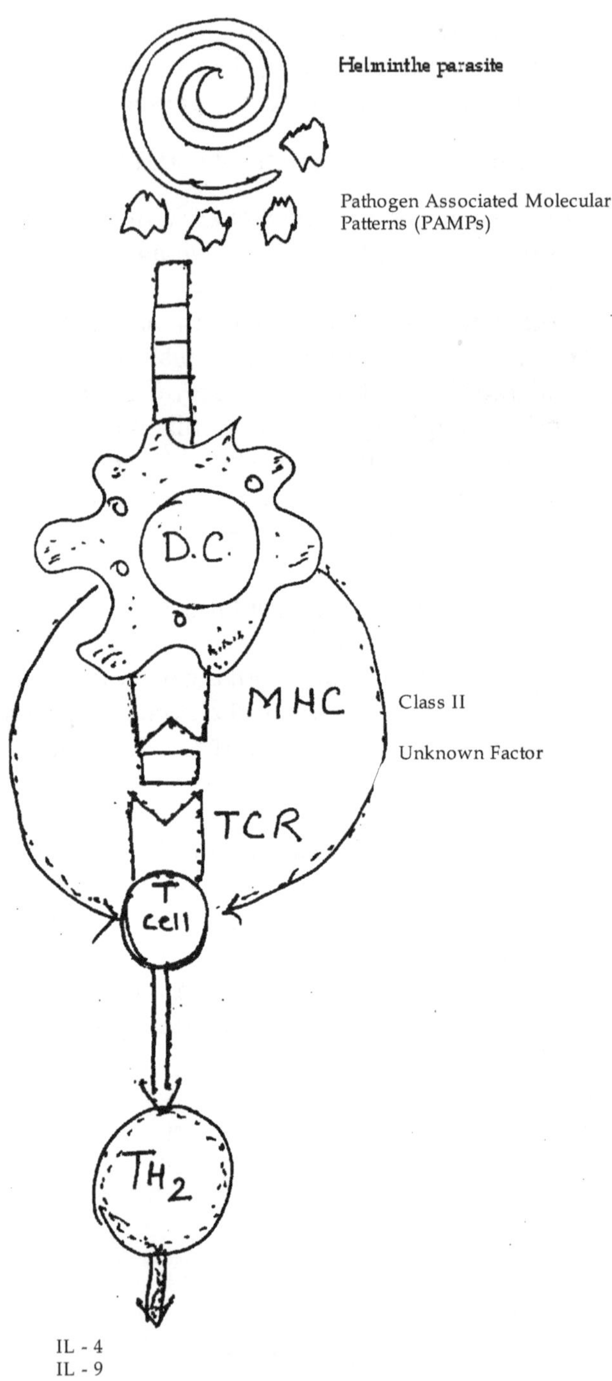

Fig. 13.8 : Pattern recognition receptors recognize pathogen associated molecular pattern.

Mucosal mast cells and eosinophils are important in determining the outcome of some helminthe infection and proliferate in response to the products of T cells and macrophage colony stimulating factor (GM-CSF) and IL-3, IL-5. IL-10 and transforming growth factor β (TGFβ) regulate the proinflammatory response and thus minimize pathological damage. In some helminthe infection, the immune system cannot eliminate the parasites but reacts by isolating the organism with inflammatory cells. The host reacts to antigen which stimulates the production of cytokines that recruit cells to the region of infection.

Eosinophils function specifically in the defense of tissue stages of the parasite which cannot be phagocytosed due to their huge size. IgE dependent mast cell activity is vital to localize eosinophils near the parasite and develop anti-parasitic activity. Major basic protein (MBP) of eosinophils crystalloid core causes damage to the parasite. Sometimes, mast cell and eosinophils act together and damage the *S. mansoni* larvae. All macrophage effector functions are enhanced soon after infection. The effector properties displayed by macrophages are observed in neutrophils. Neutrophils are phagocytic and kill by both oxygen dependent and oxygen independent mechanisms. They produce a more intense respiratory burst than macrophage and their secretary granule contain highly cytotoxic proteins. Neutrophils are present in parasite infected inflammatory lesions and remove the parasite from bursting cells.

Platelets have also capability to kill larval stage of flukes. Their cytotoxicity is augmented by cytokines such as IFN and TNFβ. In Some cases platelets become larvicidal. Platelet mediates antibody dependent cytotoxicity which is related to IgE.

Strategies of Helminthes to Evade Defense Mechanism

Different helminthe parasites have a characteristic feature to evade the immune response of host. Parasites turn out to be resistant to immune effector mechanism. Lung stage *Schistsome* larvae develop a tegument that is resistant to injury and is unaffected by antibody and cytokines. Some alter the surface coat through some biochemical process. The resistance that schistosomules acquire, as they mature, is also correlated with the appearance of surface molecules which is similar to decay accelerating factor (DAF).

Some parasites produce molecules that interfere with host immune system. Filarial worms secrete a protease inhibitor that affects proteases enzymes in processing of the antigens to peptides resulting in the reduction of class II MHC molecules.

Immunosuppression is a common feature of chronic helminthes infections. The ability of a parasite to suppress hyperactive immune response is due to the induction of regulatory T cells. Some parasites also restrain host immune response by multiple mechanisms. Nematodes have thick extracellular cuticle which protects them from toxic effect of immune response. Tapeworms prevent attack by secreting an elastase inhibitor which stops those attracting neutrophils. Some nematodes and trematodes have evolved an elegant method of disabling antibodies by secreting protease, which cleave immunoglobulins and prevent them from phagocytic process.

POINTS TO REMEMBER

- Different types of infectious organisms trigger distinct patterns of innate and specific immunity towards them.
- Antibodies, produced as a result of humoral immune response against viral antigens, are responsible for prevention of spread of virus during acute infection and against reinfection.
- The initial defense against virus invasion is the integrity of the body surface.
- Virus which cause infections are of two types: cytopathic viruses and non-cytopathic viruses.
- In case of cytopathic viruses, viruses enter into a host cell and uncoat. Their nucleic acid is released and transcription starts.
- Virus infection of a cell leads to the production of interferons IFNα and IFNβ.
- IFNα and IFNβ activate antiviral mechanisms in the adjacent cells. They activate a number of genes with direct antiviral activity.
- In most viral infections, cytolytic T cells (CTLs) activities arise within 3 or 4 days after infection. Within 7 to 10 days of primary infection, most virions have been eliminated.
- The CTLs system, operating against viruses, is highly efficient and selective. CTLs focus at the site of virus replication and destroy virus infected cells.
- The humoral immune response is the main response against extracellular bacteria.

- The antibodies act in several ways to protect the host from microorganisms.
- Extracellular bacteria produce inflammation, tissue damage and destruction at the site of damage.
- Bacterial toxins may be exotoxins and endotoxins.
- In some bacteria, complement activation directly lysis the bacteria.
- Bacteria secrete enzyme coagulase that precipitates a fibrin coat around them and protect them from phagocytic cells.
- Antibodies eliminate parasites from blood through the process of phagocytosis, opsonization and lysis,
- Many protozoans induce NK cell activity. Major functions of NK cells include lysis of target cell and the production of cytokines.
- Helminthes are resistant to cytocidal action of neutrophils and macrophages. In some helminthic infections, initial infection is not eliminated but the body develops resistance to invasion of new worms.
- Helminthes parasites turn out to be resistant to immune effector mechanism.

REVIEW QUESTIONS

1. Describe mode of viral infection and how immune response acts against it.
2. Describe different strategies of virus to evade immune response.
3. Describe immune response to extracellular and intracellular bacteria.
4. Describe immune response against protozoan parasites.
5. Give an account of different evading mechanisms of protozoa against host defense mechanism.
6. How does body protect against helminthic infection?

CHAPTER - 14

Vaccines

One of the greatest achievements in science is the development of vaccines. Vaccines have been revolutionary for the prevention of infectious diseases. They have prevented a large number of deaths caused, every year, by dreadful diseases like smallpox, polio, hepatitis etc. Diseases like smallpox and polio have been successfully eradicated by vaccination. Normally, when the body encounters a microbe, both humoral and cytotoxic immune responses are induced to neutralize its effect. In addition to eliminate the invader, body forms memory cells which help the body to protect itself from the same pathogen, in future. Vaccines not only function to kill the pathogens but they also stimulate the body to form memory cells which provide immunity against a specific pathogen

The word 'vaccine' has been derived from Latin word *vaccîn-us,* from *vacca,* cow. It was Edward Jenner who, for the first time, in 1748, deliberately inoculated people with cowpox virus, in order to protect them from smallpox. Since cowpox virus is closely related to smallpox virus, the injection of cowpox virus induced a primary immune response in the individuals, with the result of formation of antibodies and memory cells. When the smallpox virus attacked the body, memory cells were stimulated and produced secondary immune response which protected the individuals from smallpox. It was a milestone in the field of treatment of diseases. Jenner is credited to lead the foundation of vaccination.

Vaccination is process of inducing immunity through vaccines. A vaccine is a biological preparation used to enhance the adaptive immunity against some specific diseases. It is the preparations of organisms or a part of them. Vaccines may be prophylactic i.e. to prevent or ameliorate the effects of a future infection by pathogen, or therapeutic. Depending on the type of the vaccine, a vaccine may stimulate either *active* or *passive immunity* in the body.

ACTIVE AND PASSIVE IMMUNITY

Immunity can be classified into two types- innate and acquired immunity. Innate immunity is present at the time of birth and acquired immunity develops after birth. Acquired immunity can be further divided into two categories- active and passive immunity.

Active immunity: When a person suffers from a diseases, the body develops antibodies which protect the person from the same disease in future. Such type of immunity in which a person's own immune systems actively participates in the formation of antibodies is called active immunity. Depending on how the active immunity is acquired, it can be further classified as *naturally acquired active immunity* and *artificially acquired active immunity* **(Fig.14.1)**.

Naturally acquired active immunity: When a person suffers from a disease, the immunity which the body develops against the disease is called naturally acquired active immunity.

Artificially acquired active immunity: Besides suffering from a disease and acquiring immunity, active immunity can also be acquired by vaccination. When a live attenuated or killed microorganism (antigen) is inoculated in the body (immunization), the body develops antibodies and other specialized cells. This type immunity is called artificially acquired active immunity.

Passive immunity: When the body itself does not form antibodies but acquires readymade antibodies either naturally or artificially this is called passive immunity. Passive immunity is also of two types- naturally acquired passive immunity and artificially acquired passive immunity.

Naturally acquired passive immunity: A newborn acquires antibodies naturally from his mother. Maternal IgG antibodies are transported into the fetus via placenta which protect the child from a number of disease during early days of life. A newborn also receives IgA antibodies from his mother via colostrum (early milk). These antibodies protect the child until his own immune system develops.

Artificially acquired passive immunity: Passive immunity can also be acquired artificially by direct injection of antibodies. These

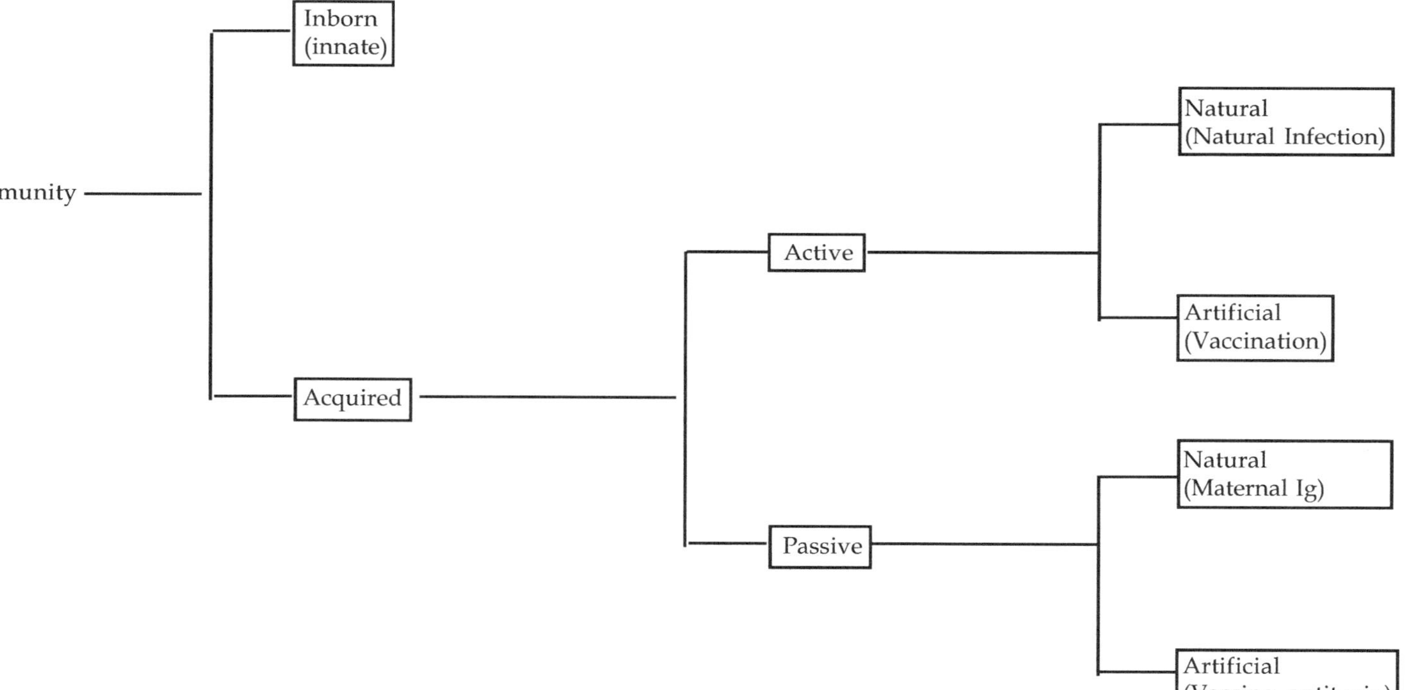

Fig. 14.1 : Types of immunity

antibodies can be obtained by several ways. These can be obtained from healthy blood donors or can be raised in suitable animals such as horse. The blood of those animals who have been vaccinated will contain high titers of antibodies for a disease for which they have been vaccinated. Antibodies to some diseases such as tetanus and diphtheria are raised in some animals such as horse by immunizing them with the toxins and by recovering the antitoxin antibodies from the immunized animals. These antibodies provide immediate protection to the person. But compared to active immunity, passive immunity is short lived and antibodies need to be administered every three weeks or so. Antibodies, obtained from animals can raise type III hypersensitivity reactions in some individuals.

HERD IMMUNITY

Vaccination is the best way of preventing diseases in a population. If in a population a large proportion of individuals are immunized for a particular disease then the chances of other individuals in the population to get infected by that disease are minimized. This is called *herd immunity.* Acquisition of herd immunity depends on the nature of the infectious organism. For example, if the pathogen is highly infectious, a large proportion of the population needs to be immunized for achieving herd immunity whereas in case of a pathogen being less infectious even a small proportion of the population can be sufficient to maintain herd immunity.

FEATURES OF AN IDEAL VACACINE

The objective of a vaccine is to protect an individual from a disease. It should not only provide protection but there should not be any side effect. There are certain features which identify an ideal vaccine. These are:

- ***Immunogenicity:*** A vaccine can be categorized as a good vaccine only if it has high immunogenicity. It should induce a fast and long lasting immune response. Most of the vaccines induce only humoral immune response. But an ideal vaccine should evoke both humoral as well as cell-mediated immune response.
- ***Safety***: Safety is the prime concern. There should not be any side effect associated with the vaccine. Even during manufacturing of the vaccine, there should not be any direct

exposure with the pathogen or its toxicants. It should be safe to administer. It should not cause any adverse reaction i.e. allergic reaction in the body.

- *Cost:* Cost is a very important factor. A vaccine should be cheap and easily available to the people who really need it, especially the poor people.
- *Stability:* An ideal vaccine should be stable and should not require any special condition (cold temperature) for storage and transport. It should be able to be stored at high temperature so that it is available at regions with hot climate with limited facilities for refrigeration.
- *Easy to administer:* An ideal vaccine is one which is easy to administer, without causing much discomfort. A needle free vaccine such as oral vaccine is the best approach in this regard.

TYPES OF VACCINES

Vaccinations can be broadly divided into two categories-those that are in current use and that which are experimental (Table 14.1).

Table 14.1 : Types of antigens used in the preparation of vaccines in use

Types of pathogen/antigen	vaccine
Vaccines in current use	
Whole Organism	
Natural intact organism	Vaccinia, Cowpox virus
Attenuated	Sabin polio, measles, mumps,
Inactivated (killed)	Hepatitis A, Influenza, Cholera
Subunit (Subcellular fragments)	
Capsular polysaccharide	*Streptococcus pneumoniae*
	Nesseria meningitides
Surface antigen	Hepatitis B
Toxoid	Tetanus, Diphtheria
Vaccines under experiments	
Anti-idiotype	
DNA vaccine	
Synthetic peptide vaccine	
Edible vaccine	

VACCINES IN CURRENT USE

Whole Organism Vaccine

Whole organism vaccines use killed organisms as vaccines or live attenuated organisms as vaccines.

The first vaccine-'vaccinia', the cowpox virus, was the only vaccine which used live organism for vaccination. But natural organisms cannot be used for vaccines as they are not safe. Strains which are virulent for another species but non-virulent in men can be used as whole organism vaccines.

Killed Whole Organism Vaccines

Pathogens are killed or inactivated by chemicals, irradiations or high temperature under controlled conditions. This destroys their ability to cause disease but they retain their antigenicity. Cholera, typhoid, pertusis and Salk polio vaccines are the examples of killed vaccines. Proper care must be taken during inactivation so that important protective antigens (proteins) are not destroyed and conformational determinants are retained. Chemical inactivation is preferred over heat because heat may destroy some of the surface proteins of the microbes. Though whole killed vaccines are safe in use, they have many drawbacks. Along with the specific antigens, killed whole organism vaccine retain many molecules which may cause allergic or other destructive reactions

Live Attenuated Vaccines

Live attenuated vaccines are much more effective than killed vaccines. They are used against many diseases. *Attenuation* refers to the weakening of a microbe's ability to cause disease. The objective of attenuation is to diminish the virulence of the microorganisms while retaining their antigenicity. Virulence is the ability of an organism to replicate and disseminate within the body without causing diseases. Thus, attenuation produces organisms which are not capable to cause a disease but can elicit an immune response. There are different methods to attenuate an organism. It can be attenuated by growing it under abnormal culture medium or modifying growth conditions. Microbes can also be attenuated by culturing them in their unnatural hosts. This modifies the microbes and they become less capable of causing disease in their natural host while retaining their antigenicity.

Louis Pasteur, for the first time, produced the live and non-virulent forms of chicken cholera bacillus and anthrax. Calmette and Guerin, for the first time, attenuated *Mycobacterium bovis* (a strain of *M. tuberculosis*)by culturing it for 13 years into a less virulent form known as BCG (Bacillus Calmette–Guerin). Attenuated vaccines are used against a number of diseases like polio, mumps, measles, rubella etc. Sabin polio vaccine was developed by growing the polio virus in epithelial cell line of monkey kidney. It consists of three strains of attenuated polioviruses which colonize intestinal areas and bring immunity against all the three virulent strains. MMR, a triple vaccine used against measles, mumps and rubella, uses live attenuated forms of viruses.

Recombinant DNA technology can also be used to remove virulence gene from the microorganisms. In general, it is easier to attenuate viruses than bacteria.

Live attenuated vaccines have many advantages over killed vaccines. In contrast to killed pathogens, live attenuated vaccines stimulate both arms of the immune system i.e. cell-mediated and humoral immune response and provide stronger immune response than killed vaccines which mainly induces humoral immune response. Live attenuated vaccines frequently confer lifelong immunity. Killed or inactivated microorganisms cannot replicate in the host and multiple doses of vaccines are required for a proper immune response. But in comparison to live attenuated vaccines, killed vaccines are much more safe and do not require refrigeration for storage. In live attenuated vaccines there is always a risk of reversal of the attenuated microbe into virulent form and causing full-blown diseases. Table 12.2 enlists some common killed and attenuated vaccines.

Table 14.2 : Killed and live attenuated vaccines

Killed vaccines	Attenuated vaccines
Virus	*Virus*
Hepatitis A	Measles
Influenza	Mumps
Rabies	Rubella
Salk polio vaccine	Sabin polio vaccine
	Varicella zoster
	Yellow fever
	Hepatitis A
Bacteria	*Bacteria*
Anthrax	Tuberculosis
Cholera	Typhoid
Plague	

Subcellular Fragments (Subunit) Vaccines

In subunit or subcellular vaccines, purified macromolecules, derived from pathogens, are used for immunization. These include inactivated exotoxins, capsular polysaccharides and recombinant surface antigens.

Capsular Polysaccharides

Some bacteria like *Haemophilus influenza* type B and *Streptococcus pneumoniae* have specific capsular polysaccharides which are used as vaccines. These bacteria are normally resistant to macrophages. Antibodies against these capsular polysaccharides provide effective protection from the bacteria. When capsule is coated with antibodies, it increases phagocytosis of the bacteria by macrophages and neutrophils. *Streptococcus pneumoniae* vaccine consists of 23 distinct capsular polysaccharides. Capsular vaccines are safer than live attenuated vaccines but capsular polysaccharides are relatively poorly immunogenic. The immunogenicity of these vaccines can be increased by conjugating them to a carrier protein. Conjugation results in enhanced immunogenicity than the polysaccharides alone. Such vaccines are also called conjugate vaccine. *Haemophilus influenza* type B (HIB) is a good example of conjugate vaccine in which capsular polysaccharide is conjugated with tetanus and diphtheria toxoid.

Toxoids

Many bacteria produce exotoxins which cause a variety of harmful effects in their hosts. For example, bacteria like *Corynebacterium diphtheriae* produce exotoxins which cause myocardial, nervous and renal damage. In such cases, vaccine is developed by inactivating the toxins. These toxins can be inactivated by treating them with some chemicals or heat. The treatment renders them inactivated. Toxins lose their toxicity but retain their antigenicity. These detoxified toxins, called, as toxoids, are unable to cause disease but can stimulate the body to form antitoxoid antibodies. These antibodies bind to the toxins and neutralizing their effects. DTaP is vary commonly used toxoid vaccine used against diphtheria, tetanus and acellular pertusis. The tetanus and diphtheria toxoid vaccines are the most successful bacterial vaccines.

VACCINES UNDER EXPERIMENTS

Anti-Idiotype Vaccines

It is another approach in vaccines which is especially useful in case where the antigen is unable to make immunogenic response. For example, polysaccharides and lipids are not immunogenic and cannot raise immune response when injected without a carrier.

In idiotype vaccines, antibodies are raised against variable regions of an effective antibody. This antibody (anti-Id) is similar to the original antigen and is immunogenic in nature **(Fig.14.2)**. This can be used in place of the original antigen which is not immunogenic.

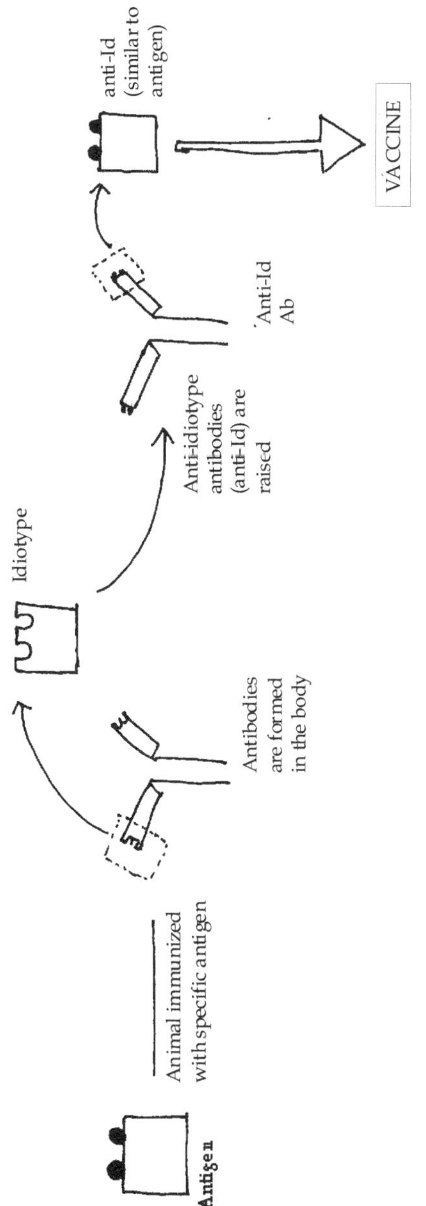

Fig, 14.2 : Development of anti-idiotype vaccine

Recombinant Vector Vaccine

Subcellular antigens can be produced by recombinant DNA technology. In this technique, genes coding for these antigens are isolated and cloned in bacteria or virus serving as a vector. This produces the required antigen in large amount which can be used for vaccine development. For example, a pathogenic gene can be introduced in attenuated vaccinia virus which would not only direct the expression of its own antigens but also cause the expression of pathogen's antigen. The technology has been used to develop vaccine against Hepatitis B surface antigens (HBsAg). The genes coding for these antigens are cloned in yeast and synthesized proteins are purified. These vaccines are much safer and less expensive. Immunogenicity of these vaccines can be further enhanced by gene manipulation.

DNA Vaccine

DNA vaccine is a new approach in the field of vaccines. They are quite different in structure from traditional vaccines.

In DNA vaccine specific genes are selected from a disease causing pathogen. These are spliced into plasmid. Plasmids are the extra chromosomal DNA, present in bacteria. They have the capacity of self replication. The genes isolated from the pathogen are spliced into the plasmid. The recombinant plasmid (plasmid containing specific genes) is then delivered into the body. There are several ways of delivering the plasmids. They may either be injected into small groups of cells by injecting into the muscle cells or by a device, called gene gun, into the cells, near the surface of the body like skin or mucous membrane. Once inside the cells, the recombinant plasmids enter the nucleus and synthesize the encoded antigenic proteins. These proteins are released by the cells or chopped into small peptide fragments. They elicit both humoral and cell-mediated immune response.

The released antigenic proteins bind to the B cells and activate humoral immune response. The B cells are activated into plasma cells to secrete antibodies which would destroy the pathogens. Some of the B cells become memory cells and will act during infection in the future. T_H cells release T_{H2} cytokines which help in the activation of B cells. During the process, the antigen processing cells take and process the antigenic proteins and display them on the surface along with MHC class I molecules. The T_H cells recognize these antigenic proteins and release T_{H1} cytokines, as shown in **figure 14.3**.

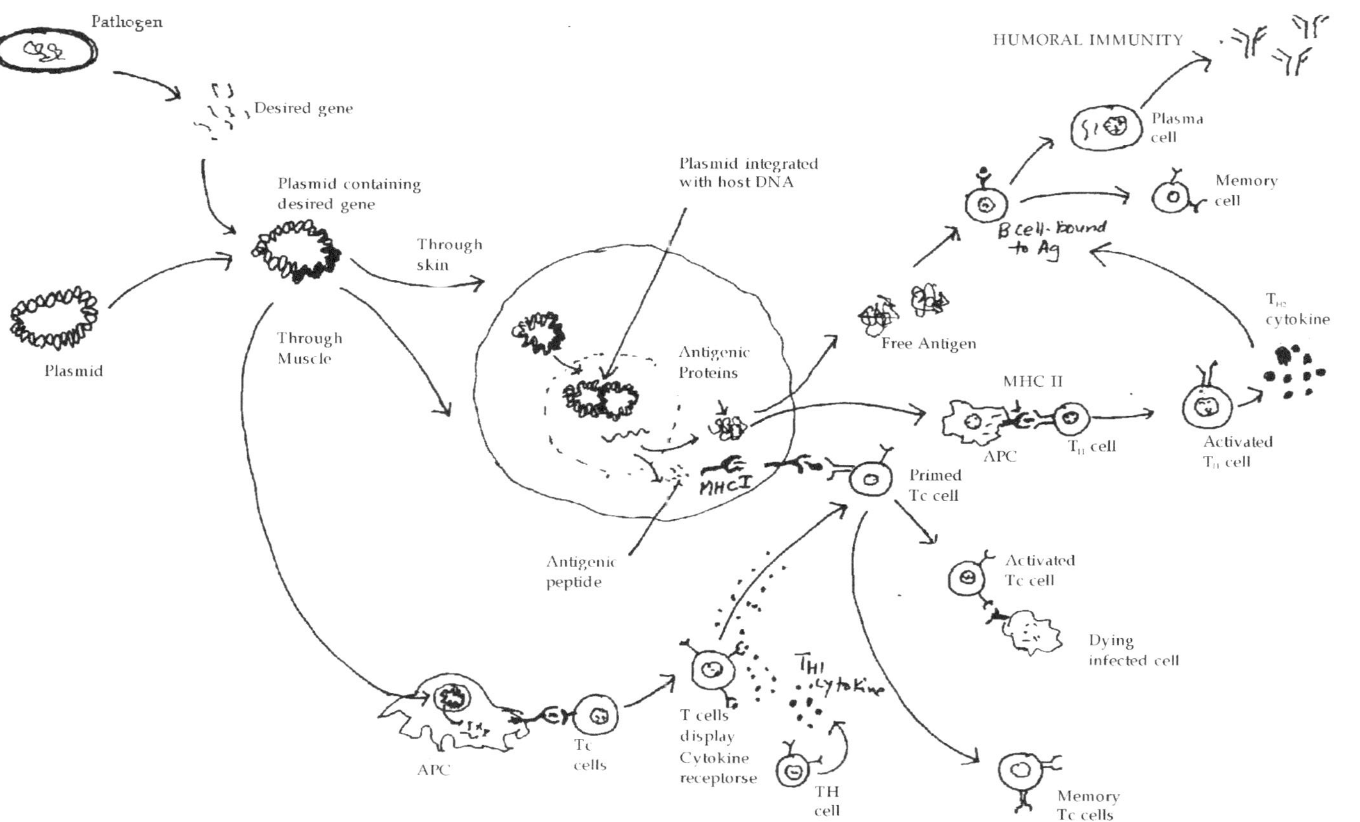

Figl. 14.3 : Mechanism of Action of DNA vaccines

The small peptides bind into the groove of MHC class II molecules and are displayed on the surface of inoculated cells. They activate Tc cells and trigger cell mediated immune response. Tc cells recognize these antigenic peptides along with MHC class II molecules. Tc cells multiply and kill the inoculated and other infected cells. Some of the activated Tc cells become memory cells. T_H cells also help in activating Tc cells by releasing T_{H1} cytokines

Advantages and limitations of DNA vaccine

DNA vaccine has several advantages over conventional vaccines. Since no pathogenic gene is used in the development of DNA vaccine, they are safe and there is no chance of reversal of the virulence form as in attenuated vaccines. DNA vaccines stimulate both arms of the immune system i.e. humoral immune response as well as cell-mediated immune response and provide better protection against a disease. A characteristic feature of a good vaccine is that it should be less expensive and easily available to a mass population. DNA vaccines can be easily produced in large quantity by recombinant technology and are less expensive. In addition, DNA vaccines produce proteins which stay longer in the body and provide immunity for longer time. These vaccines are more stable and do not require cold for storage and transport. DNA vaccines can be used both for therapeutic as well as preventive purposes.

Some limitations are also associated with DNA vaccines. Since antigens used for DNA vaccines produce proteins in the body such vaccines can be used for the protein antigens only and not for carbohydrate or any other molecule. There is also a risk of production of antibodies against self DNA and it may also affect other genes in the body.In addition, there is also possibility of inducing tolerance to the antigen (protein) produced.

Researches are also being carried out to increase the immune reactivity of the DNA vaccines. There are certain sequences of DNA in the plasmid which are known as *immune -stimulating sequences.* The presence of such sequences can amplify the immunogenicity of the genetic codes in a DNA vaccine. By increasing the number of such sequences, the immunogenicity of the genetic vaccines can be enhanced. Another approach is by incorporating genes of cytokines along with the antigen genes in the plasmid.

Synthetic peptide vaccine

These vaccines consist of synthetic peptide containing B and/or T cell epitopes thus, stimulating both arms of immune system. The peptides can be easily synthesized in bulk and hence are cheap in cost. Since they do not involve DNA or inactivation of toxins, synthetic peptide vaccines are safe. But due to their small size, peptides are less immunogenic and require some carrier proteins. It is also difficult to identify T cell epitopes.

Edible vaccines

Plant-derived vaccines comprise a new area in the field of vaccines. Several plant-derived vaccine antigens have been found to be safe and induce sufficiently high immune response. Edible plant parts can be excellent alternatives for the production of vaccines. Edible vaccine is a concept which is a cost-effective, easy-to-administer, easy-to-store, safe and socio-culturally acceptable vaccine.

Edible vaccines involve introduction of selected desired genes into plants and then inducing these altered plants to manufacture the encoded proteins. The process is called *transformation* and such plants are called *transgenic plants*. Plants like banana, potato, tomato, lettuce and rice are under study to be used as edible vaccines. The genes, encoding orally active antigenic proteins, are isolated from the microbe and integrated into the genome of the selected plant. The foreign DNA can be introduced in the plant either by gene gun method or through *Agrobacterium tumefaciens*, a naturally occurring soil bacterium which has the ability to get into plants through some kind of wound. A circular plasmid -Ti plasmid (tumor inducing) is present in the bacteria which enables it to infect plant cells and integrate into their genome. Once the genes are integrated into the genome of the selected plant, the plant produces the specific proteins in its parts or products **(Fig. 14.4)**. Production of transgenic plants is species dependent and takes 3 to 9 months. The part or product is fed raw to the human / animals to bring out immunization. Multi-component vaccines can be obtained by crossing two plant lines harboring different antigens.

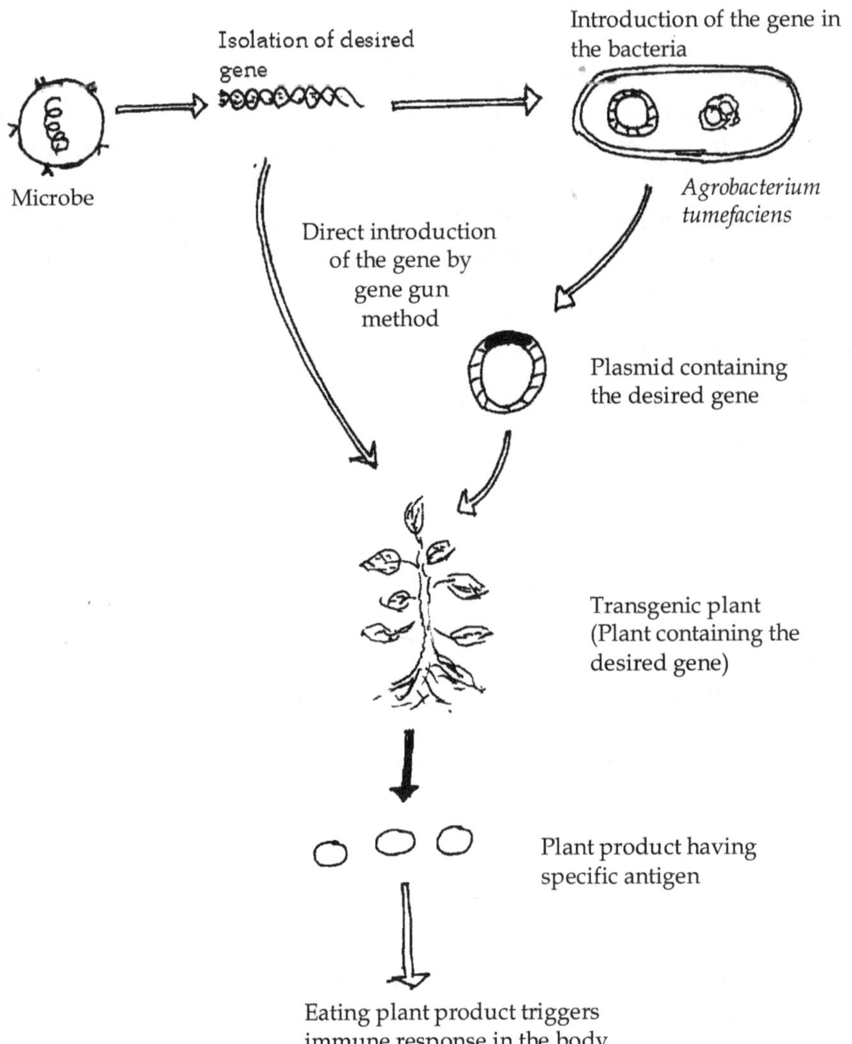

Fig. 14.4 : Development of edible vaccine

Edible vaccines have a bright future. In comparison to traditional vaccines, these vaccines have some significant advantages such as low cost, greater safety, and greater effectiveness. As they consist of antigenic proteins and do not involve any pathogenic gene, edible vaccines are safe and do not have side effects. Their production is highly efficient i.e. plants can be grown in a small area with high yield. They can be grown locally using standard methods and do not require large intensive pharmaceutical manufacturing facilities. Vaccines exhibit good genetic stability. They are heat-stable and do not require cold-chain maintenance. Vaccines can be stored near the site of use and eliminate long-distance transportation. As these vaccines do not require syringes and needles, chances of infection are very less. There is no need for skilled medical personnel. Edible vaccines also have applications in birth control, malaria and cancer therapy.

Currently, the edible vaccines are being developed and focused for a number of human and animal diseases, including measles, cholera, foot and mouth disease and hepatitis B, C and E. Improved understanding of plant molecular biology and consequent refinement in the genetic engineering techniques can lead in designing of new vaccines which can induce very specific and sufficiently high immune response.

The concept of edible vaccine seems to be a very revolutionary one. But there are many questions which need to be answered. The first problem is that the antigenic protein may be rapidly degraded in the digestive tract. The determination of the right dosage is a major problem in edible vaccines. The dose may vary according to a person's weight, age and sex. The dose may also depend on fruit/plant's size, ripeness and protein content. The amount to be eaten is very critical because too low a dose would fail to induce antibodies and too high a dose would, instead, cause tolerance. Orally administered antigen often tolerize rather than immunize.

FUTURE CHALLENGES

Though vaccines have been developed against a number of diseases, still there occur approximately two million unnecessary deaths every year, especially in the remote and impoverished parts of the world. There are many diseases for which effective vaccines are still awaited or if they exist, they are unreliable or very expensive. Vaccines are required against cancer, malaria, AIDS, hepatitis C and other parasitic diseases. With advancement in the field of biotechnology, there has been a hope that edible vaccines can be a promising one against the malaria parasite. Significant progress has been achieved in the research for edible vaccine against *Plasmodium falciparum.*

There is a need for the vaccines which are not only cost effective but are also easily available to the people. Development of a single dose vaccine which does not require booster dose; and a single vaccine against a multiple diseases are the new challenges in the field of development of vaccines.

POINTS TO REMEMBER

- A vaccine is a biological preparation, used to enhance the adaptive immunity against some specific diseases.
- Vaccines may be prophylactic or therapeutic.
- There are different types of vaccines such as whole organisms, sub cellular, anti-idiotype , DNA, and edible vaccines.
- Whole organism vaccines are either inactivated killed vaccines or live attenuated vaccines.
- In killed vaccines, killed or inactivated microorganisms are used as vaccines. Killed vaccines are used against many diseases like cholera, typhoid, pertusis and polio.
- In attenuated vaccines, microorganisms are attenuated which renders them incapable of causing disease. Attenuation can be achieved by heat, chemicals or irradiation.
- Attenuated vaccines are more effective than killed vaccines. They stimulate both cell-mediated and humoral immune response and immune response is stronger and for a longer time.
- Surface antigens such as inactivated exotoxins, capsular polysaccharides and recombinant surface antigens are also used as subcellular vaccines.
- Capsular vaccines are safer than live attenuated vaccines but capsular polysaccharides are relatively poorly immunogenic.
- Anti-tetanus and DPT are the most commonly used toxoid vaccines.
- In Anti- idiotype vaccines antibodies are raised against variable regions of an effective antibody. They are under experiment.
- DNA vaccine is a new approach in the field of vaccines which consists of plasmid- a small, circular, genetically engineered DNA, which is injected in the body to produce an immunological response.
- DNA vaccines have many advantages over conventional vaccines. They are more safe and stimulate both humoral as well as cell-mediated immune response and provide better protection against a disease. DNA vaccines are under experiment.

- Plant products can also be used as vaccines. Here, selected desired genes are introduced into the plants to produce the encoded protein in their products. Plants like banana, potato, tomato, lettuce and rice are under study to be used as edible vaccines.
- Edible vaccines are cost effective, easy to administer, easy to store and socially acceptable.

REVIEW QUESTIONS

1. What are vaccines? How do they protect the body?
2. Describe general characteristics of a good vaccine.
3. Why is a live attenuated vaccine more protective than a killed vaccine?
4. What are DNA vaccines? Describe the mechanism of their action.
5. Describe some advantages and disadvantages of DNA vaccines.
6. What are subunit vaccines? Describe different types of subunit vaccines.
7. Write short notes on:
   i. Edible vaccines
   ii. Anti-idiotype vaccines

Chapter - 15

Tolerance

Immunity is defined as the resistance to infection. This is carried out by the process of identification and removal of non-self material. In contrast to this, our body has to maintain immune tolerance to our own cells and tissues and develop the ability to distinguish these self-antigensfrom non-self (foreign) antigens. Tolerance means unresponsiveness of the immune system to an antigen. Immune tolerance is the process by which the immune system does not destroy an antigen. Tolerance can be either natural or induced. *Natural or self tolerance* prevents the body from mounting an immune attack against its own tissues. *Induced tolerance* is the process in which tolerance to an external antigen can be created by manipulating the immune system. Natural or self tolerance is a physiologically learnt process. When the immune system recognizes a self antigen and mounts a strong response against it, autoimmune disease develops but in induced tolerance, the immune system has to recognize self-MHC to mount a response against foreign antigens. So, the immune system is constantly challenged to discriminate self versus non-self and mediate the right response. Importance of induced tolerance is to protect us from foreign antigen and allergic reactions.

DEVELOPMENT OF IMMUNE TOLERANCE

Immune tolerance can be achieved at different levels and in different forms. It occurs in three forms:

- Central tolerance
- Peripheral tolerance
- Acquired tolerance

Central Tolerance

Central tolerance occurs during the development of T cells in the thymus and B cells in the bone marrow. During central tolerance the lymphocytes (T and B lymphocytes) that are reactive to self antigens are removed during their maturation in thymus and bone marrow. This can also be achieved by altering the receptors on the cells that react with self antigens. T lymphocytes are not simply effecter cells of the immune system but they also regulate the immune system.T lymphocytes develop from precursors in the bone marrow and are derived from a common lymphoid progenitor cells like B cells. When immature T cells enter the thymus they express neither CD4 nor CD8 co-receptor molecules. These cells called double–negative (DN) cells. DN cells constitute up to 3% of total lymphocytes. In the **figure 15.1**, the two stages of central tolerance, in T lymphocytes, are shown. In stage 1, thymocytes with receptor bound to class I MHC and class II MHC molecules survive, whereas those which are not bound, die by apoptosis. In stage II, those thymocytes which bind to self peptide MHC complexes die by apoptosis. Those T cells which survive recognize antigen peptide, derived from foreign antigen and displayed by self MHC molecules. These T cells mature into CD4 and CD8 and are released from thymus and circulate in the blood. T cell development involves positive and negative selection. Positive selection occurs when naïve T cells are exposed to antigen in the thymus. T cells have receptors with sufficient affinity for self MHC molecules. T cells recognize antigens only when they are associated with MHC molecules. This is very essential for self recognition. Many self antigens are processed and presented by thymic antigen presenting cells (APCs) in association with self MHC. T cells are positively selected for usefulness, so positive selection is the key for the development of self tolerance. Positive selection is important for auto reactivity. The negative selection is critical for self tolerance. The developing T cells are exposed to different combination of self antigens. Young T cells bind to self antigen and die by apoptosis in the thymus, so they do not reach in circulation. This may be the cause of auto immune disease. Negative selection process occurs during B cell development. Self reactive lymphocytes cannot be eliminated by central tolerance mechanism because most self antigens are absent in the primary lymphoid organs.

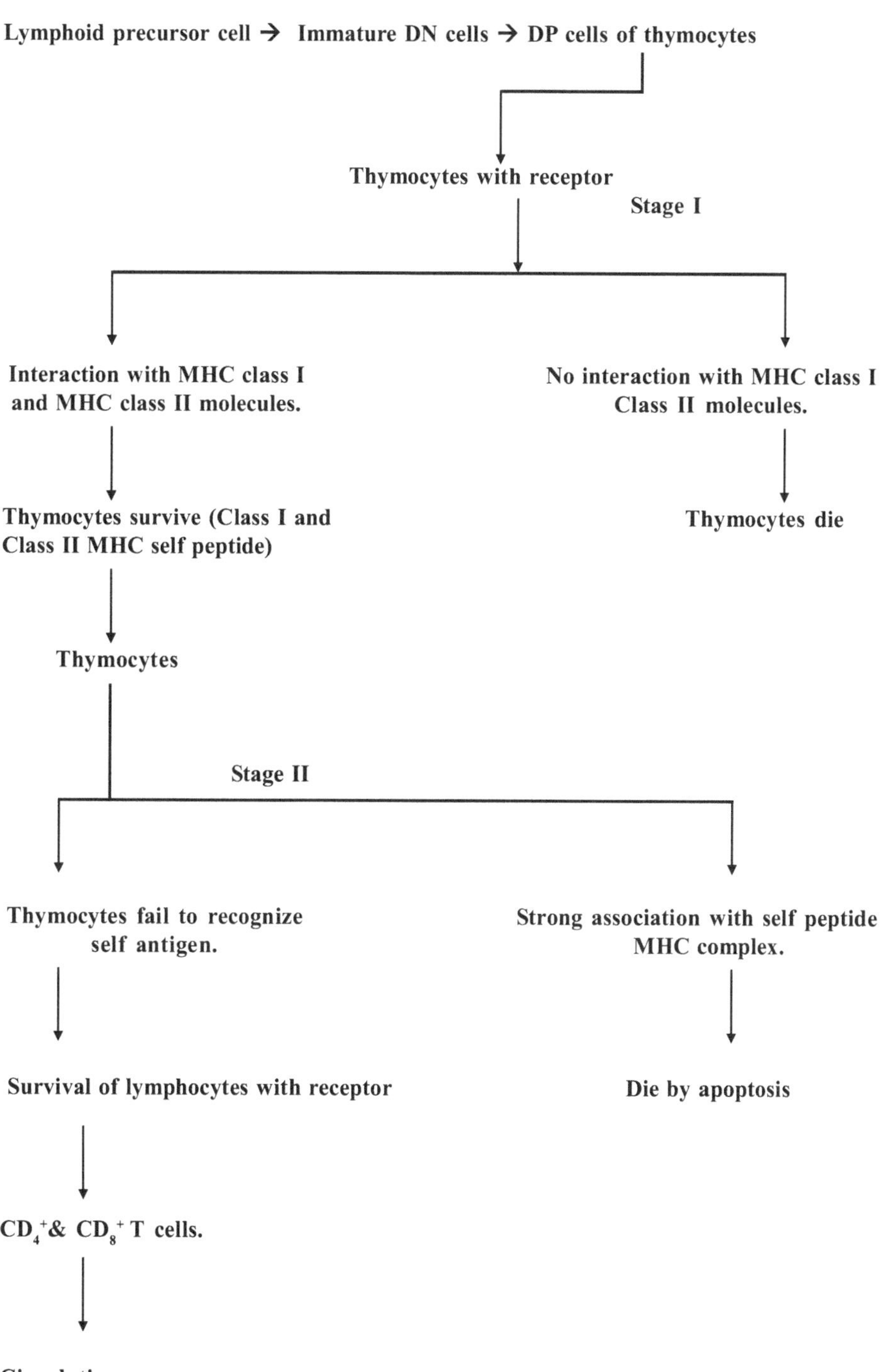

Fig. 15.1 : Central Tolerance in T Lymphocytes

Peripheral Tolerance

Peripheral tolerance is developed after the T and B cells mature and enter in the periphery. If ,the self reactive T and B cells escapes deletion in the thymus or bone marrow and appear in the periphery, they are inactivated by peripheral tolerance.

T cells that escape tolerance in the thymus also ignore self antigens if they are on privileged sites. Privileged sites include brain, anterior chamber of eye and testes. In privileged sites pro- inflammatory lymphocytes are controlled by cytokines or by apoptosis. Immune system privilege contributes to peripheral tolerance in these particular organs. Apoptosis of lymphocytes is essential for the maintenance of immune homeostasis. The lymphocytes that are deprived of co-stimulatory and cytokine stimuli are known to lose anti-apoptotic proteins and ultimately die by ignore. This process is completed by two mechanisms that are *activation induced cell death* (AICD) and *programmed cell death*(PCD). In both conditions, cell death seems to show same morphological and biochemical appearance but their molecular and physiological control is different. Many activated cells die by PCD because their antigens are eliminated. The survival rate of T cells is related with self antigens but at the time of immune response against infection. T cells will be increased in the absence of AICD. Both AICD and PCD are strongly regulated. AICD is also regulated by IL-2 cytokines.

Dendritic cells also contribute to peripheral tolerance by deleting self antigen responsive T cells.B lymphocytes occur in bone marrow. Some microorganisms have cross reactive antigens and stimulate B cells. These antigens also stimulate antibody response. B lymphocyte tolerance occurs in both stages, during their development and after antigenic stimulation. Self reactive B cells may be deleted or anergized but it depends upon the nature of antigen and affinity of B cells to antigens. At the time of development pre B cells are converted into immature B cells. B cells leave the bone marrow by expressing antigen binding receptor IgM. Immature B cells are converted into mature B cells. Mature B cells have an antigen binding receptor IgM. Self tolerance begins when IgM appears at the surface of B cells. When a B cell reacts to an antigen through IgM, the cell gets activated and divides rapidly. If the new IgM receptor does not react with a self antigen in bone marrow, B cell development can proceed. Immature B cell is resistant to apoptotic cell death but the transitional cell is sensitive. Immature B cells leave the bone marrow. These cells express the antigen and migrate to spleen. Spleen eliminates unwanted B cells. MostB cells require help from T cells to develop into plasma cells. This occurs through co- stimulatory signals which include co-stimulatory molecules and cytokines. In the absence of these signals they may undergo a state of unresponsiveness. B cells require two signals for activation by T dependent antigens. Signal 1 is delivered by antigen binding to mIgM and the second signal when CD40 L on the T cell surface interacts with CD40 expressed on the B cell. Self reactive B cell, on contact with cell antigen, do not receive the required co-stimulatory signals from T helper cells and as a result become anergic. In the absence of signal II, self reactive B cells are exposed to self antigens. B cells are tolerized by only high avidity antigens. Self reactive B cells which are present in outer T cell zone are short lived cells. These short lived B cells contribute to immune response to foreign antigen. B cells provide an additional level of protection from infection.

In addition,peripheral tolerance is regulated by a unique type of TH cells that express high levels of IL-2Rα chain. These cells called T_{reg} cells down regulate autoimmune processes.

Acquired or Induced Tolerance

Acquired or induced tolerance refers to the adaptation in immune system to a particular antigen. In these circumstances cell mediated or humoral immunity is generated. It is clearly visible in pregnancy where the fetus and the placenta must be tolerated by the maternal immune system. In adults acquired tolerance is induced by administration of antigen. The antigen is administered either intravenously or intramuscular. These antigens should be nontoxic and capable of being ingested. Induced tolerance is important in organ transplantation. Acquired tolerance is influenced by the nature of antigen and its route of administration.

TYPES OF TOLERANCE

Tolerance is classified into different types:

- Low dose tolerance
- High dose tolerance
- Classical tolerance
- Infectious tolerance
- Mutation tolerance

- Clonal tolerance

Low dose tolerance: This type of tolerance is induced by small dose of antigens. It is mediated by active suppression of immune response by T cells. This tolerance is due to the unresponsiveness of T lymphocytes.

High dose tolerance: This type of tolerance is induced by high dose of the antigen. High dose of oral antigen can also induce lymphocyte anergy. Anergy means absence of normal immune response to a particular antigen. Anergy occurs through T cell receptor (TCR). High dose tolerance is due to the unresponsiveness of B cells as well as T cells.

Classical tolerance: This type of tolerance is induced by proper administration of antigens. It is brought about by inactivation of lymphocytes. In this type of tolerance antibody is not produced and the recovery of T and B cell become very slow.

Infectious tolerance: This type of tolerance is due to blocking of lymphocyte receptors by antigens and also by antigen-antibody complex. In this type of tolerance antibodies are produced. Recovery of lymphocytes is fast.

Mutation tolerance: In this type of tolerance, the gene which is responsible for the differentiation of immature and mature lymphocytes may be defective. In this process self antigens do not react with T cells, B cells and macrophages.

Clonal tolerance: The production of T and B cells during the development of immune response against an antigen is a clonal selection. These cells interact with an antigen and proliferate into many clones and only one cell is selected for immune system. Complete elimination of a group of lymphocytes from immune system is called clonal deletion. In clonal deletion antibodies are not produced. Inactivation of a particular group of lymphocytes refers to clonal abortion and here also antibodies are not produced.

MECHANISM OF TOLERANCE

The immune system identifies an antigen as self antigen and develops tolerance. Tolerance is easier to achieve before birth or during natal life. Tolerance is induced to certain antigens under appropriate conditions. It is important to immune defense. Acquired tolerance is primarily associated with tolerance to non-self antigen and involves anergy, deletion and active suppression by T_{H2} cells. Tolerance is influenced by the nature of antigen. An antigen that induces tolerance is termed as *tolerogen*. Size and molecular complexity are the most important factors determining whether a substance is antigenic or tolerogenic. Large or complex antigens are usually not tolerogenic because they are phagocytosed by macrophages but small or soluble antigens may not be taken up by the macrophages. This will favor the development of tolerance. Tolerance can be induced with antigen doses, both low dose and high doses. Low dose tolerance affects T lymphocytes and high dose tolerance affects both T and B lymphocytes. T cell unresponsiveness develops after one or two days of administration of antigen, while B cell unresponsiveness develops after a week. Tolerance also develops as a result of mutation under sudden change in genes. Tolerance is induced by the production of T suppressor cells. They inhibit the T helper cells. So T_H cells can not cooperate with B cells to elicit an immune response leading to the development of tolerance.

CONDITION FOR INDUCTION OF TOLERANCE

The substance capable of stimulating antibody production can also induce tolerance and are called *tolerogens*. Tolerogens are proteins, polysaccharides and haptens. Tolerance is specific to particular antigenic determinant. All body components, during early stage of development are exposed to immune system. Tolerance is more easily induced in the fetus and in the newly borne, when the immune system remains relatively immature. The form of antibody is an important factor for the induction or non-induction of tolerance. Aggregated molecules are immunogenic but monomers are tolerogenic. It is usually difficult to induce tolerance to virus, bacteria and other pathogens which are highly immunogenic. The soluble antigen in its monomeric form is tolerogenic and in polymeric form it is immunogenic. Route and intake of antigens also influence tolerance. Immunogenic antigens areinjected into tissues whereas tolerogenic antigens are injected intravenously. Tolerance is achieved in T cells from spleen and thymus after low antigen dose within few hours and tolerance of spleen B cells require high dose of antigen and more time. High dose of orally administered antigen can cause anergy. Low dose of orally administered antigen induce T cells in gut. At this time antigen antibody response develops. Antibody response in gut requires IgA. Low dose of tolerance is mediated by active suppression of immune response by T cells.

Regulatory T cells are divided into subgroups such as $CD4^+$, $CD25^+$ regulatory T cells, T_{H3}cells and T_{R1} cells. These cells produce cytokines IL-10 and transforming growth factor β (TGF-β) and serve to inhibit inflammatory responses mediated by T_{H1} cells. TGF-β inhibits the proliferation and function of B cells, cytotoxic T cells and NK cells.Cytokine production in lymphocytes antagonizes the effects of TNF. Subgroups of T cells can suppress immune response in surrounding areas. This effect is known as *bystander suppression*.

Tolerance is a normal feature of the immune system at mucosal surfaces. Tolerance can also be generated through nasal administration of antigen.

Anti-idiotypic response also induces tolerance. In this reaction, antibody combines with an antigen and produce anti-idiotypic antibody. These antibodies can block B cell responsiveness of the cell. The idiotype of T cell is represented by the polymorphic regions of the TCR α and β chains. TCR specific regulatory cells are induced after immunization with self reactive T cells, TCR peptides and TCR molecules.

An animal is recovered from tolerance through many ways, such as elimination of antigens from immune system, production of antibodies and by the addition of immunocompetent cells. Tolerance could be valuable to the damaging immune response in hypersensitive state and autoimmune disease.

POINTS TO REMEMBER

- Tolerance is a state of antigen specific immunological unresponsiveness.
- Tolerance is antigen-specific which involves the T and B lymphocytes.
- Tolerance or unresponsiveness of the immune system occurs in three forms: central, peripheral and acquired.
- Central tolerance is the process where immature T and B cells acquire tolerance to self antigens during maturation within the primary lymphoid organs.
- The immature T cells enter the thymus through two types of selection processes such as positive and negative selection.
- In positive selection, T cells recognize MHC peptide which is essential for self recognition.

- In negative selection, T cells are exposed to different combination of self antigens.
- Peripheral tolerance is the process where mature lymphocytes acquire tolerance to self antigens in the peripheral tissue.
- Central and peripheral tolerance are associated with anergy as well as deletion of lymphocytes.
- Acquired tolerance is associated with non-self antigens and involves anergy and deletion.
- Tolerance is classified into different categories such as low and high dose tolerance, classical tolerance, infectious tolerance, mutation tolerance and clonal tolerance.
- Tolerance is induced by both tolerogenic and immunogenic antigens.
- AICD and PCD pathways of apoptotic death show same terminal effectors' phase.

REVIEW QUESTIONS

1. What is tolerance? Enlist various kinds of tolerance.
2. Explain different forms of tolerance.
3. Give an account of the mechanism of tolerance.
4. Describe various conditions for induction of tolerance.
5. Describe the role of antigens in tolerance.

CHAPTER - 16

Transplantation Immunology

Transplantation refers to the implantation of a tissue or organ from one site to another site in an individual or from one individual to another.It involves implanting of organs or grafting of tissues to repair defects and to stimulate regeneration. Auto-transplantation or autoplasty of the skin, cartilage, bone, muscle, tendon, nerve and fatty tissue is used in plastic and cosmetic surgery. The replacement of a diseased organ by a transplant of a healthy tissue is called *transplantation.* The animal from which the graft is taken is called *donor* and the animal in which the graft is placed is called *recipient.* The transplant leads to various complications in the host which are mediated through host body's immune response and very often the transplant gets rejected. *Transplantation immunology* refers to the study of the response of the immune system where a graft is implanted in an animal. Transplantation immunology is a general term for the complex phenomenon of the pathobiology of the immune system in transplantation. Skin and tumor grafts are rejected mainly by T lymphocytes that enter the tissue and release lymphokines. In kidney, heart or other organ transplantation, a different rejection mechanism is processed. The blood vessels of the organ remain functional in the recipient's body and antigens stimulate host B lymphocytes to produce antibodies. The antibodies enter the organ, where they combine with the antigens on the surface of the cells. This interaction activates complement and the lysed due to cytotoxic hypersensitivity.

TYPES OF GRAFTS

- Autograft or Autogenic graft

- Isograft or Syngenic graft
- Allograft or Allogenic graft
- Xenograft or xenogenic graft

1. **Autograft:** It is a self tissue, transferred from original donor site to another site of the donor. It is a common feature in plastic surgery and orthopaedic surgery. For example, skin graft from thigh to face in deformed cases or in a burnt patient. The antigen present in the autograft is same as present in the body. An autograft survives throughout life.
2. **Isograft** : It is also called *syngraft.* It is a graft that takes place between two genetically identical individuals of same species. In isograft, the histocompatibility antigens are identical so the graft survives and it is not rejected.
3. **Allograft:** This graft is transferred between genetically different members of the same species. This is called *homograft.* In allograft, the histocompatibilty antigens are dissimilar.
4. **Xenograft:** This graft is transferred from one species to another. In Xenograft, the histocompatibility complex antigens are so different that the graft is more vigorously rejected than in an allograft. This is also called *heterograft.*

MECHANISM OF GRAFT REJECTION

The major problem of rejection is the immune system of recipient. The graft area shows increased population of polymorphonuclear lymphocytes which invade the area. Thrombosis and acute cell destruction can be seen after 3 or 4 days after grafting. T cells are involved in allograft rejection. T cell sub population $CD4^+$ and $CD8^+$ both are involved in graft rejection. The role of humoral antibody is in the destruction of lymphocytes through cytotoxicity in allogenic graft. Tissues that are antigenically similar, that are histocompatible, do not induce immune response to graft rejection. Tissues that display significant antigenic differences are non histocompatible and induce immune response to graft rejection. The various antigens that determine histocompatibility are encoded by more than 40 different loci. These loci are responsible for the allograft rejection and are located within major histocompatibility complex (MHC). MHC loci are closely linked; they are usually inherited as a complete set called haplotype, from each parent. MHC provokes the most powerful rejection of grafts between members of same species. The intensity of MHC mismatched rejection is a consequence of the alloreactive cells which are present in normal individuals. Allogenic MHC differs from the recipient. The allogenic MHC will be able to bind to a number of peptides which are derived from proteins, common to donor and host, which might be unable to fit the groove in the host MHC and fail to induce self tolerance.

Differences in blood groups and major histocompatibility antigens are responsible for intense graft rejection. The blood group antigens are expressed on RBCs, epithelial cells and endothelial cells. Antibodies produced in the recipient to any of these antigens are present on transplanted tissue and will induce antibody mediated complement lysis of the donor cell. Repeated blood transfusion develops antibody titer against MHC antigens on WBCs in the transfused blood.

If the MHC antigens are same as those on subsequent grafts then hyper acute rejection develops due to the action of antibody on graft. Sometimes hyper acute rejection of graft may occur due to presence of antibodies against blood group antigen in the graft. Therefore, ABO blood group matching and tissue typing are carefully performed prior to transplantation to avoid rejection. Graft rejection is developed up to 10 days of transplantation. It is due to infiltration of the graft mononuclear cells and rupture of peritubular capillaries and appears as cell mediated hypersensitivity.

Major histocompatibility complex (MHC) is the major cause of graft rejections. MHC that encodes transplantation antigens includes human leukocyte antigens, HLA-A, HLA-B and HLA-C. MHC was originally studied in leukocytes. Every person has a unique antigenic profile, which is a very big hurdle in successful transplantation. If the MHC identity of donor and recipient are almost same, there is possibility of rejection due to differences at various minor histocompatibility loci. MHC in human exists on chromosome 6 and close to immune response (Ir) genes. The MHC consists of four clusters of genes, each gene having number of versions. MHC genes and antigens are the key factors in acceptance or rejection of a transplant. When a tissue is transplanted between genetically different individuals, even if the MHC antigens are identical, the transplanted tissue might be rejected because of differences at various minor histocompatibility loci. MHC antigens are generally recognized by T_H and T_C cells which are called alloreactivity. The mechanism of graft rejection can be divided into two phases or stages-*sensitization stage* and *effector stage.*

Sensitization Stage

This is the stage of recognition of alloantigens (foreign protein) through direct or indirect pathways. In this stage T cells (CD4 and CD8) of the recipient, recognize alloantigens, which are expressed on cells of foreign graft.

In the direct pathway of allorecognition, the recipient T_H cells can recognize donor MHC molecules and associated peptide ligand in the cleft of the MHC molecule. The peptide is derived from processing of antigens that occurred in the donor before transplantation. A host cell becomes activated when it interacts with an antigen presenting cell (APC) that both expresses MHC molecule complex and provides the requisite co- stimulatory signals which depends upon the different tissues and different cells within a graft and work as antigen presenting cells (APCs).

Dendritic cells are found in different tissues. In many cases,these cells act as APCs. Dendritic cells express high level of MHC molecules, and serves as APCs in grafts. Host APCs are also known to migrate into a graft and endocyte the foreign alloantigens and present them as processed peptides with self MHC molecule. In some organs like kidney, pancreatic islets and thymus, donor APCs are called *passenger leukocytes.* These cells migrate from the graft to the regional lymph nodes and called dendritic cells. Passenger leukocytes express the allogenic MHC antigens of the donor graft and they are recognized as foreign protein. They stimulate T cells in lymph nodes. Direct recognition provides a stronger stimulus to rejection than the indirect pathway.

Indirect pathway is more important in chronic rejection. Dendritic cells and vascular endothelial cells express allogenic antigens and induce host T cells proliferation. The major proliferating cell is the $CD4^+$ T cell. $CD4^+$ T cell recognizes alloantigens presented by APCs. T_H cells play a central role in inducing the effector mechanism of allograft rejection.

Effector Stage

Effector mechanisms play important role in allograft rejection. Different types of effector mechanisms are: cell mediated reactions, antibody dependent cell mediated cytotoxicity (ADCC) and antibody plus complement lysis. In cell mediated reactions delayed type hypersensitivity (DTH) and CTL mediated cytotoxicity are involved.

During cell mediated graft rejection reaction, an influx of T cells and macrophages in graft occurs. In cytotoxic T lymphocyte mediated toxicity, Interleukin 2 (IL-2) and interferon gamma (IFN-λ) are released by T_H cells. IL-2 is responsible for promotion of T cell proliferation and is necessary for the production of cytotoxic T lymphocyte. IFN-λ plays a vital role in delayed type hypersensitivity (DTH) response, promotion of influx of macrophages into graft and their subsequent activation into destructive cells. Tumor necrosis factor (TNFα), released by activated macrophages , has cytotoxic effect on graft cells.Various cytokines like IFNα, IFNβ, IFNλ, TNFα and TNFβ all increase class I MHC expression. IFNλ also increases class II MHC expression. At the time of rejection, the levels of cytokines increase and induce the different cell types, within the graft, to express class I or class II MHC molecules.

CLINICAL INDICATION OF GRAFT REJECTION

Transplantation rejection occurs when transplanted tissue is rejected by the recipient immune system which destroys the transplanted tissue. Graft rejection reactions depend upon the type of tissue or organ grafted and on immune response of host. *Hyperacute rejection reaction* occurs within 24 hours of transplantation. *Acute rejection reaction* develop after few weeks of transplantation and *chronic rejection reaction* develop after few months of transplantation. These rejection reactions are caused by pre-existing host serum antibodies for antigen of the graft tissue. Graft rejection is mediated by both cell mediated and humoral immunity. Hyperacute rejection is mediated by pre-existing circulating antibodies in the recipient to antigens of the donor. In the acute rejection, the graft shows infiltrates of activated lymphocytes and monocytes.

Acute rejection is divided into two subgroups: early acute rejection and late acute rejection. *Early acute rejection* is due to the stimulation of T lymphocytes and cell mediated immunity. *Late acute rejection* is due to the stimulation of B lymphocytes and humoral immunity. In the chronic rejection, the gradual loss of the function of the grafted organ occurs.

Suppression of the immune response is important to prevent antibody response in autoimmune diseases, allergies, as well as allograft rejection. The Immunosuppression therapy is used to decrease reactivity to antigens.

GENERAL IMMUNOSUPPRESSIVE THERAPY

Nonspecific interference with the expression of immune response is called *immunosuppression.* Immunosuppression involves an act that reduces the activation of the immune system. In some cases a part the immune system itself has immunosuppressive effects on other parts of immune system. A person who is undergoing immunosuppression or whose immune system is weak is said to be immunocompromised. Induced immunosuppression is performed to prevent the body from rejections.There are three methods of immunosuppression: surgical ablation, irradiation and drugs. All three methods inhibit the antibody formation. Immunosuppressive drugs are targeted only at hyperactive component of the immune system and would not cause any immunodeficiency. Immunodeficiency is the root cause of many infections. Many immunosuppressive drugs are used for chemotherapy. These drugs are highly toxic for dividing cells. Most common immunosuppressive drugs are, azathioprine, cyclophosphamide, methotrexate, corticosteroids and cyclosporine. Azathioprine or Imuran is a mitotic inhibitor. It is used just after transplantation. Imuran acts on cell in the S phase of the cell cycle to block the synthesis of inosinic acid. Inosinic acid is a precursor of purines adenylic acid and guanylic acid. This drug has advantageous effect on T cell mediated reactions. It also inhibits the synthesis of nucleic acid. Methotrexate and cyclophosphamide are also mitotic cell division inhibitor. They are used with other immunosuppressive drugs as a conjoint treatment. Methotrexate is an inhibitor of folic acid. So it inhibits the synthesis of nucleic acid. Cyclophosphamide is an alkalytic agent which is responsible for disruption of DNA chains. These drugs exert toxic effect on cell during mitosis and are most effective after grafting, when cell division takes place.

Corticosteroids are anti-inflammatory agents which affect the immune response at different levels. It affects the lymphocyte, neutrophil adherence to endothelium, suppression of macrophage and also directly influences the gene expression. In human beings, steroids have their own side effects. After the use of steroids there is a decrease in the numbers of neutrophils, basophils and eosinophils. Corticosteroids also suppress the production of interlukin-1 by macrophages and interlukin-2 by T_H cells.

SPECIFIC IMMUNOSUPPRESSIVE THERAPY

Nonspecific immunosuppressive therapies produce generalized immunosuppression during transplantation and this type of immunosuppression increases the risk of infection.So, at this stage, ideal antigen specific immunosuppressive therapy against alloantigens of graft is required. In some cases, monoclonal antibodies can suppress graft rejection response. Use of monoclonal antibodies suppress the T cell activity in general or suppress the activity of sub-population of T cells. Rapid depletion of T cells in blood is brought out by CD3 monoclonal antibodies. Use of monoclonal antibodies to CD3 receptors in human being is to reverse acute rejection reaction. The depletion of T cells from blood is caused by binding of antibody coated T cells to Fc receptor on phagocytic cells. So cell become phagocytosed and clear the T cells from blood. Anti CD4+ monoclonal antibodies are known to prolong graft survival and it does not deplete the population of T cells. Monoclonal antibody therapy is also used in the bone marrow transplantation.

In modern trends, monoclonal antibodies are used against different cytokines because cytokines play an important role in allograft rejection. TNFα, IFNλ and IL-2 cytokines have been studied in this therapy. To prolong bone marrow transplantation in mice, monoclonal antibodies are used against TNFα and to prolong cardiac transplantation in mice, monoclonal antibodies are used against IFNλ and IL-2. These experiments have not been tried in human transplantation.

IMMUNE TOLERANCE TO ALLOGRAFTS

There are some examples in which allograft may be accepted without the use of immunosuppressive measures. There are two cases in which allograft may be accepted. First is, tissues transplanted at certain sites are protected from response and survive indefinitely. These are the privileged sites which include the anterior chamber of the eye, uterus, testes and brain. Second is a state of tolerance which is induced biologically by exposing to the antigens of the donor in the manner that causes immune tolerance rather than sensitization in the recipient. Each of these privileged sites includes poor lymphatic vessels. There is a failure of alloantigens to sensitize lymphocytes of the recipient as dendritic cells are few in number. The privileged location of the cornea has allowed corneal transplant to be highly successful. Corneal grafts survive without the need for immunosuppression. Because they are avascular, they tend not to sensitize the recipient. This privileged protection is boosted by the local production of immunosuppressive factors.

In an experiment in mice, the transplant of pancreas in diabetic mice was accepted because pancreatic islet cells were covered with semipermeable membrane.The islet cells survived and produced insulin. The transplanted cells were not rejected because the recipient's immune cells could not penetrate the membrane. Today Type 1 diabetes is a major problem in human beings. Diabetes is the result of autoimmune destruction of insulin secreting β cells of pancreas. Diabetic persons require insulin to support life but many complications occur. So, transplantation of islet cell is a better remedy for diabetic patients.

CLINICAL TRANSPLANTATION

In many cases, a transplant is the only means of therapy. Liver, kidney and pancreas are the organs which are commonly transplanted. The first kidney transplant was performed in 1950s. Since then thousands of kidney transplants have been performed. The next most frequently transplanted solid organ is the liver followed by heart, lung and pancreas. The frequency with which a given organ or tissue is transplanted depends on different factors like clinical position of patient, availability of tissues or organs and difficulty in transplantation and caring of post transplantation patients.

Kidney is the most commonly transplanted organ. Most common diseases like blood pressure, diabetes, nephritis are the cause of kidney failure. For transplantation, kidney can be obtained from relatives or volunteers because after donation of one kidney donor can live a normal life with remaining kidney. The major problems faced by kidney patients, are the shortage of kidney and increasing number of sensitized recipients. Patients are partially immunosuppressed at the time of transplantation. The combination of azathioprine and prednisolone was commonly used in the long term management of kidney grafts. If kidney function is poor during a rejection crisis, renal dialysis can be used. After a kidney transplant, patients have to depend upon many drugs for their entire life but these drugs create many complications like hypertension, bone diseases, risk of cancer and many other infections.

Bone marrow transplantation is a common and most frequent transplantation after kidney transplantation. Bone marrow transplantation has been increasingly adopted as a therapy for a number of malignant and nonmalignant hematologic diseases including thalassemia, anemia, lymphoma, leukemia and immunodeficiency diseases. Bone marrow contains not only hematopoietic but also mesenchymal stem cells, which can give rise to cartilage, bone and tendons. Bone marrow transplantation is an excellent treatment of *osteogenesis imperfecta. Osteogenesis imperfecta* is a genetic disease in children. Chemotherapy in patients suffering from malignant diseases of the bone marrow destroys not only cancerous cells but also normal cells of bone marrow. So in these cases bone marrow transplantation is used to save patients from side effects of chemotherapy. Nowadays, before the treatment, bone marrow of the patient is stored for future. The stored marrow can be re-infused into the patient. Normally, in the routine procedures, the host is immunologically suppressed before carrying out graft. The immune suppressed state of the recipient makes graft rejection rare. But because the donor bone marrow contains immunocompetent cells, sometimes, the host may reject the graft due to *graft versus host disease* (GVHD). The immunocompetent T cells, present in the donor bone marrow recognize alloantigens on the host cells. The T cells are proliferated and activated to produce certain cytokines such as TNFβ and IFNλ. These cytokines generate inflammatory reactions. These reactions are called *Graft versus host reaction* (GVHR) and are responsible for GVHD.

GVHD is a serious problem in bone marrow grafting which affects 50 to 70 percent bone marrow transplantations. The disease targets many organs which include skin, gastrointestinal tract and liver.In some cases GVHD produces anemia, retarded growth and damage bile duct.Various treatments are used to avoid GVHD in bone marrow transplantation.The best method of prevention is by proper HLA matching.

Liver performs many functions in the body which are related to clearance and detoxification of chemical and other waste substance. Liver can be damaged by viral disease (hepatitis), malaria, chronic alcoholism and exposure to harmful chemicals. In case of liver failure, liver transplantation can be used as therapy.

Diabetes and its associated problems are caused by malfunctioning of insulin producing islet cells in the pancreas. Transplantation of pancreas maintains the level of insulin though transplantation of complete pancreas is not necessary. Only islet cells can restore the function and avoid the major surgery.

POINTS TO REMEMBER

- The term 'transplantation' means removing something from one location and introducing it in another location.

- The implanted tissue or organ is known as 'graft'.
- Graft rejection is an immunological reaction as a result of self recognition between the grafted tissue and host.
- MHC is the major cause of graft rejection.
- The rate of rejection depends upon different tissues.
- T cells and sub division of T cells like CD4+and CD8+are involved in allograft rejection.
- Graft rejection is divided into two stages: Sensitization stage and effector stage.
- Sensitization stage is the stage of recognition of alloantigens.
- A variety of effector mechanisms e.g. cell mediated reactions participate in effector stage.
- Graft rejection reactions are of three types: hyperacute rejection reaction, acute rejection reaction and chronic rejection reaction.
- Immunosuppression therapy is used to prevent graft rejection.
- Kidney, liver, bone marrow, pancreas and cornea are most commonly transplanted organs.
- Bone marrow transplantation is extensively adopted in malignancy, hematologic diseases and in *osteogenesis imperfecta.*
- Graft versus host disease is a major problem of bone marrow transplantation.

REVIEW QUESTIONS

1. What is transplantation? Describe various types of transplantations.
2. Describe graft rejection and its mechanism.
3. Explain immunosuppressive therapy.
4. Write a brief note on immune tolerance to allografts.
5. Give an account of bone marrow transplantation.
6. Explain *graft versus host diseases.*

CHAPTER - 17

Tumor Immunology

A tumor is said to be an abnormal growth of cells to form a lump or mass of transformed cells. This growth occurs due to loss of regulatory control of cell division. The majority of cells and organs have a balance between cell renewal and cell death. A cell has a particular life span;when it dies a new cell is generated. In normal condition, this system is regulated through cell cycle and cell division but at the time of any abnormality this growth mechanism is totally disturbed and failed. These cells give rise to clone of the cells and eventually produce tumor. The tumors are of two types: *benign tumor* and *malignant tumor*. Benign tumor is usually encapsulated and confined to a single site and does not invade the healthy tissue. In malignant tumor, the cells break off from the main tumor mass and travel through the blood system and lodge themselves elsewhere and initiate new tumor formation at the site where they have lodged. Malignant tumor continuously grows and becomes progressively invasive. Malignant cells have such ability to change the surface molecules and have receptors on the membrane. The term 'cancer' refers particularly to malignant tumor. Malignant tumors exhibit metastasis, where cancerous cells dislodge from tumor and invade the blood or lymphatic vessels and are carried to other tissues, so a primary tumor gives rise to a secondary tumor. The tumor is also referred as *neoplasm.*Tumor cells are parasites on parent cells for food and energy. This condition is called cancer. The development, from a normal cell to a cancerous cell, is usually a multistep process of clonal evolution, driven by a somatic cell mutation from normal growth to a precancerous state and finally cancerous state **(Figure 17.1)** The development of malignant tumors is called *cancer*. Tumors are usually named by appending the suffix "oma" to the name of the tissue in which the tumor has arisen. Thus, a *lymphoma* is a cancer of lymphoid tissue, *sarcoma*, a cancer of fleshy, non epithelial tissue, and *adenocarcinoma*, a cancer of glandular tissue. The exception is the leukemia of white blood cells. Cancer causing substances are called *carcinogens*. The leukemia, tumors contains undifferentiated cells, often with large nuclei and nucleoli.

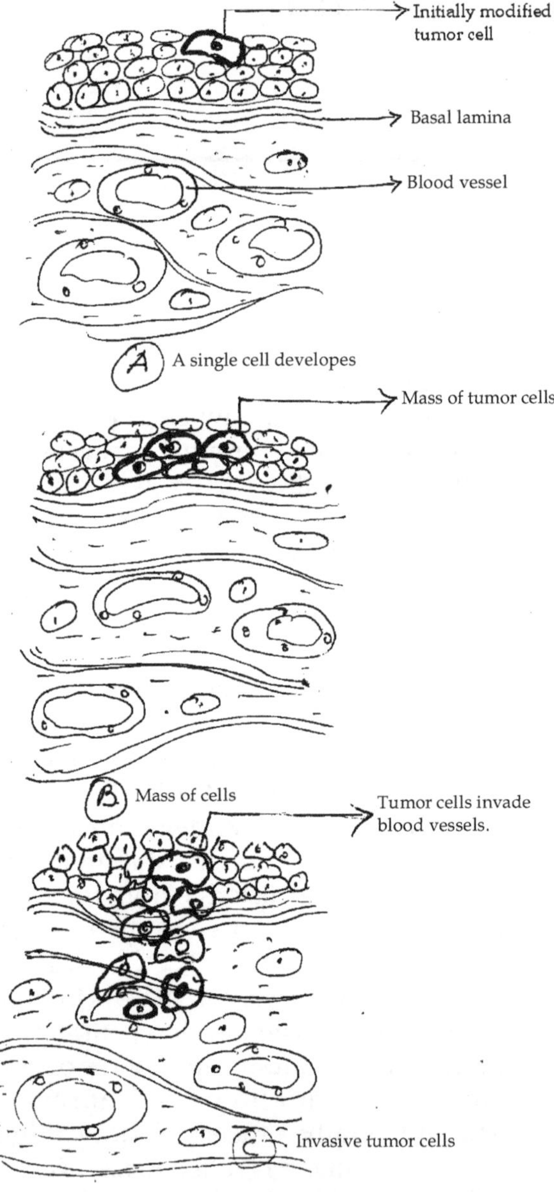

Fig. 17.1 : Tumor Growth and metastasis.

TYPES OF CANCER

Cancerous tumors are mainly of four types:*carcinomas, sarcomas, lymphomas* and *leukemia*.

Carcinomas are tumors made up of epithelial cells of ectoderm or endoderm.They form solid tumors in nerve tissue and on body surfaces. Cervical, skin, brain and breast cancer are common *carcinomas*. 85 to 90 % tumors are *carcinomas*.

Sarcomas are the tumors made up of connective tissues of mesodermal cells. They are solid tumors of connective tissue, bone and muscle. 2% human cancers are *sarcomas*.

Lymphoma tumors develop due to excessive production of lymphocytes of lymph nodes and spleen. 5% human cancers are*lymphomas*.

Leukemia is neoplastic growth of white blood cells and increased production of the cells. 4% human cancers are *leukemia*.

FEATURES OF TUMOR CELLS

Almost all types of differentiated cells can become neoplastic. These cells lose their ability to control their rate of division and they get converted into tumor.

Some important features of cancerous cells are as follows:

- Cancerous or tumor cells are immortal and can grow indefinitely.
- Tumor cells do not stop dividing after forming a monolayer i.e. several layers of cells are formed. These cells develop a change in the property of their cell membranes which become less adhesive. These cells dissociate from neighboring cells and infiltrate to other organs, where they form tumors.
- These cells have reduced cellular adhesion.
- These cells show invasiveness due to change in plasma membrane.
- In tumor cells, the receptors are more mobile within the membrane.
- Tumors cells are thinner as compared to normal cells. Micro filaments and microtubules disaggregate and disappear. Myosin filaments also disappear.
- Tumors cells have increased negative charge on cells surface.

- Tumor cell consumes more glucose so that rate of sugar transport across the cell membrane is high.
- Plasma membrane of tumor cells contain specific antigens.
- Tumor cells secrete proteolytic enzyme *protease* which converts plasminogen into plasmin. Plasmin participates in a wide variety of physiologic and pathologic processes including tumor growth, invasion and metastasis, through their effect on angiogenesis and cell migration.
- In tumor cells, *aldolase* enzymes A, B and C cannot be differentiated.
- Anaplasia, autotomy, increased number of golgi body and mitochondria are also observed in cancerous cells.

MALIGNANT TRANSFORMATION OF CELLS

Cell transformation means change in the genetic qualities that modify the phenotype of a cell in a specific manner. The genetic changes are brought about by the exogenous genetic material. In transformation of cells, two types of changes mainly occur, *continuous cell division* and *biochemical alterationin cell.* Majority of the normal cells require a solid support for anchorage. They will never grow if growth factors are not present. The normal cells stop proliferation when they come in contact with each other. On the contrary, the transformed cells are anchorage independent, serum independent and are not inhibited by contact or high density of cells. It may be possible that the path of malignancy is via transformed cells. Many changes develop in the cell, before a cell become malignant. Tumorigenic cell have the transformed phenotype as well as the capacity to proliferate indefinitely and divide faster. Recent research has proved that tumors have some immortal cells called *continuous cell lines.* Some genetic changes convert normal cell into immortal cells. Cell lines originate from malignant tissues and they are called *semi transformed cell.* Generally, cell lines possess anomalous numbers of chromosomes. *HeLa* cells may carry 76 chromosomes instead of 46 normal diploid cells.

Transformed cells may be recognized by some special features like immortalization, growth on the top of normal cell, different cell colony pattern and agglutination by lectins. The transformed cells cause tumor through a long process.Tumor formation involves complex interactions with the immunological system of the host. Transformation provides a base for the development of malignant potential by infected cells.

The deregulated and uncontrolled cell division of a cancer cell creates a milieu in which the products of normally silent genes may be expressed. Sometimes these genes encode differentiation of antigens which are normally associated with fetal stage. Thus, tumors, derived from the same cell type are often found to express such *oncofetal* antigens, which are present on embryonic cells. Although tumors are derived from tissues of one's own body, the malignant transformation process leads to the expression of molecules on the tumor cells which are recognized by one's own immune system as foreign bodies.These molecules are known as *tumor antigen*. Treatment of normal cells with chemical carcinogens, irradiation and certain viruses can alter their morphology and biochemistry. In some cases, the cells are able to produce tumors.This is called *transformation*. Both cancer cells and transformed cells can be sub cultured indefinitely i.e. they are immortal. A variety of chemical agents and ultraviolet rays cause mutations which induce transformation. Some genes are usually silent and are not expressed in normal cells but when these genes are switched on, as a result of malignant transformation of a cell, and are expressed in appropriately in the wrong tissue, they may behave like tumor. Most of the tumors express those genes whose products are required for malignant transformation as well as for the maintenance of the malignant phenotype. Mostly these genes have been found to be altered or mutated form of normal cellular genes and they regulate cell division and differentiation process. Carcinogens induce point mutations and deletions and alter the proto-endogenes to oncogenes, whose products have transforming activity.

Both DNA and RNA viruses have been shown to induce malignant transformation. Virally induced tumors usually contain integrated proviral genomes in their cellular genomes and these tumors often express viral genome coded proteins. DNA viruses such as *papova viruses, semian virus (SV-40)* and *adenoviruses* induce malignant tumors. There are many genes in these viruses which interact to bring about malignant transformation. Most RNA viruses replicate in the cytosol. Retroviruses transcribe their RNA into DNA by means of *reverse transcriptase* enzymes and then integrate the transcript into the host's DNA. RNA tumor viruses (*retroviruses)* induce tumors in days to week ,after infection and are called *acute transforming retroviruses*. *Murine leukemia virus* causes tumors in months after infection. In some cases,*retrovirus* induced transformation is related to the presence of oncogenes. *Rous sarcoma virus* is a transforming *retrovirus* and carries oncogene which is called *v-src*. When this oncogene is cloned and transferred into normal cells, the cells are converted into malignant.

ONCOGENES AND CANCER INDUCTION

A substantial minority of tumors arise through infection with oncogenic viruses. After infection, the viruses express genes homologous with cellular oncogenes which encode factors, affecting growth and cell division. Howard Temin suggested that oncogenes, unique to transforming viruses,are found in normal cells.A virus might acquire oncogenes from the genome of an infected cell.These cellular genes called *proto-oncogenes* or *cellular oncogenes*. Viral oncogene sequences were almost identical to sequences found in the normal cellular DNA. Proto-oncogenes are named after their viral homologies because their sequences were first discovered in viruses. Viral oncogenes are designated by a three letter word that relates the oncogenes to the virus. e.g. *feline sarcoma virus*oncogeneis referred to as *V-fes*, *simian sarcoma* virus oncogene is called as *v-sis*. Thus oncogenes are referred to as *fes* and *sis*.

Cellular oncogenes consist of exons and introns. There viral counter parts consist of uninterrupted coding sequences. The actual coding sequences of viral oncogenes and the corresponding proto-oncogenes exhibit a high degree of homology. Both viral and cellular oncogenes are derived from cellular genes that encode various growth controlling proteins. The conversion of proto-oncogenes into oncogenes occurs due to a change in the level of expression of growth controlling protein. In normal tissues homeostasis is maintained through the process of cellular proliferation to cell death. If there is any imbalance, the cell is converted into cancerous cell. Cancer associated genes are divided into three categories:-

i. Genes associated with induction of cellular proliferation.

ii. Genes associated with inhibition of cellular proliferation.

iii. Genes associated with regulation of programmed cell death.

Induction of Cellular Proliferation

In this category proto-oncogenes and their oncogenic counterparts encode protein that induce cellular proliferation. This protein plays an important role in growth.

Inhibition of Cellular Proliferation

In this category genes are anti-oncogenes and are called *tumor suppressor genes*. These genes inhibit excessive cell proliferation.

Regulation of Programmed Cell Death

In this category genes regulate programmed cell death. These genes produced apoptosis. These genes play important role in regulating cell survival during hematopoiesis and in the survival of selected B and T cells during maturation.

There are two types of oncogenes, *viral oncogenes* and *cellular oncogenes. Cellular oncogenes* have been observed in various cells and cause various types of cancer. Mutation in proto-oncogenes has been associated with cellular transformation **(Figure 17.2).** Many tumors show increased levels of growth factor. In breast cancer, increased growth factor is responsible for causing cancer. The presence of *myriad chromosomal abnormalities* in precancerous and cancerous cells supports the role of multiple mutations in the development of cancer. Most *lymphomas* and *leukemia* have visible chromosomal abnormalities. Lymphoma genesis is a multistep process. The lack of proliferative control,stimulated by deregulation of chromosome and other similar events induced by chromosomal translocation is permissive for the vulnerability to the induction of neoplasia.But it is not sufficient to bring about malignant transformation.

TUMOR ANTIGENS

Tumor cells are entirely different from normal cells and they lose all their original antigens and new set of antigens are developed on cell surfaces **(Figure 17.3)**. Cancerous tumors behave abnormally due to expression of mutated or viral genes or deregulated expression of normal genes of a cell. Cancerous cells express proteins that are either not expressed at all or present in much lower quantities in normal cells. These proteins may appear foreign to the host body because they were never there on the tissue surface, prior to tumor development. These proteins can stimulate immune response to tumor cells.

According to immunological probe tumor antigen can be classified into two main groups:

i. Tumor antigen recognized by T lymphocytes.

ii. Tumor antigen defined by xenogenic antibodies.

TUMOR ANTIGEN RECOGNIZED BY T LYMPHOCYTE

Some tumor antigens, recognized by T lymphocytes, are unique and specific to certain tumors and some are present on many cancers

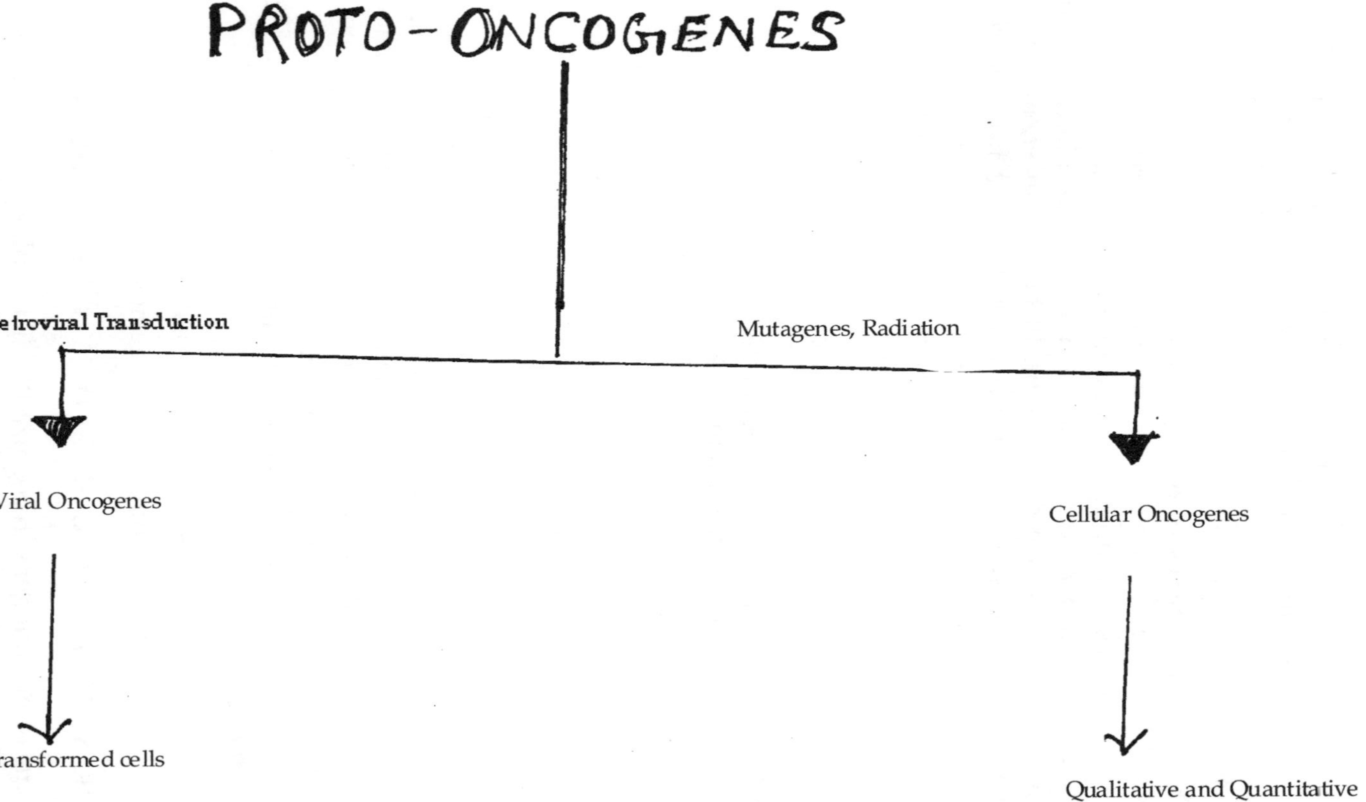

Figl. 17.2 : Conversion of proto-oncogenes into oncogenes

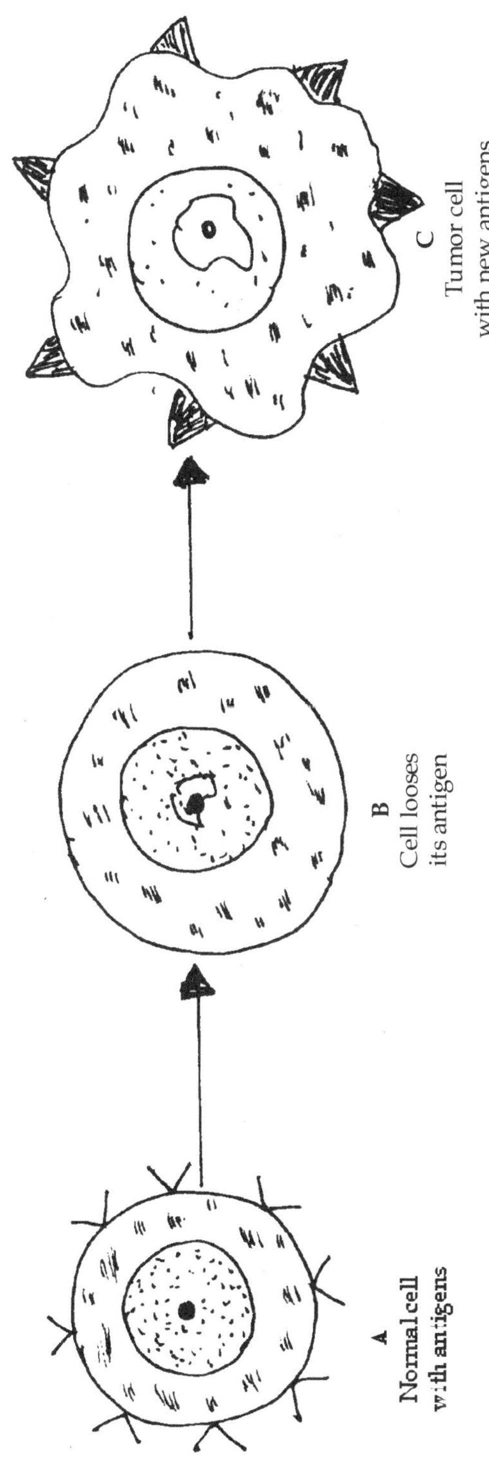

Figl. 17.3 : A normal cell converted into tumor cell with new antigens

of different types. Tumor antigens, recognized by T lymphocyte, include cellular and viral proteins. They are divided into different groups:-

- Tumor specific transplantation antigens (TSTAs)
- Tumor antigens expressed on *in vitro* mutagenized tumor cells.
- Tumor antigens encoded by normally silent cellular genes in spontaneous tumor.
- Tumor antigens induced by oncogenes.
- Tumor antigens induced by genome of oncogenic viruses.
- Tumor antigens resulting from mutations.

Tumor specific antigens are unique to tumor cells and do not occur on normal cells in the body. They may result from mutations in tumor cells that generate altered cellular proteins. Some tumor antigens, expressed on experimental tumors, are mutant cellular proteins with single amino acid substitution. This has been shown by tumor specific CTL clones which are used to isolate genes, encoding tumor antigens. Tumor antigens were derived from the non-tumorgenic line and these genes were transfected into parental tumorigenic line. The transfectants were screened for their sensitivity to lysis by the tumor specific CTL clones. The proteins produced by these genes are endogenously synthesized. If these proteins are normal self proteins, they do not induce immune response because of the absence of T cells.If these proteins are altered proteins, they will be recognized by CTLs and will be lysed. Some genes are usually silent and are not expressed in normal cells or they are expressed in the early stage of development during differentiation and embryogenesis. Some antigens are identified on tumors induced by virus. Virally induced tumors usually contain integrated proviral genome in their cellular genomes and these tumors express viral genome coded proteins. Different kinds of tumors are induced by viruses in different organs (Table 17.1).

Table 17.1 : Different types of tumors caused by viruses

VIRUSES	TUMORS
HTLV: *Human T cell lymphotrophic virus*	*Lymphoma/ T cell leukemia*
EBV: *Epstein Barr virus*	*Hodgkins lymphoma, nasopharyngeal carcinoma, B cell lymphoma*
HPV: *Human Papillomavirus*	*Cervical carcinoma*
HBV: *Hepatitis B virus*	*Liver carcinoma*

Growth factor and growth factor receptor are increased significantly in tumor cells. Many types of tumor cells express more amount of epidermal growth factor (EGF) receptor than as that expressed by normal cells.

TUMOR ANTIGENS DEFINED BY XENOGENIC ANTIBODIES

In tumor cells, surface molecules are identified by antibodies. Tumor antigens are found in different types of tumors. These antigens do not stimulate immunological response in the host. These are called *tumor associated transplantation antigens* (TATAs). TATAs are not unique to tumor cells;they are proteins and are expressed on normal cells, during fetal development. Tumor antigens are important in the diagnosis and treatment of cancers. These tumor antigens are divided into different categories such as *oncofoetalantigens, glycoprotein antigens* and *differentiation antigens.*

Oncofoetal tumor antigens are found on cancerous cells as well as on normal cells. These antigens appear early in embryonic development before the development of immune system. Oncofoetal antigens are expressed on tumor cells due to derepression of genes. They are antigenic in the host. Two oncofoetal antigens are: *alpha-fetoprotein* and *carcinoembryonicantigen* (Table 17.2).

Alpha fetoprotein (AFP) is synthesized and secreted in foetal life by the yok sac and liver. *AFP* levels are increased in liver cirrhosis and liver cancer pateint. *Carcinoembroynic antigen(CEA)* is a membrane glycoprotein, present on gastrointestinal and liver cells of 2 to 6 month old fetuses. *CEA* is expressed by gastrointestinal (GI) derived tumors including *colon carcinoma*, pancreatic *carcinoma*, liver *carcinoma, gallbladder carcinoma* as well as *breast tumor. Colorectal cancer* have increased level of *CEA*. Recent researches have shown that *CEA* promotes the binding of tumors to one another through adhesion molecules.Glycoprotein antigen is expressed on several tumors. Abnormal glycoprotein and glycolipid synthesis is associated with tissue invasion and malignant behavior of tumor cells.

Table 17.2 : Antigens produced by different tumors

ANTIGENS	TUMOR
CEA:*Carcinoembryonic antigen*	*Gastrointestinal and breast Tumor*
AFP: *Alpha-feto protein*	*Hepatomas*

IMMUNE RESPONSE TO TUMOR ANTIGENS

Experientially it has been proved that tumor antigens induce both cell mediated and humoral immune responses which result in the destruction of tumor cells. Many tumors induce tumor specific cytotoxic T lymphocytes (CTLs) which recognize tumor antigen, presented by class I MHC molecules on tumor cells. NK cells and macrophages are important in tumor recognition. NK cells can bind to antibody coated tumor cells. NK cells are increased in certain types of cancer. The antitumor activity of NK cells is enhanced by cytokines including interferon tumor necrosis factor α (TNFα). Activated macrophages play very significant role in immune response to tumors. Macrophages are not MHC restricted. They are often observed in different ways to destroy tumors which include cluster around the tumors and release of lysosomal enzymes. Macrophages secrete tumor necrosis factor (TNFα) which is capable to select tumor cells. TNFα bindsonthe cell surface receptors and destroy the tumor. TNFα is also involved in disruption of cytoskeleton protein and gap junction protein. If TNFα is injected into tumor, it has been found to induce hemorrhage and necrosis of tumor cells.

Cytotoxic T lymphocytes (CTLs) play important role in antitumor activity. Anti-tumor activity of CTLs is similar to CTLs antiviral activity. CTLs can be isolated from tumors. Mononuclear cells, found in tumors as infiltrates, are called *tumor infiltrating lymphocytes (TILs)* which include CTLs with the capacity to destroy the tumor. Helper T cells (T_H cells) also show antitumor activity. CD_4^+ Tcells (T_H cells) are also known to secret TNF and interferon which help in antitumor activity.

Tumor bearing host do produce antibodies against tumor antigen. Human cancer patients produce antibodies against the tumor. The antibody mediated destruction of tumor cell depends upon complement activation or antibody dependent cell mediated cytotoxicity.

TUMOR EVASION OF THE IMMUNE RESPONSE

An individual has innate ability to build an immune response against cancer.It is fact that many people die due to cancer because of failure of immune response. Tumor has the ability to evade immune response. There are many mechanisms by which tumor cells adopt to evade the immune system.

Antibodies can be produced against tumor specific antigens. A lot of attempts have been made to protect animals against tumor growth by active immunization with tumor growth or by passive immunization with antitumor antibodies. Occasionally, antibodies enhance the growth of tumor. In many circumstances antitumor antibodies are used as a blocking factor. The antibody binds to tumor specific antigens. In some occasions antibody develop antibodies complexes with tumor antigens. These complexes block the CTL response and inhibit antibody dependent cellular cytoloxicity (ADCC). Some tumor antigens have been observed to disappear from the tumor cells in the presence of antibody. If the antibodies are not present in serum these tumor antigens reappear. This process is *antigenic modulation*. Malignant transformation of cells expresses low levels of class I MHC molecules. It is related with tumor growth. CD_8^+ CTLs recognize antigen only when it associated with class I MHC molecules. Any alternation in the expression of MHC class I molecules leads to failure in immune response. The decrease in expression of class I MHC molecules can be correlated with tumor growth. T-cell activation requires an activating signal but some tumor cells provide poor signals. The poor immunogenicity of many tumor cells is a cause of cancer in the individuals.

IMMUNE THERAPY OF CANCER

Immunotherapy of tumors requires the activation of multiple immune defense mechanisms. Cytokine therapy can augment immune responses to tumors. A number of cytokines are involved in cancer immunotherapy such as IFNα, IFN β, IL-1, IL-2, IL-4, IL-5, IL-12 and TNF. Interferon mediated anti-tumor activity may involve several mechanisms. All three types of interferons have been shown to increase the expression of class I MHC molecules ontumor cells. IFN λ has been shown to increase MHC class II molecules on macrophages. MHC molecules increase the activity of CTL againsttumors.Some anti-tumor effects of interferons are related to the ability of interferon to inhibit tumor growth. Interferon λ directly or indirectly increase the activity of Tc cells, macrophages and NK cells. These cells play important role in anti-tumor activity. *Tumor necrosis factors* (*TNFα* and *TNFβ)* have been shown to exhibit anti-tumor activity, killing some tumor cells and reducing the rate of proliferation of cancer cells into normal cells. TNFα inhibits tumor induced vascularization, decreasing the flow of blood and oxygen which is necessary for growth of tumor. Studies have demonstrated that lymphocytes mediate more effective tumor destruction. Researchers have proved that in the presence of IL-2, lymphoid cells activate and kill tumor cells called *lymphokine-activated killer cells(LAK cells)*. These cells are derived from NK cells. *Tumor infiltrating lymphocytes (TIL)* therapy is also used in anti-tumor reactivity. *TIL* cells have cytolytic activity against tumor cells. Monoclonal antibodies are also used as an immunotherapeutic agent for cancer. Anti-idiotypic antibodies and monoclonal antibodies are used in B cell lymphoma. Antibody dependent cell mediate toxicity (ADCC) kill the lymphoma cells. Monoclonal antibodies can also be used to prepare tumor specific immunotoxins. A variety of tumor receptors which are targets for anti-tumor monoclonal antibodies are also used to prepare tumor specific antitumor agents. Antitumor antibodies, associated with radio isotopes and chemotherapy drugs, have also been tried for immunotherapy of cancers. An antibodywhich is specific for cancer antigen is associated to an antibody, directed against a surface protein of NK cells or CTLs cells targeted against tumor cells. This is called *heteroconjugate.*

POINTS TO REMEMBER

- A tumor is an abnormal growth of cells.
- The tumors are of two types: benign tumor and malignant tumor.
- Malignant tumors are those grow continuously and become progressively invasive.
- Tumor cells display tumor specific antigens and the more common tumor associated antigens. Tumor antigens are induced by chemicals and virus. Virally encoded tumor antigens are shared by all tumors induced by the same virus.
- On the basis of the nature of the tissue affected, cancerous tumors are classified into many types: carcinoma, sarcoma, lymphoma and leukemia.

- Tumors cells have uncontrolled growth and fail to respond to the regulatory signals which cause orderly growth and tissue pattern of normal cells.
- The transformed cells are anchorage independent and are not inhibited by contact of cells.
- Proto-oncogenes encode proteins involved in the control of normal cellular growth. The conversion of proto-oncogenes to ontogenesis is one of the key steps in the induction of most human cancers.
- Cell transformation means change in the genetic qualities that modify the phenotype of the cell in a specific manner. It involves genetic changes brought about by the exogenous genetic material.
- Tumor cells can be differentiated from normal cells by quantitative and qualitative differences in their antigens.
- Tumors antigens can be classified into two main groups i.e. tumor antigen recognized by T lymphocytes and tumor antigens defined by xenogenic antibodies.
- Tumor antigens recognized by T lymphocytes include a wide variety of cellular and viral proteins.
- Tumor antigens are defined by xenogenic antibodies. These antigens do not stimulate immunologic response in the host and are called tumor associated transplantation antigens (TATAs).
- Oncofetal antigens such as carcinoembryonic antigen (CEA) and alpha-fetoprotein (AFP) are highly associated with gastrointestinal (GI) derived tumors and liver tumors.
- Tumor antigens induce both humoral and cell-mediated immune response.Tumor antigens can stimulate CTLs and $T_H$cells to kill the tumor cells.
- Tumor antigens-antibodies complexes block the CTLs response and inhibit antibody dependent cellular cytotoxicity (ADCC).
- Multiple immune defense mechanisms play important role in immune therapy of cancer.

REVIEW QUESTIONS

1. What is cancer? Enlist some important characteristics of a cancerous cell.
2. What is a malignant tumor? Describe malignant transformation of a cell.
3. Describe tumor antigens.
4. What are oncogenes? Describe their role in cancer induction.
5. Describe tumor evasion of the immune system.
6. Explain : Immune therapy of cancer.

Chapter - 18

Immunodeficiencies

The well-being of a person depends on his immune system. Any deficiency in any component of immune system may lead to defects in the normal functioning of his body and makes him more susceptible to diseases. The immunodeficiency may be either in specific components (B cells or T cells) or nonspecific component (phagocytes and complement) of the immune system.

Immunodeficiency can be classified into two types- primary and secondary. ***Primary immunodeficiency*** refers to genetic defects in the components of immune system. The diseases caused by such defects are congenital i.e. present at birth or may manifest later in life. Primary immunodeficiency may affect both adaptive as well as innate immunity. ***Secondary immunodeficiencies*** are those defects which are acquired during life. These are not genetic in nature but may be a consequence of infection or other diseases. Acquired immunodeficiency syndrome (AIDS), caused by HIV (Human immunodeficiency virus) is the most common and well known secondary immunodeficiency.

PRIMARY IMMUNODEFICIENCIES

Primary immunodeficiencies include deficiencies of B cells, T cells, complement and phagocytes. Deficiency of B and T cells affect the adaptive immunity while deficiency of complement and phagocytes impairs innate immunity. In case of B cell deficiencies there may be complete absence of B cells, plasma cells or selective absence of immunoglobulins. It affects the humoral immunity. The persons are more susceptible to pyogenic bacterial infections but immunity to viral and fungal infection remains unaffected.

T cells are involved in both humoral and cell mediated immunity. Deficiency of T cells affects both branches of immune response and individuals are susceptible to a broad spectrum of infections. Table 18.1 summarizes some primary immunodeficiency diseases.

Table 18.1 : Some Primary Immunodeficiency diseases

Disease	Immunecomponent involved	Defects
Severe Combined immunodeficiency (SCID)	B and T cells	Toxic metabolites in T and B cells, no proper development of thymus
Wiskott- Aldrich syndrome	T cells and platelets	Defective T cells and low levels of IgM and increased levels of serum IgA
X-linked agammaglobulinemia	B cells	No mature B cells, low IgG
Common variable immunodeficiency(CVI)	B cells	B cells do not produce antibodies
X-linked hyper-IgM syndrome(X-HIGM)	T cells	Defect in T cell surface molecules
Di George syndrome	Development of B and T cells	Thymus does not develop normally, development of T cells affected
Chediak-Higashi syndrome	Defective intracellular transport protein	Defective phagocytes, processing and presentation of antigens are impaired
Chronic granulomatous disease	Phagocytes	Defective oxidative pathway, phagocytes are unable to phagocytose the microbes
Ataxia Telangiectasia	IgA and IgE	Low level of IgG and IgA, low number of T cells
Leukocyte Adhesion deficiency	Leukocyte extravasation	Leukocyte extravastion
Barr lymphocyte syndrome	MHC class II molecule	Deficiency of MHC class II molecules
Complement deficiency	C1,C4,C2,C3, C5,C6,C7,C8,C9	Impairs the immune system, immune complex diseases

Severe Combined Immunodeficiency (SCID)

Severe Combined immunodeficiency (SCID) is due to the deficiency of T cells or both T and B cells and natural killer cells. The condition occurs in infants and is usually fatal, if not treated. The affected infants have very low number of lymphocytes. The thymus does not develop properly and remains fetal in appearance. The infants develop recurrent infections and suffer from prolonged diarrhea, pneumonia and skin, mouth and throat lesions and a host of other opportunistic infections. The only treatment for the disease is bone marrow transplant. Gene therapy can also be used to transfer the correct gene in the affected children. The disease is caused either due to a defective gene on X chromosome or recessive genes found on some other chromosomes. About 50% of the SCID cases are due to a defective gene on a X chromosomes which encodes γ chain of interleukin IL-2. Rest of the cases are due to recessive gene, present on other chromosomes which cause deficiency of adenosine deaminase (ADA) or purine nucleoside phosphorylase(PNP). Due to the absence of these enzymes some toxic metabolites accumulate which are toxic to lymphoid stem cells.

Common Variable Immunodeficiency (CVI)

Common variable immunodeficiency (CVI) is an immunodeficiency in which B cells do not mature to plasma cells and do not function properly. It affects both males and females and usually occurs after puberty. They are unable to produce immunoglobulins. B cells are not defective but they do not receive proper signal from T cells to mature into plasma cells. There may be several causes of CVID. Normally, B cells may synthesize antibodies but are unable to secrete them. In some cases, auto antibodies to B cells may also be present. Thus, there are low levels of most of the immunoglobulins. The patients suffer from recurrent infections. They are more susceptible to pyogenic microorganisms. The disease can be controlled to some extent by administration of gammaglobulins. CVI is associated with other diseases like rheumatoid arthritis, hemolytic anemia, neutropenia and chronic lung disease.

X-linked hyper-Immunoglobulin Syndrome (X-HIGM)

It is an X-linked recessive disease due to a defective gene encoding CD40 ligand. This results in defective CD40 ligand on T_H cell surface which is required for B cell activation and isotype switching. B cell response to T cell dependent antigen is impaired but B cell response to T cell independent antigens is unaffected. There is deficiency of IgG,IgA and IgE though the level of IgM is increased. The children suffer from recurrent infection, especially that of respiratory infections.

X-Linked Agammaglobulinemia

X-linked agammaglobulinemia is a B cell deficiency disease which occurs due to a defective gene *btk,* present on the long arm of X chromosome . The disease usually occurs in male infants between 9 months and 2 years. The gene is involved in B cell maturation. Due to mutation in the gene, maturation of B cells is arrested at pre-B cell. Hence, in the affected individuals, the number of pre-B cells is normal but the number of B cells is reduced. The males have a few or very less number of B cells in their blood or lymphoid tissues. The serum level of IgG is very low while other isotypes i.e. IgA, IgD, IgE and IgM are usually absent. The patients suffer from pyogenic recurrent infections such as pneumococci, streptococci, staphylococci and meningococci due to very low number of IgG, usually less than 100 mg/dl. The disease is usually fatal.

Wiskott - Aldrich Syndrome

It is a X-linked immunodeficiency disease which is mainly present in males. In this disease T cells are defective. The affected individuals have low levels of IgM and increased levels of serum IgA and IgE. The level of IgG remains normal. The patients catabolize their immunoglobulins faster than normal individuals. Individuals with Wiskott-Aldrich syndrome suffer from recurrent bacterial infection, thrombocytopenia (increase in number of thrombocytes), eczema and pyogenic and opportunistic infections. The disease can be treated by bone marrow transplantation.

Ataxia Telangiectasia

Ataxia telangiectasia is a progressive neurologic disease in which there is deficiency of IgG and sometimes also of IgA. It is an autosomal recessive disorder in which the number and functions of circulating T cells are greatly reduced. The disease is named so as the affected infants have difficulty in maintaining balance (ataxia) and appearance of dilated capillaries in eyes (telangiectasia) and in skin.

Barr Lymphocyte Syndrome

Barr lymphocyte syndrome results from the deficiency of MHC class II molecules, normally expressed on antigen presenting cells (APC) such as macrophage, B cells and dendritic cells. Deficiency of these molecules leads to a deficiency of CD4+ T cells (T_H) cells. Since T_H cells have very important role in antibody formation, deficiency of T_H cells leads to deficiency of antibodies as well. Thus, it affects both cellular and humoral immunity. The disease is characterized by recurrent infections and may cause early death. In absence of MHC class II molecules antigens cannot be presented to CD4+ cells.

Di George Syndrome

There are many immunodeficiencies in which thymus does not develop normally. This has effects on normal functioning of different types of T cells such as T_H cells, Ts cells and T cytotoxic cells.The defect in T cells may vary from person to person. Di George Syndrome is caused by genetic deletions found on the long arm of one of the two 22nd chromosomes. Deletion syndrome may involve migration defects of neural crest-derived tissues, particularly affecting development of the third and fourth branchial pouches (pharyngeal pouches).The disease can be treated by thymus transplantation in case of absence of the thymus in the rare, so-called "complete" DiGeorge syndrome.

Chronic Granulomatous Disease

It is an X-linked disorder which is caused due to defective oxidative pathway by which phagocytes generate hydrogen peroxide. Thus, phagocytes are unable to phagocytose the microbes. As a result, microorganisms remain alive. Due to the persistence presence of bacteria in the phagocytes, cell mediated responses are activated and granulomas form in skin, lungs and lymph nodes. The disease is characterized by hepatosplenomegaly, pneumonia, osteomyelitis, abscesses and recurrent bacterial infection.

Chediak-Higashi Syndrome

This condition occurs due to a mutation in the protein LIST, involved in the regulation of intracellular trafficking. This results in the defective phagocytes which are unable to kill bacteria. The symptoms include hepatosplenomegaly, abnormalities of central nervous system, recurrent bacterial infection and partial *oculocutaneous albinism*. Other than the general defect in the functioning of the phagocytes, the ability of mononuclear cells to serve as APCs is also affected. Both processing and presentation of antigens are impaired. The disease is usually fatal and most children do not survive their childhood.

Leukocyte Adhesion Deficiency

There are several cell surface molecules which function as adhesion molecules. Such adhesion molecules are involved in the adhesion of leucocytes to endothelial cells. Dysfunction of adhesion molecules results in the impairment of adhesion of leucocytes to vascular endothelium. This causes impaired neutrophil recruitment to the site of infection and inflammation. The patients are more susceptible to bacteria and fungi. Immunity to virus is also affected.

Reduction in Neutrophil Count

Decrease in neutrophil count or total absence of neutrophils may occur either due to genetic defect or may be acquired through extrinsic factors. When there is total absence of neutrophils, it is called agranulocytosis, in which myeloid stem cells are unable to differentiate into neutrophils. Reduction in the concentration of neutrophils in the blood is called granulocytopenia or neutropenia. Decrease in neutrophils affects the phagocytosis of bacteria and children become more susceptible to bacterial infection.

Selective IgA Deficiency

IgA deficiency is the most common type immunodeficiency in which the patients have very low (usually less than 5 mg/dl) or absence of serum IgA. The levels of other immunoglobulins are normal. The deficiency of IgA is usually associated with allergies, recurrent sino-pulmonary infection and type III hypersensitivity. Some IgA deficient patients are also deficient in IgG2 and IgG4. They are highly susceptible to pyogenic infections.

Complement Deficiency

Deficiency of any component of complement system impairs the immune system. The patients are more susceptible to bacterial infections or immune complex diseases. Deficiency of C1q, C1r, C1s, C4 or C2 may cause immune complex disease and that of C3, factor H or factor

P may result in increased susceptibility to pyogenic infection. Deficiency of MBL (mannose binding lactin) leads to increased susceptibility to bacteria and fungi. Deficiency of CI inhibitor results in hereditary angioneurotic edema (HAE), an autosomal dominant disease which involves edema of various organs such as intestine or respiratory organs.

TREATMENT OF IMMUNODEFICIENCIES

The best option for an immunodeficient person is the total isolation of him from the exposure to any kind of microbial agent. But practically this is not possible. There are some other options of treatment:

i. ***Injection of immunoglobulins*** : Injection of antibodies can be an effective way in case of the diseases where there is deficiency of particular types of immunoglobulins. For example, Ataxia telangiectasia can be treated by injection of IgG antibodies. Injection of IgA can be useful in the treatment of selective IgA deficiency.

ii. ***Administration of cytokines*** : Administration of some cytokines can be effective in certain diseases. Many cytokine genes have been cloned and their products can be used for the treatment.

iii. ***Hematopoietic stem cell transplantation*** : In those diseases which involve deficiency of hematopoietic system, hematopoietic stem cell transplantation can be an useful approach. Impaired immune functions, in case of Severe Combined immunodeficiency (SCID) which is characterized by deficiency of T cells or both T and B cells, can be restored by hematopoietic stem cell transplantation.

iv. ***Gene therapy*** : Gene therapy can be a very useful, especially in the diseases which involve a single gene defect. It involves replacement of missing gene or cell type or that of protein with a proper gene.

SECONDARY IMMUNODEFICIENCIES

In addition to inborn immunodeficiency diseases, immunodeficiencies may also develop due to various agents such as drugs, nutritional deficiency or pathogens. Such immunodeficiencies which develop after birth are called secondary immunodeficiencies.

Nutritional Immunodeficiencies

Nutritional deficiency may affect various aspects of immune system. The effects vary from disease to disease. For example, diseases like measles, diarrhea and tuberculosis are highly affected by nutritional deficiency while there is no significant effect of nutritional deficiency in diseases such as tetanus and encephalitis. Lymphoid tissues are highly vulnerable to nutritional deficiency. Malnutrition can lead to lymphoid atrophy. It also affects thymus severely. In case of protein - energy malnutrition, the number of CD4+ and CD8+ is greatly reduced. PEM also affects phagocytosis. Production of some cytokines such as IL-2 and TNF and complement components like C3, C5 and factor B are decreased following the deficiency some nutrients. Similarly, deficiency of some mineral elements causes a variety of effects on immune system. Deficiency of zinc, for instance, causes a decrease in CD4+/CD8+ ratio and T cell dysfunction. Iron deficiency leads to impaired functions of neutrophils and NK cells. It reduces the ability of neutrophils to kill bacteria and fungi. Other elements like selenium and copper are also important for proper functioning of immune system. Experiments have shown that selenium deficiency can also alter the virulence of viruses.

Vitamins play important role in proper functioning of immune response. Deficiency of certain vitamins such as vitamin B6 and folate diminishes cell mediated immunity and can influence antibody production.

Drugs

Immune system is severely affected by some drugs. Steroids cause change in number of leukocytes. Administration of steroids induces leukocytopenia. The number of T cells (especially, CD4+ T cells) is greatly reduced. Steroids inhibit activation and proliferation of T cells. The number of neutrophils is significantly increased after steroid treatment. Steroids also cause monocytopenia and inhibit the synthesis of IL-1 and TNF by monocytes. Other than their effects on blood cells, steroids also inhibit the synthesis of some cytokines like IL-1,IL-2,IL-4, IL-6,IL-10,TNF and IFNα.

Acquired Immunodeficiency Syndrome (AIDS)

Acquired immunodeficiency syndrome (AIDS) is the most common example of secondary immunodeficiency. AIDS has become a big problem which is responsible for more than three million deaths per year.

Human immunodeficiency virus (HIV), the virus that causes AIDS, was first identified in 1983, by Francois Barre-Sinoussi and Luc Montagnier. HIV is a reterovirus which contains RNA as its genetic material. A second strain of HIV, HIV-2 was identified in 1986. Both HIV-1 and HIV-2 can cause AIDS but HIV-1 is more common cause of the disease. HIV-1 and HIV-2 differ in their virulence and geographical locations. HIV-2 is less pathogenic and is primarily found in West Africa. Both HIV-1 and HIV-2 are natural viruses of primates and have jumped species to infect humans. HIV-1 has come from chimpanzees though it does not cause AIDS in them. HIV-2 came from sooty mangabey. In addition to these two types(strains) of HIV i.e. HIV-1 and HIV-2, there are many different subtypes (clades) of HIV-1, designated by letter A through K. HIV has a very high capacity to mutate, giving rise to different forms of virus known as variants. Although HIV-1 differs from HIV-2 in genomic structure and antigenicity, both HIV-1 and HIV-2 have a similar life cycle.

Mode of Transmission of HIV

There are several ways of transmission of HIV. Major ways of transmission are (i) sexual contact,(ii)sharing of infected needles,(iii)transfusion of infected blood and blood products, and (iv) transmission from infected mother to her children. The virus normally does not transmits by hand shaking, kissing or hugging.

Structure of Virus

HIV is an enveloped retrovirus which contains RNA as its genetic material. It contains two copies of single stranded RNA(ssRNA), associated with various enzymes such as *reverse transcriptase, integrase* and *protease*. Genome is surrounded by nucleocapsid. p^{24} protein forms the chief component of inner shell of nucleocapsid and outer layer consists of p17 protein. The outer envelope of the virus consists of phospholipids with several virus specific proteins. There are two viral proteins gp^{120} and gp^{41} which form knob like projections **(Fig. 18.1).** Here gp refers to glycoprotein and the number denotes the molecular weight of the protein. For example, gp^{120} is a glycoprotein with molecular weight 120000 Daltons. HIV mainly infects CD4+ T_H cells of the immune system. Macrophages and dendritic cells which also express CD4 , although at low levels than T cells, are also infected by the virus.

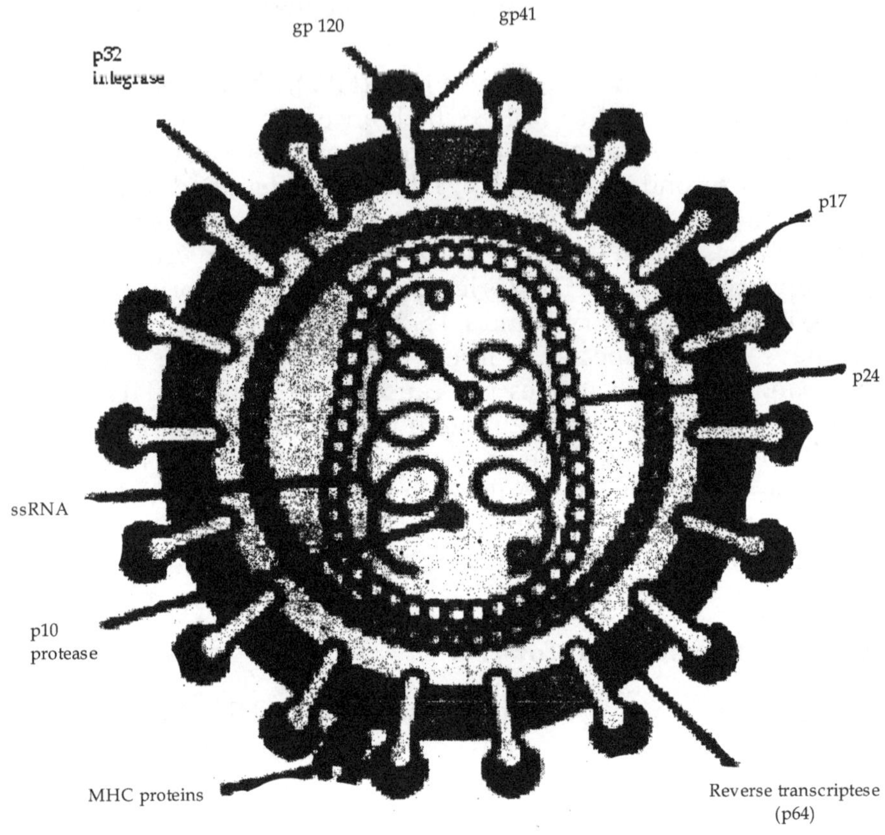

Fig. 18.1 : Structure of HIV

Stages of AIDS

From infection to the development of AIDS, the progression of HIV can be divided into several stages **(Fig. 18.2)**. *Early or acute phase* generally lasts for 2 to 8 weeks, following exposure to HIV. During this stage, virus multiplies and disseminates rapidly throughout the body. Initial viraemia (increase in HIV titre in the blood) is followed by a decrease in viral load. About 70% patients may experience flu-like illness. *Latency or asymptomatic phase* lasts for 6 months to 10 years. The virus continues to replicate, especially in the lymphoid tissue. T cells start decreasing in number due to their gradual destruction. *Chronic symptomatic stage* is characterized by development of various symptoms due to progression of AIDS. The symptoms include weight loss, persistent diarrhea and fever,lymphadenopathy and certain cancerous and precancerous conditions of cervix. Patients are more susceptible to opportunistic infections such as candidiasis, herpes simplex and shingles. Virus replication continues and there is further decline in the number of CD4+ cells. The balance between different types of T cells is disrupted. The virus may also destroy dendritic cells. The final stage of AIDS is characterized by infection of esophagus, bronchi and lungs. The commonest infection is pneumonia, caused *by Pneumocystis carinii.* Tuberculosis, toxoplasmosis, Kaposi sarcoma; and cryptococcalmeningitis also occur. Central nervous system may also be infected, leading to autoimmune neuropathies, cerebro vascular disease and brain tumors. More advanced stages of the disease are characterized by dementia and severe sensory and motor changes.

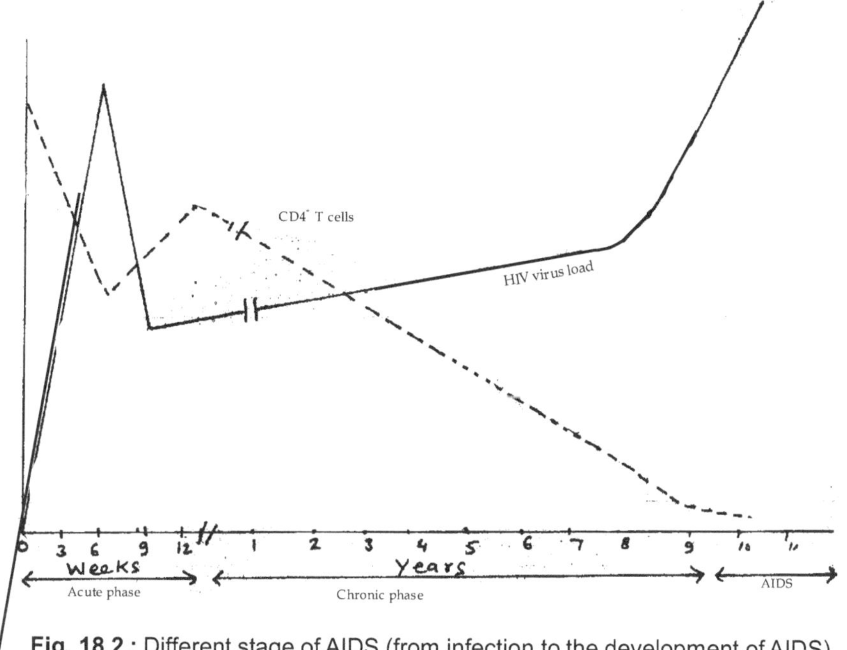

Fig. 18.2 : Different stage of AIDS (from infection to the development of AIDS)

Life Cycle

The infection of the virus begins with the binding of the virus to the host cells. gp120 protein , present on the surface of the HIV, binds to the CD4 molecules, present on the host cell surface. In addition to gp 120, binding of gp41 to a second protein (co-receptor) is also required for the entry of the virus inside the cell. These co-receptor / second proteins include CCR-5, present on the surface of CD4+T cells, macrophage and dendritic cells and CXCR-4 which is present only on CD4+ T cells but not on macrophages and dendritic cells. Accordingly, different cells are infected by different variants of the virus. Binding to the CD4+ receptor and co-receptor makes some conformational changes which mediate the fusion of HIV with the host cells membrane. This allows the viral nucleocapsid to enter the cell.

Inside the cell, the virus releases its RNA which is transcribed into cDNA by the viral enzyme *reverse transcriptase***(Fig. 18.3)**. This cDNA enters the nucleus and integrates into the DNA of the host cell with help of enzyme *viral integrase*. The DNA is now known as *provirus*. Within the cell, provirus stays hidden from the immune system, in latent form, till T cells are activated. This latency may last for months or years depending on the state of activation of T cells. On activation of T cells, proviral DNA is transcribed into viral RNA. Some of this RNA is translated into viral proteins. The viral proteins and RNA assemble into viral particles that bud from the infected cell. These viral particles can now infect other cells.

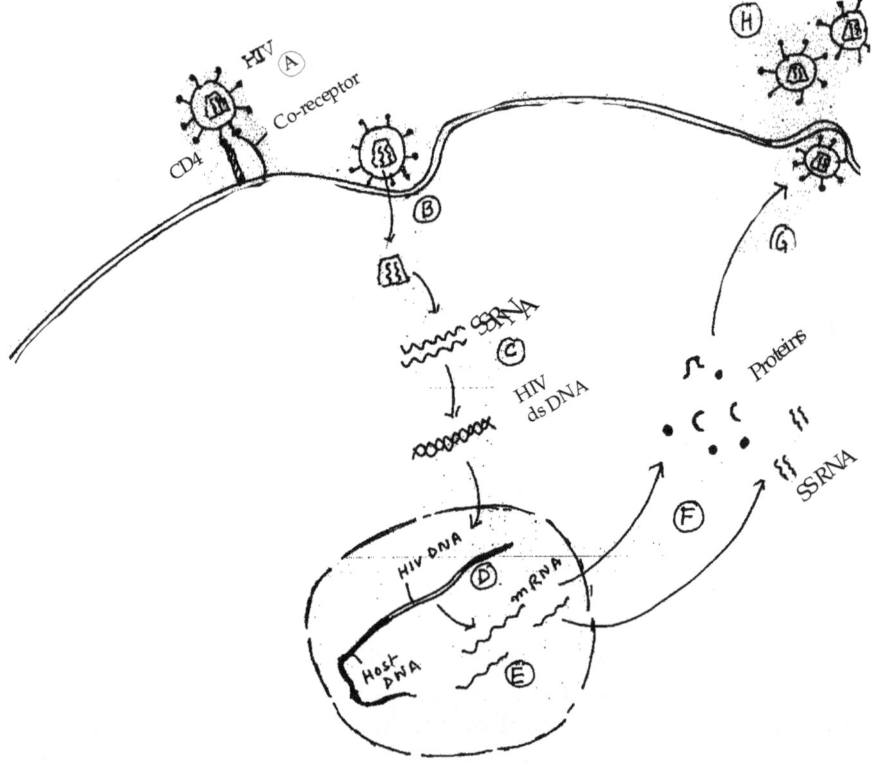

Fig. 18.3 : Life Cycle of AIDS Virus

A = Binding of HIV to the CD4 and receptor molecules on the target cell
B = Entry of the nucleocapsid containing virall genome into the cell
C = Reverse transcription - forming ds DNA from RNA
D = Entry and integration of HIV DNA with the host cell DNA
E = Formation of mRNA
F = RNA comes out of the nucleus, new viral proteins synthesize
G = Assembly of RNA and viral proteins
H = HIV exits by budding

Treatment of AIDS

At present, there is no cure for AIDS. The best way is to prevent the spread of the disease. Some antiviral drugs are used for the treatment but these drugs do not kill the virus but they only reduce the virus load. Some drugs are used to treat opportunistic infections and malignancies. Drugs like zidvudine (AZT or Retrovir) , didanosine (Videx) , tenofovir(Viread) and lamivudine(Epivir) act as nucleoside reverse transcriptase inhibitors and interfere with reverse transcription of viral RNA into DNA. Though AZT inhibits replication of the virus, but the virus mutates very quickly and become resistant to the drug.

Other targets for the drugs are enzyme *viral integrase* and *viral protease*. Inhibition of *integrase* prevents integration of HIV genome into the genome of the host cell. Drugs like indinavir (Crixivan), nelfinavir (Viracept) and saqinavir (Invirase) are used as *protease* inhibitors. These drugs interfere with the assembly of new virus particles. There are some other drugs like enfuvirtide (Fuzeon) which inhibit binding of the virus to the receptor and co-receptor and fusion of the virus to the cell membrane and thus prevent the entry of the virus into the cell.

Another approach of treatment is to use a combination of drugs. This is called highly active anti retroviral therapy (HAART). In this therapy several drugs are used which act at different levels such as inhibitors of *reverse transcriptase*, viral *integrase* and *viral protease*.This reduces the viral load at a very low level and prevents the infection of new cells. Some virus may remain dormant inside the cell and may be reactivated. This therapy is also associated with several serious side effects and needs to be evaluated.

Many vaccines have been developed and some are under trial but development of an effective and safe vaccine has remained a challenge till now.

There are many difficulties in the development of a proper vaccine. The ability of the virus to mutate rapidly and change the antigenic properties of its envelope proteins is a major hindrance. The presence of large number of strains is another problem. A vaccine can be regarded successful if it can provide protection from all variants of the virus. Lack of a suitable animal model is another obstacle in the development of vaccine.

POINTS TO REMEMBER

- Immunodeficiency refers to any deficiency in any component of the immune system.
- Immunodeficiency can be classified into two types- primary and secondary immunodeficiency.
- Primary immunodeficiency refers to genetic defects in the components of immune system.
- Secondary immunodeficiencies are not genetic in nature but may be a consequence of infection or other diseases.
- Primary immunodeficiencies include deficiencies of B cells, T cells, complement and phagocytes.
- Some primary immunodeficiencies are SCID, Wiskott-Aldrich syndrome, X-linked agammaglobulinemia, CVI, X-linked hyper-IgM syndrome(X-HIGM), Di George syndrome, Chediak-Higashi syndrome, Chronic granulomatous disease, Ataxia Telangiectasia, Leukocyte Adhesion deficiency, Barr lymphocyte syndrome and Complement deficiency.
- There are different ways for the treatment of immunodeficiencies. These are: injection of immunoglobulins, administration of cytokines, hematopoietic stem cell transplantation and gene therapy.
- Secondary immunodeficiencies may develop due to various agents such as drugs, nutritional deficiency or pathogens.
- AIDS is the most common example of secondary immunodeficiency which is caused by HIV.
- HIV is an enveloped retrovirus which has may subtypes and variants.
- HIV acts on CD4+ cells in the body.
- There is no cure for AIDS but some drugs such as zidvudine(AZT or Retrovir) , didanosine (Videx) , tenofovir (Viread) and lamivudine (Epivir) are used in the treatment of the disease.
- Though some vaccines have been developed but an effective and safe vaccine is yet to be developed.

REVIEW QUESTIONS

1. What is immunodeficiency? Differentiate between primary and secondary immunodeficiencies.
2. Describe different types of T cellimmunodeficiencies.
3. What are the various causes of secondary immunodeficiencies?
4. Describe the structure of HIV.
5. Explain the life cycle of AIDS virus.
6. "In case of AIDS, prevention is better than cure." Explain.
7. What are various approaches for the treatment of immunodeficiencies?
8. Write short notes on :
   i. Common variable immunodeficiency (CVI)
   ii. Chediak-Higashi syndrome
   iii. HAART

CHAPTER - 19

Autoimmunity

One of the characteristics of the immune system is to recognize and discriminate self from the non-self antigens. But, sometimes, there is a failure to discriminate between self and non-self antigens in the body and an autoimmune response occurs. Body starts producing antibodies against its own cells/antigens. These antibodies are called *autoantibodies* and this state is called *autoimmunity*. Thus, autoimmunity can be defined as the failure of an organism in recognizing its own constituent parts as self, thus leading to an immune response against its own cells and tissues. Autoimmunity leads to various diseases called *autoimmune diseases*. Autoimmune diseases afflict both men and women. At least 40 diseases are considered to be autoimmune in nature. Some common examples of autoimmune disease are: Diabetes mellitus type 1, systemic lupus erythematosus (SLE), Hashimoto's thyroiditis, Graves' disease, Addison's disease, rheumatoid arthritis, Polymyositis, dermatomyositis and allergies.

Many of the autoimmune diseases are strongly linked with MHC class I or MHC class II antigens. In general, about 5% of individuals are affected by autoimmune disease. Though the diseases affect both males and females, they are more prevalent in females and have a bimodal age distribution i.e. a first peak of incidence occurs around puberty and the second peak is in forties and fifties.

CAUSES OF AUTOIMMUNITY

Several factors are responsible for the development of autoimmune diseases. Thus, autoimmune diseases are multifactorial in origin. These factors include:

✓ Genetic factors
✓ Environmental factors
✓ Age and gender factors
✓ Drugs

Genetic Factors

Antigen specific autoimmune diseases tend to occur in clusters in certain families. Thyroid reactive antibodies are common in genetically related family members. If, in a family, an individual has autoantibodies against thyroid, it is very likely that other members will also have autoantibodies against thyroid. Several autoimmune diseases have been found to be associated with class II MHC genes. The role of the MHC is evidenced by a strong connection between Human leukocyte antigen (HLA) and incidence of certain autoimmune diseases (Table 19.1). Certain microbial antigens bind preferentially to particular MHC molecules.

Table 19.1 : Autoimmune disease showing HLA association

Auto immune diseases	HLA allele
Ankylosing spondylitis	B27
Diabetes mellitus	DR3/DR4
Multiple sclerosis	DR2
Myasthenia gravis	DR3
Psoriasis vulgaris	DR4
Reiter's disease	B27
Rheumatoid arthritis	DR4
Systemic lupus erythematosus	DR3

Polymorphism or mutations of genes also play a role in activation or suppression of lymphocytes. This may also lead to the development of autoimmune disease. Complement deficiency, due to mutations in genes for C2, C4, C5, and C8, results in autoimmune diseases.

Environmental Factor

Environmental agents such as diet, sunshine and petrol fumes can also cause autoimmune disease. Sunshine is a trigger of the skin lesions in systemic lupus erythematosus. Exposure to organic solvents can initiate the basement membrane autoimmunity and develops

Goodpasture's syndrome. Petrol fumes also induce autoimmune disease is some individuals.

Main environmental agents that cause autoimmunity are infectious microorganisms. Some microorganisms grow very slowly in host and it is very difficult to culture them. But they are associated with some hypersensitive state. These microbial antigens enter in the immune system because of the failure of cellular barrier and destroy the cell through necrosis. Activation of T lymphocytes and macrophages also destroy peripheral tolerance.

Microbes may contribute to development of disease by stimulating an increase in expression of B7 co-stimulatory molecules on APCs. Microbial infection leads to breakdown in T cell tolerance to self antigens. Cross reactivity between microbe and self antigens is also a cause of autoimmunity. Group A hemolytic streptococci have several epitopes that cross react with tissue antigens. One of them cross reacts with an antigen, found in cardiac myosin. The normal immune response to such a strain of streptococcus will generate lymphocyte clones that will react with myosin and induce myocardial damage, long after the infection has been eliminated. Rheumatic fever is an infection of streptococci. Rheumatic heart disease occurs due to an epitope that is common both in heart muscles and group A streptococci bacteria. Target antigens are highly conserved proteins such as heat shock proteins (HSPs), stress proteins, enzymes or other substances. The primary immune response to microbial infections includes a strong response to HSPs. Thus an immune response to microbial HSPs may induce a cross reactive response to human HSPs. These antigens can be cell surface, cytoplasmic or, nuclear molecules. Table 19.4 enlists some autoimmune diseases produced by antigens.

Table 19.2 : Autoimmune diseases produced by antigens.

Antigens	Diseases
Adrenal cortical cells	Addison's disease
RBC membrane proteins	Hemolytic anemia
Nuclei	Chronic hepatitis
Connective tissue , IgG	Rheumatoid arthritis
TSH receptor	Grave's disease
Basement membrane of kidney and lung	Goodpasture's syndrome
Nuclear antigen DNA	Systemic lupus erythematosus

Infectious agents such as viruses also produce autoimmune diseases. Viruses have the capacity to display viral antigens on the surface of host cell. Viral antigenic expression could act to trigger autoimmune diseases. Some viruses can have a latent state. This is the time when they remain in the host cell in a non-infective state but they may influence the cell surface for autoimmunity. Some latent viruses do not cause cytotoxicity but they can interfere directly or indirectly with several functions of infected cells. Viral protein may interfere with the functions of proteins involved in the control of cell death and survival. Viruses can stimulate autoimmune diseases by polyclonal activation of lymphocytes. An unknown non-lytic virus causes T lymphocyte activation in the thyroid gland, followed by increased expression of MHC II and thyroid derived peptides. This leads to an anti-thyroid immune reaction. Many viruses and bacteria may possess antigenic determinants which are similar to host cell components.

Age and Gender Factor

The development of autoimmunity shows a gender bias. Women have a greater risk than man. Autoimmunity is more prevalent in older people. In systemic lupus erythematosus and Graves' disease, there is a female/male 10: 1 and 7 : 1, respectively. Some hormones play an important role in autoimmune disease. Hormones secreted from hypothalamus, thyroid and adrenal glands have effects on the homeostasis of the lymphoid system and response to antigens. Systemic lupus erythematosus and Rheumatoid arthritis afflict women and myasthenia gravis afflict men. Sex steroids i.e. estrogens, progesterone and testosterone influence function and development of various lymphocytes. Pregnancy is often associated with increase in the severity in some diseases, particularly in Rheumatoid arthritis, and there is, sometimes, a striking relapse after giving birth.

Drugs

Certain drugs can also induce autoimmunity e.g. patients receiving procainamide develop systemic lupus erythematosus like symptoms and have antinuclear antibodies which disappear following discontinuation of the drug. Hemolytic anemia can be caused in susceptible people taking the antibiotic penicillin. Penicillin binds to erythrocytes membrane these membrane bound penicillin can lead to the triggering of lymphocytes. Antibody formation against erythrocyte antigen and complement activation can lead to anemia.

Cross reactivity of several antibodies to bacterial antigen is also a factor for autoimmunity. In a patient of syphilis, antibodies raised against *Treponema* antigen, can cross react with erythrocyte and produce anemia. Thrombocytopenia (low platelet count) and anemia are common examples of drug induced autoimmunity.

INITIATION OF AUTOIMMUNITY

Autoimmune diseases arise as a result of a break-down in self tolerance. Generally, some antigens are concealed away from the immune system. They are not accessible to the immune system. These antigens are sequestered or hidden antigens. Such antigens include lens antigens, sperm antigens and thyroglobulin.

Lens protein remains in its capsule and does not circulate in the blood. So this antigen is not exposed to immune system and a person does not develop immunological tolerance against lens protein. In case of autoimmune sympathetic ophthalmia, damage to one eye leads to the production of autoantibodies for lens antigen and damage the other lens.

Some antigens are modified by physical as well as chemical agents. These altered or modified antigens are also called *neoantigens*. Neoantigen is produced when a small molecule (virus or drug) binds to a protein and alters an MHC binding peptide so that it gets converted into neoantigen. The altered antigens provoke an autoimmune response and produce dermatitis, anemia, leucopenia and thrombocytopenia. The MHC haplotype can influence autoimmune disease by enhancing the presentation of peptide epitope in the periphery.

Leukocyte activation and elicitation through cytokine is associated with autoimmunity. Elevated levels of chemokines are reported in rheumatoid arthritis and multiple sclerosis.

T cells play a direct role in tissue inflammation. Cytotoxic T lymphocytes (CTL) are the major effectors of damage to the organ localized autoimmune disorders such as diabetes, chronic thyroiditis and multiple sclerosis.

B cell activation can lead to autoimmunity. Autoantibodies can damage immune complexes. AIDS patients show high levels of nonspecific antibodies and autoantibodies to RBCs and platelets.

TYPES OF AUTOIMMUNITY DISEASES

In human, autoimmune diseases can be broadly classified into two types: *organ specific autoimmune diseases* and *non organ specific* or *systemic autoimmune diseases*.

Organ specific autoimmune diseases are related to a particular organ. This category includes inflammation or dysfunction caused by immunological reactions against a specific antigen, corresponding to a single organ. The cells and tissues of the target organ may be directly affected by humoral or cell mediated effector mechanisms. In these diseases autoantibodies respond to self antigens, situated in a particular organ. These are also called *localized autoimmune diseases.*

Non organ specific autoimmune diseases affect the whole body or many organs. In this category the immune response is directed against a broad range of target antigens that are distributed systemically. These are also called *systemic autoimmune diseases.* These disease reflect a general defect in immune response. Table 19.3 and 19.4 enlist some organs specific autoimmune and systemic autoimmune diseases, respectively.

Table 19.3 : Organ specific autoimmune disease in humans

Autoimmune Disease	Immune Response	Self Antigen
Direct cellular damages		
Hemolytic anemia	Autoantibodies	RBC membrane proteins
Pernicious anemia	Autoantibodies	Gastric parietal cells.
Hashimoto's thyroiditis	T_{DTH} cells, autoantibodies	Thyroid proteins and cells.
Diabetes mellitus	T_{DTH} Cells, autoantibodies	Pancreatic β Cells
Goodpasture's syndrome	Autoantibodies	Renal and lung basement membrane
Mediated by stimulating antibodies		
Myasthenia gravis	Autoantibodies	Acetylcholine receptors
Graves's disease	Autoantibodies	Thyroid stimulating hormone receptor
Addison's disease	Autoantibodies	Adrenal cells.

Table 19.4 : Systemic autoimmune diseases in human

Autoimmune Disease	Immune Response	Self Antigen
Multiple sclerosis (MS)	T lymphocytes and auto-antibodies	Brain
Rheumatoid arthritis (RA)	Autoantibodies	Connective tissue
Systemic Lupus Erythematosus (SLE)		Autoantibodies Nuclear Protein, RBC and Platelet

ORGAN SPECIFIC AUTOIMMUNE DISEASES

Organ specific autoimmunity is mediated either by *direct cellular damage* or *by stimulating antibodies.*

Autoimmunity by Direct Cellular Damage

It occurs when lymphocytes bind to cell membrane antigen and cause inflammatory response in the organ. The cellular structure of the organ is lost and replaced by connective tissue. In direct cellular damage the whole organs as well as its function are damaged. Autoimmune anemia, Insulin-dependent diabetes mellitus, Goodpasture's syndrome and Hashimoto's diseases are examples of diseases involving direct cellular damage.

Autoimmune Anemia

It is a drug induced hemolytic anemia which is caused by sulfa drugs and antibiotics. In this disease, RBCs are destroyed prematurely and bone marrow production can no longer compensate for their loss. This condition because antibodies are formed against the drug antigenic determinants and the association of the drug with erythrocytes causes binding of antibody to erythrocytes. The autoantibodies bind to erythrocytes membrane protein and causes complement mediated lysis and phagocytosis of RBCs by macrophage. It ultimately leads to hemolysis. The clearance of RBCs or platelets in spleen produces anemia **(Fig. 19.1).**

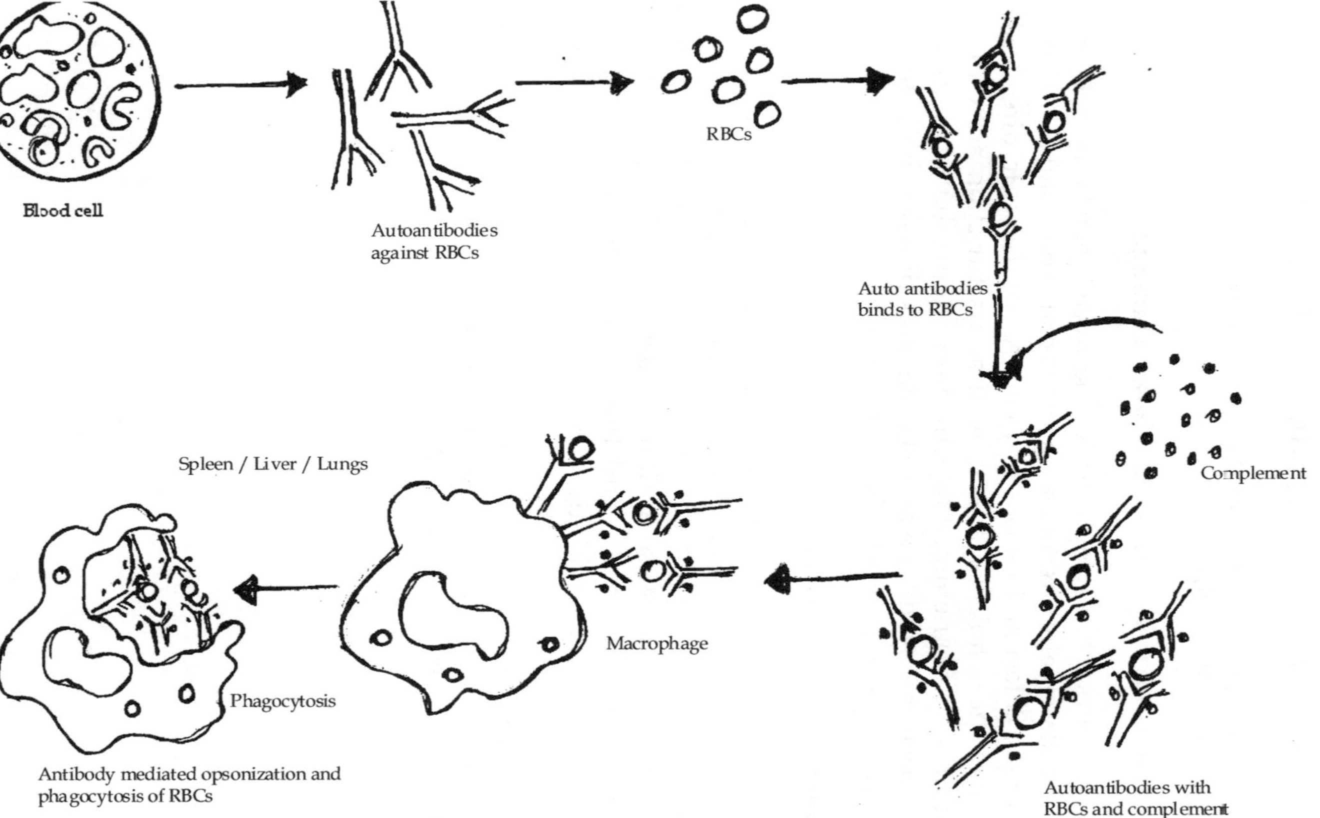

Fig. 19.1 : Antibody mediated removal of RBCs in autoimmune

Pernicious anemia is also an autoimmune disease which is caused by the binding of antibodies to intestinal intrinsic protein factor which is required for the uptake and transportation of vitamin B_{12}. This vitamin is necessary for proper development of erythrocytes. Vitamin B_{12} cannot be synthesized by the human body but it can only be intake through diet. When antibodies are present against this intrinsic factor, they neutralize it. Absence or reduced amount of intrinsic protein factor affects the absorption of vitamin B_{12} in the body **(Fig. 19.2)**. Decreased concentration of vitamin B_{12} in the body affects the maturation of erythrocytes and cause pernicious anemia.

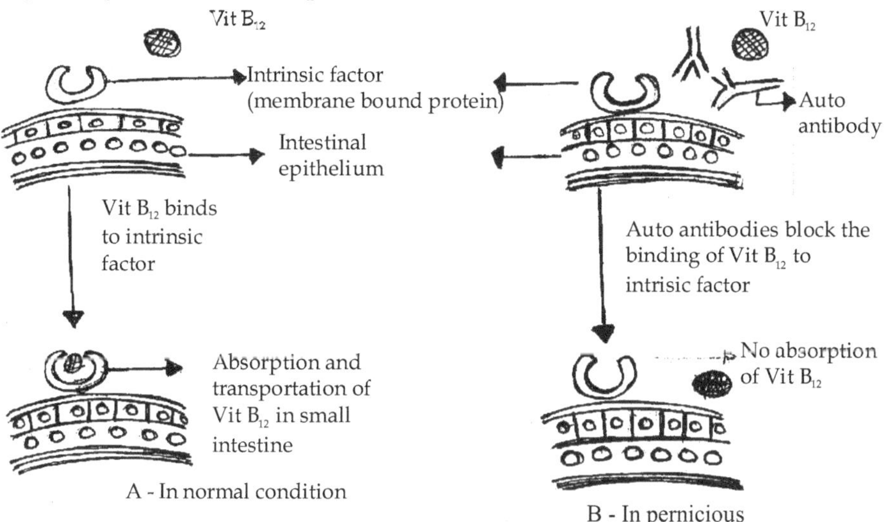

Fig. 19.2 : Auto immunity in Perni

Goodpasture's syndrome

In this disease autoantibodies attack the basement membrane of the kidney glomeruli and lung alveoli. This disease generally occurs in young men. Kidney is a vital organ which is essential for survival. The autoantibodies damage the glomerular and alveolar basement membrane and develop progressive glomerulonephritis and pulmonary hemorrhage. The IgG autoantibodies fix complement which causes necrosis in kidney glomeruli, causing disturbed renal function. The self antigen-autoantibodies complex forms a layer over the basal membrane and leads to hypersensitivity. The immunoglobulins are also deposited over the basal lamina of lung alveoli. This leads to hemorrhage and ultimately the patient dies. The condition develops due to sharing of antigenic epitopes between the epithelial cells of kidney and lung.

Insulin dependent Diabetes mellitus

Insulin dependent diabetes is an autoimmune disorder of pancreas. In pancreas, there are clusters of cells called islets of Langerhans. In the islets of Langerhans, there are three types of cells α, β and δ cells. â cells secrete hormone insulin which regulate glucose level in the blood. In insulin dependent diabetes, autoimmune response against β cells destroys these cells and level of insulin is decreased. This cause an increase in the blood sugar level. Several factors are involved in the destruction of β cells. These β cells may be destroyed by cytotoxic T lymphocytes or by autoantibodies. CTLs attack on β cells and destroy them. Autoantibodies destroy β cells either through complement lysis or antibody dependent cell mediated cytotoxicity. Reduced levels of insulin also result in ketoacidosis and increased urine production. The disease may lead to renal failure and blindness, if remains untreated.

Hashimoto's thyroiditis

This is a disease in which thyroid gland enlarges due to the infiltration of lymphocytes and phagocytes in the gland. The disease is mostly seen in middle aged women. Autoantibodies react to the thyroid proteins such as thyroglobulin and thyroid peroxidase. Both these proteins are responsible for uptake of iodine. Binding of autoantibodies to these proteins interferes with iodine uptake and leads to decrease the production of thyroid hormone. This causes hypothyroidism.

Myxedema

Myxedema is also a disease of thyroid. In this disease thyroid is totally destroyed and thyroxine hormone is not produces. Thyroxine is an important hormone for metabolism. In the absence of thyroxine, digestive functions, cardiovascular system, reproductive system as well as immune system are badly affected.

Disease Mediated By Stimulating Antibodies

In some autoimmune disease autoantibodies can modulate cell functions. Autoantibodies act as agonists and bind to the receptors of hormones and act as a hormones. This results in increased cell growth. Conversely, autoantibodies also act as antagonists and block the function of receptor for hormonal or neurotransmitter action. This causes atrophy of the cells. Myasthenia gravis, Grave's disease and Addison disease are the examples of this category of autoimmune diseases.

Myasthenia gravis

Myasthenia gravis is characterized by weaken and tired muscles. In this disease, speaking, eating and walking become tiresome. This disease is more frequently observed in young adult females and older males. The disease is caused by autoantibodies against muscle antigens and acetylcholine receptor antigens. Autoantibodies block the receptors of neurotransmitter, acetylcholine. The autoantibodies destroy the cells bearing receptors and acetylcholine cannot be produced **(Fig. 19.3).** The function of acetylcholine is severely affected. The nerve impulse cannot be transmitted from the nerves to muscles. It is an example of types II hypersensitivity.

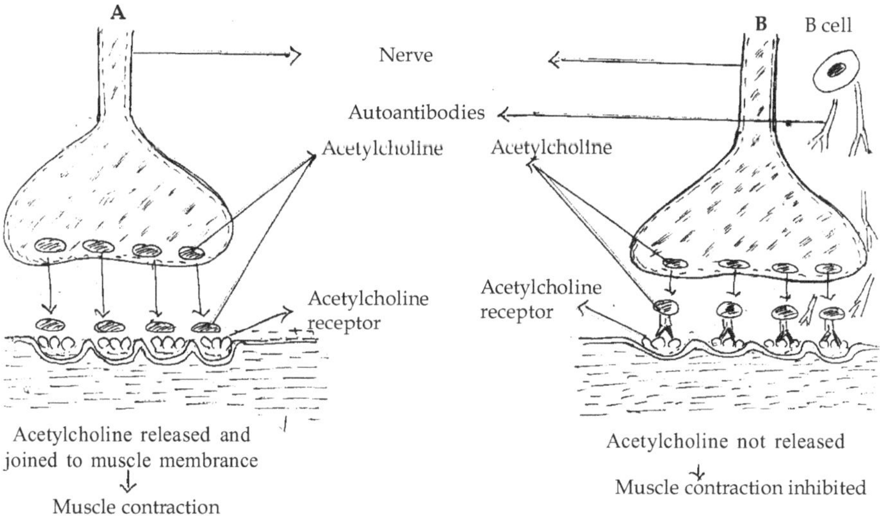

Fig. 19.3 : A - Normal stimulation of muscle : acetylcholine binds to acetylcholine receptor and muscle contracts.

B - Autoantibodies bind to acetylcholine receptors and acetylcholine is not released : muscle contraction is inhibited.

Grave's disease

It is an autoimmune thyroid disease. It is also called *thyrotoxicosis.* In this disease, autoantibodies are produced against TSH receptors. Binding of auto antibodies to these receptors mimics the action of TSH resulting in the production of excess amount of thyroid hormones **(Fig. 19.4).** If these autoantibodies are not regulated, they over stimulate the thyroid for the secretion of hormone. These autoantibodies are called long-acting thyroid stimulation (LATS) antibodies. These are IgG types of antibodies. Thyrotoxicosis is a condition of hyperthyroidism caused by generalized over activity of the entire thyroid gland. This disease is also called *diffuse toxic goiter.* The disease is associated with inflammation and protrusion of the eyes. Since IgG autoantibodies can cross the placenta, they can cause transient-hyperthyroidism in the newborn of the women who have Grave's disease.

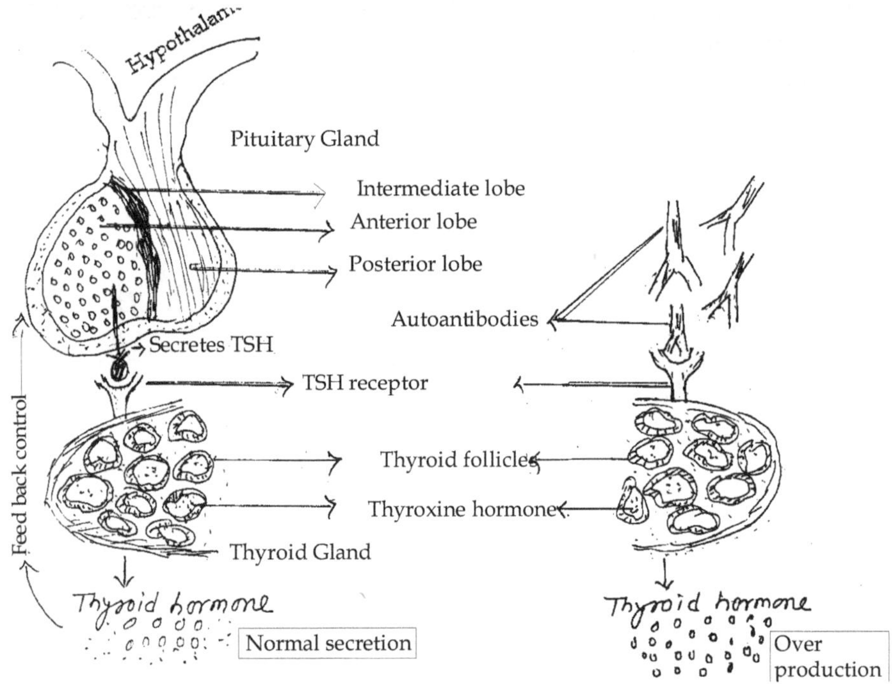

Fig. 19.4 : Binding of autoantibodies to TSH receptor

Addison Disease

This disease is caused due to adrenal hyperplasia and progressive malfunctioning of to the adrenal gland. In this disease, adrenocortical tissue is damaged so the production of adrenal hormones is affected. Adrenocortical tissue is damaged is by autoantibodies against zona glomerulosa cells of adrenal cortex. Adrenal hormones maintain the sodium and potassium level. The disturbances in the hormones cause increased excretion of water and dehydration. The symptoms are fatigue, change in skin color and gastrointestinal disorders.

NON ORGAN SPECIFIC AUTOIMMUNE DISEASES

These type of autoimmune diseases can attack many parts of the body. The immune response is directed towards a broad range of target antigens and involves many organs. The most common cause of such diseases is defective immune regulation of T and B lymphocytes. Rheumatoid arthritis, multiple sclerosis and systemic lupus erythematosus are some examples of this category of autoimmune diseases.

Rheumatoid Arthritis

It is a common autoimmune disease of joints of skeletal system. Rheumatoid arthritis is an inflammatory disease of connective tissue which is most common in women and afflicts them between 40 to 60 years of age. It is a chronic autoimmune disease of unknown etiology. The disease usually affects freely movable joints, joints capsule and synovial membrane. In this disease individuals produce autoantibodies called *rheumatoid factors* that are reactive in the F_C region of IgG. Another rheumatoid factor is IgM antibody. The IgM reacts with IgG and make a complex that is IgM- IgG complex. This gets deposited in the joints. These immune complexes initiate and augment inflammatory reactions, causing synovial membrane damage and cell lysis. These complexes activate complement reaction, leading to type III hypersensitivity. The synovial fluid of the patients contains increased number of T-cells and macrophages.

Multiple Sclerosis

Multiple sclerosis is the most common degenerative demyelinating disease of the central nervous system which causes neurological disability. The symptoms are numbness in limbs, paralysis and loss of vision. People are effected from this disease mostly at the age of 20 to 40. Myelin is an insulating substance that forms an intermittent sheath around certain neurons. It is involved in conduction of nerve impulse. Demyelination affects the neurons and conduction of nerve impulse. Neurological malfunction develops after demyelination. Patients suffering with this disease produce auto reactive T cells that participate in the formation of inflammatory lesions along the myelin sheath of nerve fibers. Cerebrospinal fluid of the patients contains auto activated T lymphocytes and autoantibodies, directed against myelin basic proteins, which destroy myelin sheath. Different factors develop the anti-myelin sheath response. In viral infections (adenovirus, Epstein bar virus and hepatitis B virus infections), the chance of disease becomes more.

Systemic Lupus Erythematosus

In the disease systemic lupus erythematosus, the word systemic refers to multiple organ connection, the word lupus is used to describe a characteristic butterfly facial rash resembling the colony of a wolf and erythematosus to red skin rashes. It is most common in middle aged females at the age of 20 to 40. Systemic lupus erythematosus results in skin rashes, joints pain, fever and weakness. The disease may cause hemolytic anemia, lymphopenia, thrombocytopenia and bleeding deformity. The affected individuals may produce autoantibodies to different tissue sell antigens like DNA, RBC, thromobocytes, leukocytes, histones and clotting factor. Interaction of autoantibodies to self antigens can initiate complement reaction as well as immune complex mediated type III hypersensitivity.

Complement activation leads to membrane attack complex formation, which ultimately results in cell damage. Viral infections and MHC gene region HLA-DR3 is a suspect of systemic lupus erythematosus. The main autoantibodies, found in systemic lupus erythematosus, are anti-DNA antibodies which can bind with free DNA to form immune complexes. The immune complexes circulate and get deposited at different tissues. After deposition, there occurs the activation of complement. In systemic lupus erythematosus, distinct types of cells appear in the blood and bone marrow. These cells are mature neutrophils that have phagocytosed the nuclear material, in the presence of antinuclear autoantibodies.

THERAPY FOR AUTOIMMUNE DISEASES

The ideal treatment of autoimmune diseases is to reinstate specific immune tolerance to self antigen. However, when more than one autoantigens are involved and induction of tolerance is difficult to achieve, some other methods are used. There are two commonly used methods- *replacement therapy* and *antigen non specific therapies.*

In ***replacement therapy*** autoantigens are removed directly through autoimmune response. To restore the damaged immune system, patients are administered normal cells or hormones. For example, patients are given platelets in thrombocytopenia, thyroid hormone in thyroid autoimmunity, insulin in IDDM and vitamin B_{12} in pernicious anemia.

In ***antigen non specific therapy,*** immunosuppressive drugs such as corticosteroids, azathioprine, cyclophosphamide and cytotoxic drugs are given. These drugs are often give responses to slow down the proliferation of lymphocytes. These drugs suppress the immune response but at the same time increase the risk of infection and malignancy. Non specific steroidal anti-inflammatory drugs like aspirin, celecoxibs and ibuprofen etc. are often used to dampen inflammation. Some corticosteroids e.g. predinisolone also reduce the inflammatory reactions. Cyclosporine, an immune suppressive drug blocks the production of lymphokines by T cells and is used in the treatment of some autoimmune diseases. Cyclophosphamide, a cytotoxic drug, blocks cell division and inhibits antibody production and thus prevents autoimmunity. Azathioprine is a cell division inhibitor and suppresses T cells. Cytotoxic drugs are used in severe case of autoimmune diseases to eliminate the autoantigen specific T and B cells.

Plasmapheresis therapy is one more method which is also used in some autoimmune diseases like Myasthenia gravis, rheumatoid arthritis and systemic lupus erythematosus. In this therapy plasma is removed from a patient's blood and red blood cells are given back by resuspending them in a suitable medium.

Several experimental therapeutic approaches are there. T cell vaccination has been used in experimental autoimmune encephalomyelitis(EAE).When rats were injected with low dose ($<10^4$) of cloned T cells specific for myelin basic protein (MBP), they did not develop symptoms of EAE. The experimental rat became resistant to the development of EAE. Further findings proved that T cell vaccine effect could be enhanced if the T cells are cross linked with formaldehyde. When cross- linked T cells were injected into rat with active EAE, permanent remission was observed.

Monoclonal antibodies are used to treat autoimmunity, in experimental rats. Oral antigen could also induce the state of immunologic unresponsiveness or tolerance. Oral antigens are used in multiple sclerosis.

POINTS TO REMEMBER

- Autoimmunity is the state in which body makes antibodies against its own cells. These antibodies are called autoantibodies.
- Diseases, associated with autoantibodies, are called as autoimmune diseases.
- Several factors are involved in the development of autoimmune diseases. These include genetics, gender, age, environment and drugs.
- Autoimmune diseases can be divided into two broad categories: organ specific and systemic autoimmune diseases.

- The organ specific diseases involve autoimmune response, directed primarily against a single organ or gland.
- The systemic diseases are directed against broad range of antigens and target a broad range of organs.
- Organ specific autoimmunity is mediated either by direct cellular damage or by stimulating antibodies.
- Systemic autoimmune diseases occur due to defective immune regulation of T and B lymphocytes which leads to widespread tissue damage.
- Organ specific autoimmune diseases include Pernicious anemia, Goodpasture's syndrome, Diabetes mellitus, Hashimoto's thyroiditis, Myxedema, Myasthenia gravis Grave's disease and Addison's disease.
- Systemic autoimmune diseases are Rheumatoid arthritis, multiple sclerosis and systemic lupus erythematosus.
- Two main approaches, used in the treatment of autoimmune diseases, are-replacement therapy and use of immunosuppressive drugs.

REVIEW QUESITONS

1. What is autoimmunity? Describe different types of autoimmune diseases.
2. Explain various factors which cause autoimmunity.
3. Write an account of organ specific autoimmune diseases.
4. What is a systemic autoimmune disease? How is it different from organ specific autoimmune disease?
5. Describe the autoimmune mechanism behind diabetes mellitus.
6. What are the various approaches for the treatment of autoimmune diseases?
7. Write short notes on:
   i. Myasthenia gravis
   ii. Rheumatoid arthritis

CHAPTER - 20

Immunotechniques

Many techniques are used in immunology to detect specific proteins (antigens) or antibodies in biological samples. These techniques are called *immunotechniques*. These methods are used for research and diagnostic purpose in many fields such as microbiology, biochemistry, endocrinology, hematology and in pathological laboratories. The techniques can be used to estimate proteins, drugs, hormones and toxic substances in a sample. Immunological techniques are highly specific, precise and reliable. These methods are simple and safe. Antigen and antibody reaction is highly specific in nature and this forms the base of the immunological methods. There are many immunotechniques which are used for the qualitative and quantitative estimation of antigens and antibodies in a sample. Some of commonly used techniques in immunology are:

- Immunodiffusion
- Immunoelectrophoresis
- Immunofluorescence
- Immunohemagglutination
- Radioimmunoassay
- ELISA
- Complement fixation test

IMMUNODIFFUSION

Immunodiffusion methods are used for the qualitative and quantitative estimation of antigen/antibodies in a sample. The methods are based on precipitation reaction. When antigen-antibody reaction is carried out in a gel, a precipitin line is formed at the equivalence point, the point where the concentration of antigen and antibody is optimal. If there is excess of antigen or antibody in the solution, antigen-antibody complexes are soluble and no visible precipitation is formed. The concentration of antibody in serum is expressed as *titre* of the antibody. Precipitation reactions can be demonstrated in test tubes**(Fig. 20.1)** or in agar plates. There are many variants of immunodiffusion.

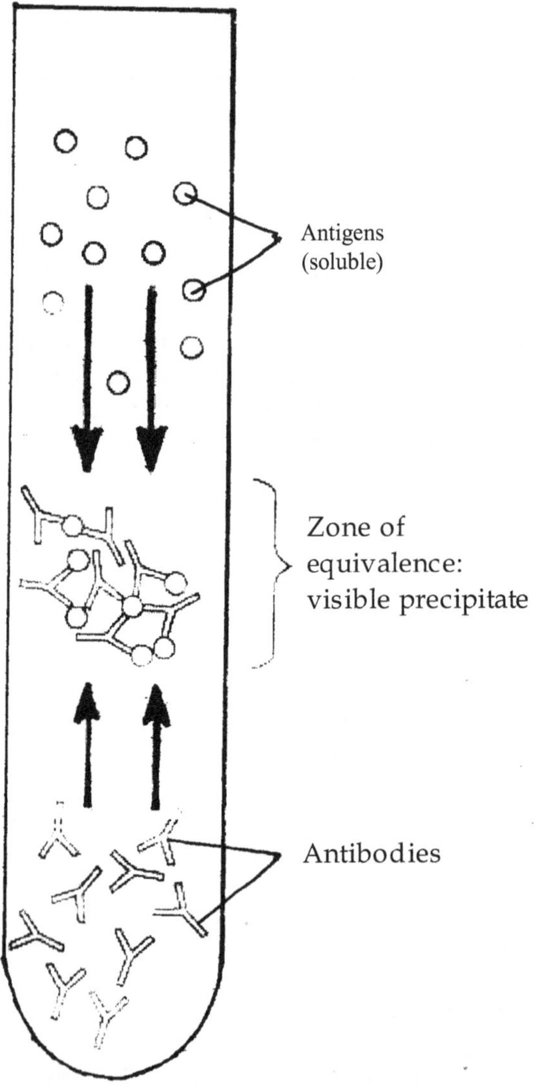

Fig. 20.1 : Precipitation reaction in a test-tube

Double Immunodiffusion (Ouchterlony Method)

Double immunodiffusion method is widely used to demonstrate precipitin reaction in antigen-antibody reaction. In this technique, agar solution is prepared and it is poured either on a slide or into the petridish. The solution is allowed to solidify. After the agar has solidified, wells are cut in the gel, using a punching device. Antigen and antibody are placed in the wells opposite to each other. The plate is left for some time to allow antigen and antibody to diffuse into the agar. Antigen and antibody diffuse radially through a concentration gradient. As the equivalence is reached, a visible line is formed at the junction of antigen-antibody complex **(Fig.20.2)**.

Fig. 20.2 : Double Immunodiffusion

If there are multiple antigens, multiple lines are formed. If there are two antigens which share identical epitopes against a single antibody, a single curved line is formed which indicates the identity between the antigens. If the two antigens are unrelated to each other and do not share any epitope, two independent precipitation lines are formed which cross each other and do not form a smooth curve. Such lines indicate that antigens are dissimilar and non cross-reacting. If two antigens share a particular epitope, the precipitin lines do not cross each other and a characteristic spur is formed **(Fig. 20.3)**.

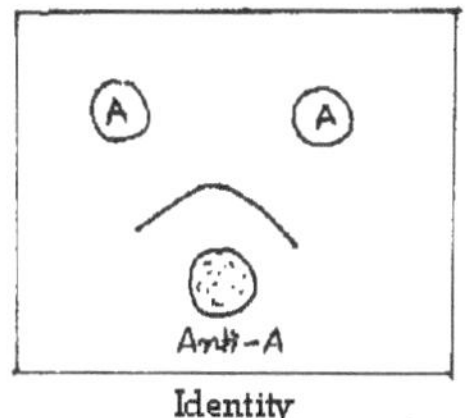

Identity

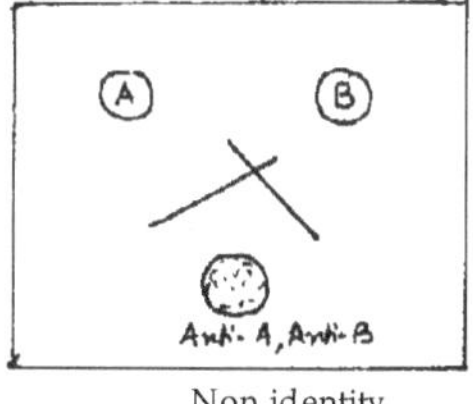

Non identity

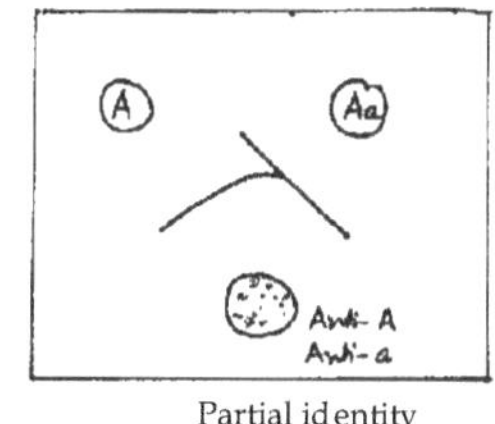

Partial identity

Fig. 20.3 : Reaction patterns with double diffusion in agar

Radial Immunodiffusion (Mancini Method)

This technique was introduced by Mancini, in 1965, for quantitative estimation of antigens. In this technique, as in the double immunodiffusion, agar plates are prepared and wells are cut in the agar. But here, in this method, antibody is incorporated in the gel in a suitable dilution, during preparation of the agar solution. The antigen is loaded into the wells. The plates are left for some time and antigen is allowed to diffuse through the gel. As the antigen diffuses into the gel, the region of equivalence is established and a precipitin line is formed around the well **(Fig. 20.4).** The area of the ring is proportional to the concentration of antigen. By using known concentrations of antigen and measuring the areas of the rings a standard curve is prepared. Concentration of unknown antigen can be determined by comparing the areas of the precipitin ring with the standard curve.

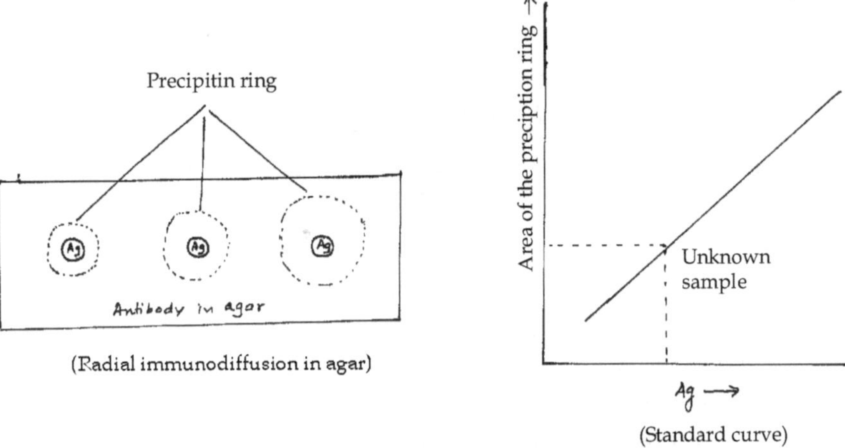

Fig. 20.4 : Radial immunodiffusion (standard curve is plotted by plotting areas of the preciptin rings against concentration of antigen)

IMMUNOELECTROPHORESIS

It is a combination of electrophoresis and immunodiffusion. First, there is electrophoresis of complex mixture of antigens and subsequent identification of the separated components by double immunodiffusion with the antiserum in the same gel. The technique is used to detect the presence of specific proteins in the serum. Both identification and quantification can be made by this technique. An agar gel slab is prepared and a trough and a well are cut into the gel. The trough is cut parallel to the well as shown in **figure 20.5.** The antigen mixture is loaded in the well and electrophorsed to separate its components. After separation of antigens, antibody is loaded in the trough and allowed to diffuse. The separated antigens and antibody diffuse towards each other and precipitation archs are formed, where antibody meets the antigens at equivalence. The technique is also used to determine the abnormality in concentration of antigen and/or the presence of abnormal proteins in a particular serum, urine or spinal fluid. A disadvantage of this technique is relatively poor resolution of antigen mixture.

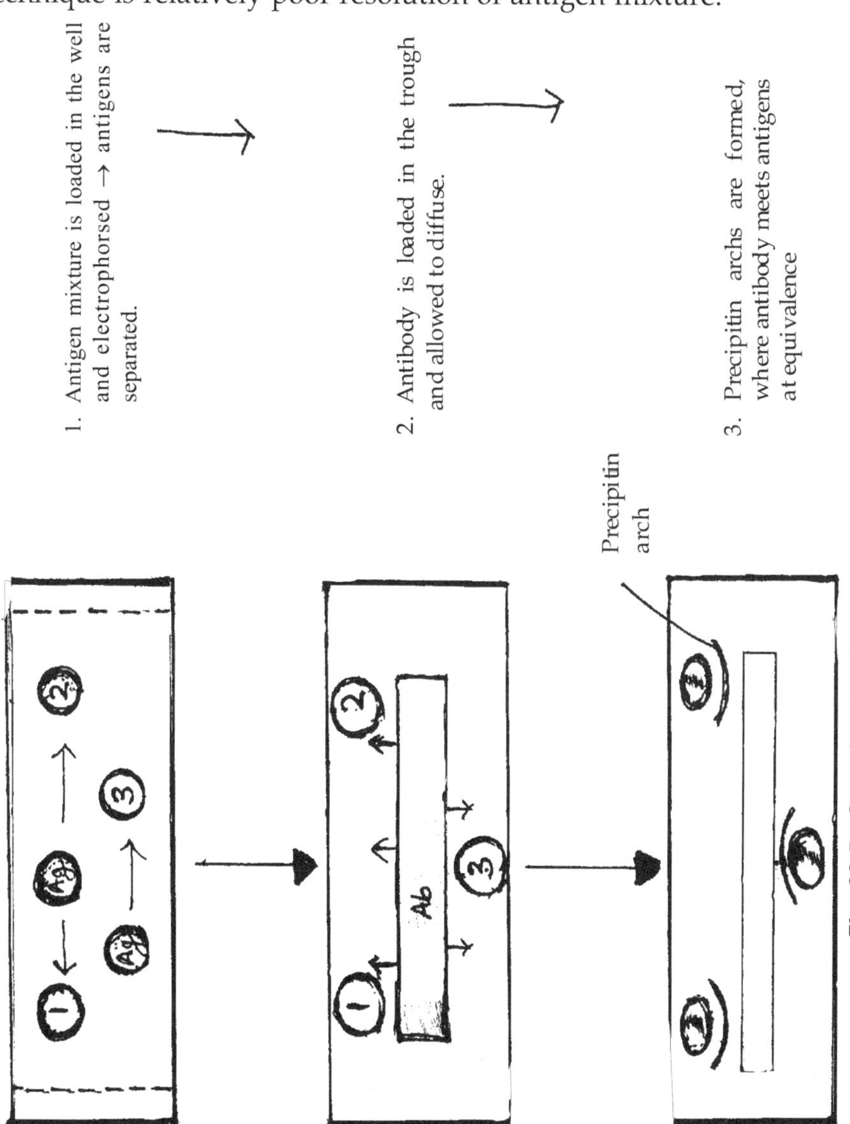

Fig. 20.5 : Steps involved in immunoelectrophoresis

There are two common variants of immunoelectrophoresis. These are:

1. Rocket immunoelectrophoresis
2. Counter current immunoelectrophoresis

Rocket immunoelectrophoresis

In this technique, a known amount of antibody is incorporated into the gel. The wells are cut on one corner of the plate. Antigen is loaded in the wells. When electrophoresis is carried out, negatively charged antigen moves towards the positive pole. Since the gel already contains antibody, when the antigen starts migrating towards positive pole, rocket shaped precipitin lines are formed in the gel. The heights of the rockets are proportional to the antigen concentration in the well. By using known concentration of the antigen rockets of different heights are formed and a standard curve is prepared by plotting the heights of the rockets against known antigen concentration. Concentration of an unknown antigen sample can be determined by comparing the rocket height of the unknown sample with the standard curve **(Fig. 20.6).**

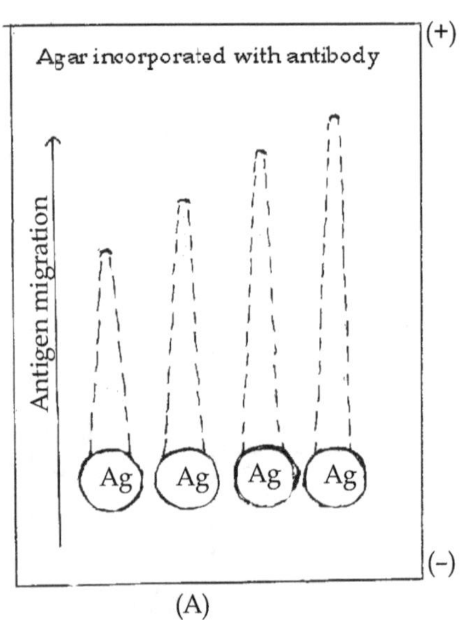

(A)

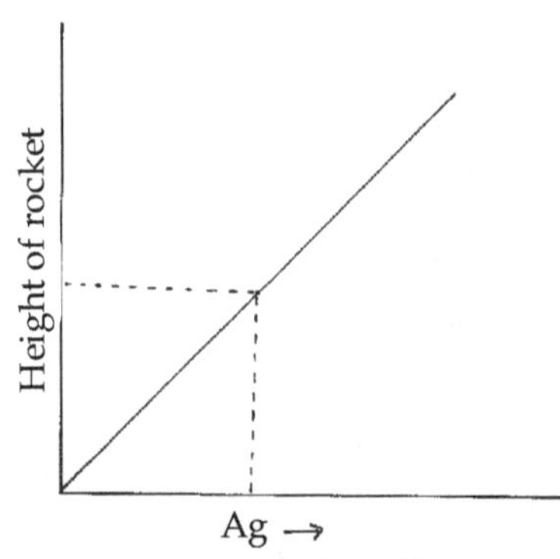

(B) Standard curve plotted between heights of rockets formed and Ag concentration.

Fig. 20.6 : Rocket electrophoresis.
(Concentration of Ag in an unknown sample can be determined by standard curve)

Counter current immunoelectrophoresis

Counter current immunoelectrophoresis is just like double immunodiffusion except that in this technique, antigen and antibody move towards each other in the presence of electric field. But in this technique, because of the presence of electric current, antigen and antibody move at faster rate.

Agrose gel plate is prepared and wells are cut just as in double immunodiffusion. Antigen and antibody are loaded in the opposite poles. Antigen is loaded in the well at the negative pole whereas antibody is loaded in the opposite positive pole. When electric field is applied antigen and antibody move towards each other and a precipitation line is formed at equivalence **(Fig. 20.7).**

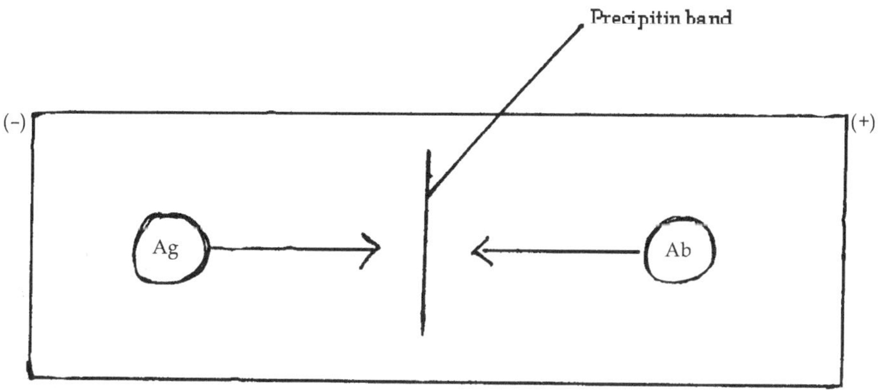

Fig. 20.7 : Counter current immunoelectrophoresis.

Crossed immunoelectrophoresis

It is slightly modified form of rocket immunoelectrophoresis. Crossed immunoelectrophoresis is used for the analysis of several antigens in a complex mixture simultaneously. In this method, antigens are first separated by electrophoresis. When the antigens get separated they are laid over another gel containing antibody and electrophoresis is carried out but now at the right angle to the direction of the previous one. Overlapping peaks of different heights, like rockets, are formed **(Fig. 20.8).** The area of the rockets corresponds to the concentration of antigen.

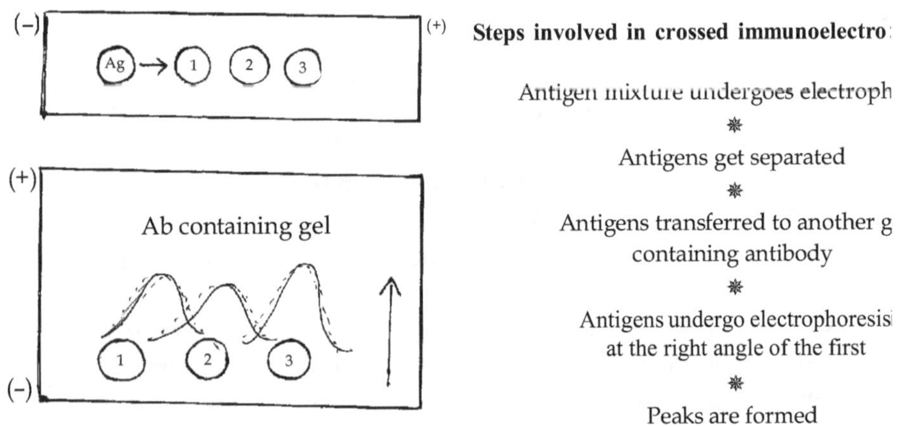

Fig. 20.8 : Crossed immunoelectrophoresis

IMMUNOFLUORESCENCE

It is an immunohistochemical technique, used for localization of antigens in the cells and tissues. The technique is widely used in research and clinical laboratories for many purposes such as evaluation of cells in suspension, cultured cells and tissues and for the detection of specific proteins. In this technique, antibodies are conjugated with fluorescent dyes such as fluorescein and rhodamine. The conjugated antibodies are allowed to combine with antigens, present in a cell or a tissue **(Fig. 20.9)**. The antibodies combine with specific antigens and bound antibodies can be visualized under fluorescence microscope. The fluorescence can also be amplified using a flowcytometer or automated imaging instrument.

To detect the presence of a specific antibody in the serum, antigens may be coupled to a fluorescent dye. There are two methods of immunofluorescence: direct and indirect.

Direct Method

In this method, antibody is directly conjugated to a fluorescent dye. It is a simple procedure which requires less time. If we wish to observe the distribution of a specific antigen on the tissue, antibody against those antigens are isolated and conjugated with fluorescence dye. These labeled antibodies are directly applied to a section of a tissue on the slide and observed in a fluorescence microscope in the form of bright spots. The technique can also be used to detect two antigens simultaneously in a sample. For this, antiserum is conjugated with two different fluoresce dyes which emit fluorescent at different wavelengths.

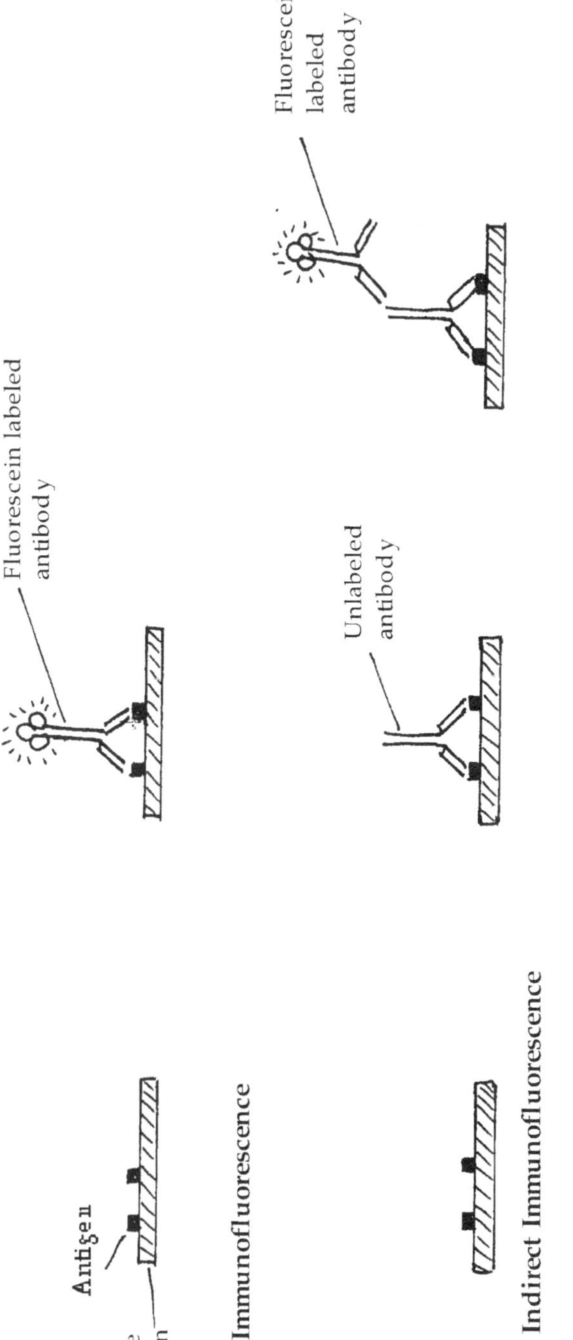

Fig. 20.9 : Direct and Indirect immunofluorescence techniques

Indirect Method

In this method, first unlabelled antibodies are applied to the tissues. The antibodies bind to the tissue antigens. Antibodies against these unlabelled antibodies i.e. anti-antibodies, are conjugated with fluorescent dye. These fluorescent conjugated antibodies are then applied to the tissues.

The technique is used to detect the presence of antibodies, developed against a specific antigen in the serum. If antibodies are present they will bind to fluorescent conjugated anti-antibodies which can be visualized in a fluorescence microscope. The technique can also be used to detect complement fixation of tissue section by first treating the tissue section with a mixture of antibody and complement, followed by a fluorescent conjugated anti-complement reagent as a second layer. This technique has better sensitivity i.e. fluorescence is higher than that with direct method.

IMMUNOHEMAGGLUTINATION

This method is used for determination of blood groups. As mentioned earlier in the book (Chapter 5: Antigen-antibody interaction), the four types of blood groups A,B,AB and O are based on the specific inheritable antigens, called *agglutinogen* which are present on the surface of erythrocytes. Blood group A possesses agglutinogen A, blood group B possesses agglutinogen **B**, blood group AB has both A and B agglutinogens and blood group O does not possess any agglutinogen. In addition to these agglutinogens, blood also contains corresponding antibodies, called *agglutinin*. These agglutinins are present in the plasma and react with the corresponding agglutinogens.

For determining blood groups, a drop of each antiserum A and antiserum B are placed on a slide at two different points. A small amount of blood is transferred to drop A and mixed gently with antiserum A. Similarly, a small amount of blood is mixed with antiserum B. The slide is left for a few minutes. After a few minutes, agglutination is observed as coarse granules. Agglutination can be seen by naked eyes or under low power of microscope.

The technique is based on agglutination reaction between antigen(agglutinogen) and antibody(agglutinin). When antigen-antibody reacts with each other agglutination occurs due to lattice formation. This can be observed in the form of clumping. Agglutination is observed, as given in table no. 20.1

Table 20.1 : Agglutination observed in different blood groups

Blood group	Antiserum A	Antiserum B
A	Clumping	No clumping
B	No clumping	Clumping
AB	Clumping	Clumping
O	No clumping	No clumping

Landsteiner and Wiener, in 1940, discovered a specific protein on the erythrocytes of the Rhesus monkey, called Rh factor. Rh factor is also present on the surface of red blood cells in some human beings. Rh factor is a mosaic of several antigens which are determined by three pairs of genes called C,D and E. Antiserum, used for determination of Rh factor is called anti D.

As in blood grouping, a small amount of blood is mixed with anti D. If agglutination occurs, this indicates the presence of Rh factor (Rh positive) and absence of agglutination indicates absence of Rh factor (Rh negative).

RADIOIMMUNOASSAY (RIA)

Radioimmunoassay is commonly used for quantification of hormones, drugs and steroids in a biological sample. This is one of the most reliable, precise and sensitive test which allows the measurement of antigen up to picogram (10^{-12}gm).

The principle of RIA involves competitive binding of radio labeled antigen (antigen is generally labeled with radioactive iodine ^{125}I or tritium ^{3}H) and unlabeled antigen to a highly affinity antibody. A fixed amount of antibody and labeled antigen react in the presence of known (standard) or unknown amount of antigen. Both labeled and unlabeled antigens compete for antigen binding sites on the antibody. With the increase in the amount of unlabeled antigen the binding of labeled antigen to the antibody is competitively inhibited. This results in the lowering of percentage binding **(Fig. 20.10).** A standard curve is prepared by recording the amount by which the binding of labeled antigen to the antibody is competitively inhibited by increasing the amount of known unlabeled antigen. The amount of antigen in an unknown sample can be detected by comparing the binding of labeled antigen from the standard curve.

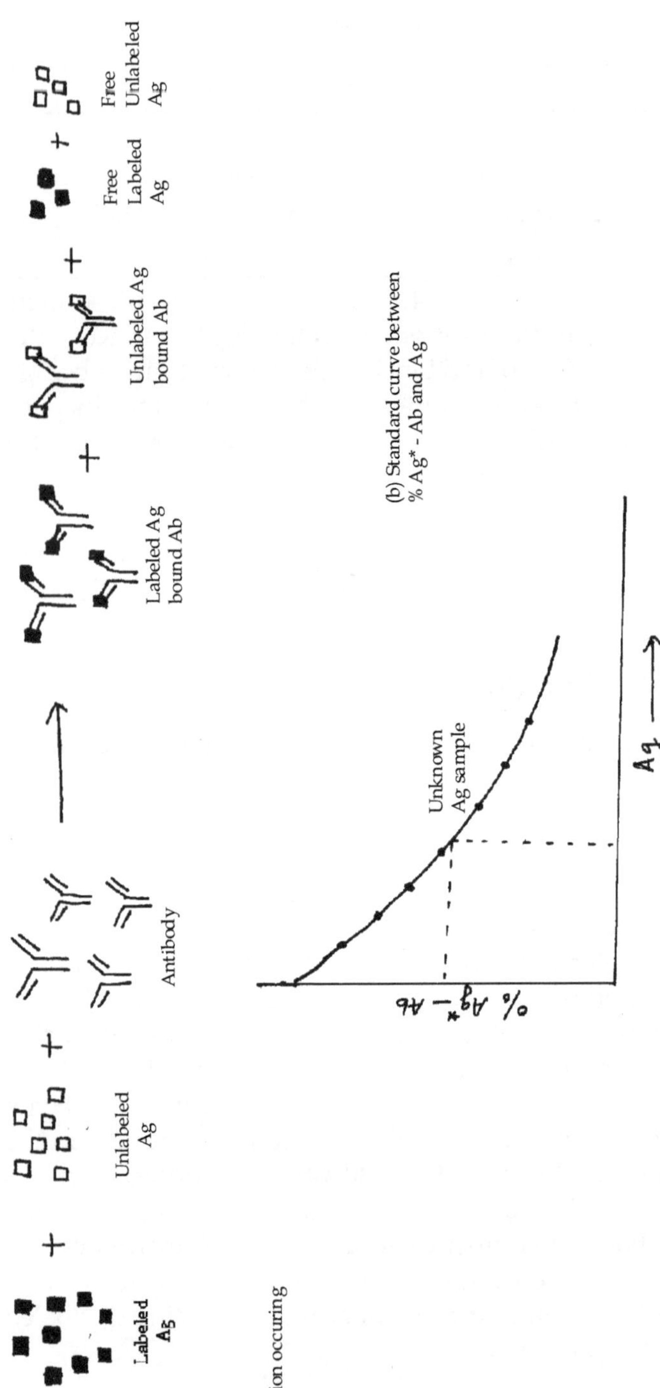

Fig. 20.10 : (a) Radio immunoassay (b) Standard curve between% of labeled Ag bound to Ab and concentration of unlabeled Ag.

$$Ag + Ag^* + Ab \longrightarrow Ag^*Ab + Ag$$

For measuring the amount of labeled antigen bound to antibody, first, the bound labeled antigen is separated from free labeled antigen. These can be separated by ion-exchange chromatography. Radioactivity is measured through gamma counters.

Though RIA is a very precise and sensitive technique but there is always a risk for exposure to the radioactive substances, if proper measures and procedures are not adopted while handling the radioactive isotopes.

ENZYME LINKED IMMUNOSORBENT ASSAY (ELISA)

Enzyme linked immunosorbent assay (ELISA) is a widely used technique for detection of antigen- antibody reaction. The technique is much safer and less expensive than RIA. ELISA test kits are commercially available and are easy to use. The tests are faster and do not require high skill. The procedures are highly automated and the results can be read by a computer scanner. There are two basic methods of ELISA: Direct ELISA and Indirect ELISA.

Dircct ELISA

This test is used to detect antigens. In this assay, specific antibodies (antibodies specific to an antigen to be detected, say a drug, if we are detecting the presence of a drug in the urine or body fluids) are placed on to the wells on the microtiter plate. The antibodies are adsorbed to the walls, coating the plate. The patient's sample is added. If specific antigens are present in the sample, it will bind with the antibodies. The well is washed to remove any unbound antigen. Antibodies, specific to antigens and linked to an enzyme, are added. These enzyme-linked antibodies will react with the Ab-Ag complex and a 'sandwich' of Ab-Ag-Ab^E is formed. Finally, a colorless substrate which reacts with the enzyme, is then added. The enzyme reacts with the substance and a colored product is formed. Such substances which are colorless and when acted on by the enzymes are converted into a colored product are called *chromogens*. The intensity of the product is measured by spectrophotometer. If specific antigens are present in the sample a final colored product is formed **(Fig. 20.11).**

Indirect ELISA

Indirect ELISA detects antibodies rather than antigen in a patient's serum. In this test, specific antigens are adsorbed onto the walls of the

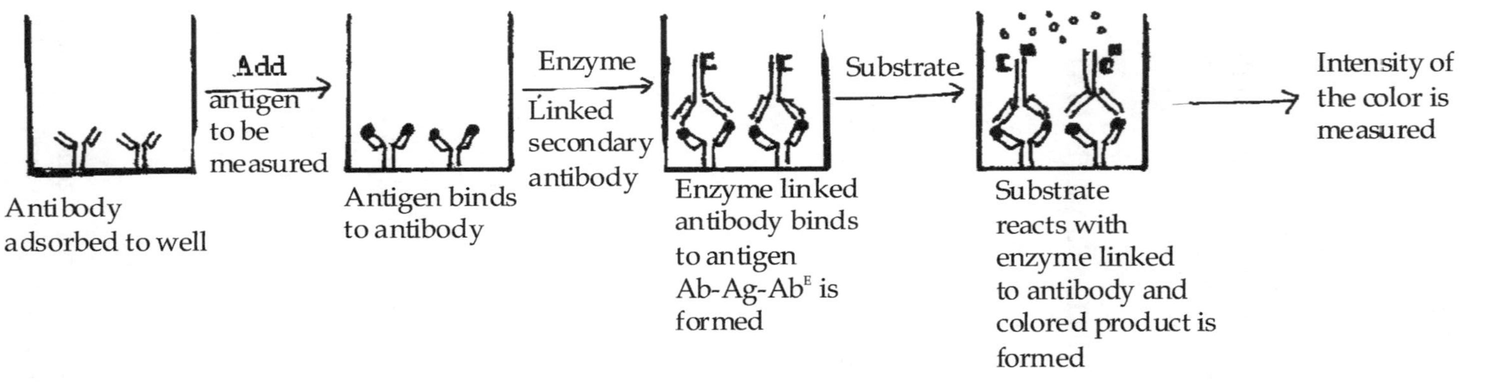

Fig. 20.11 : Steps involved in direct ELISA

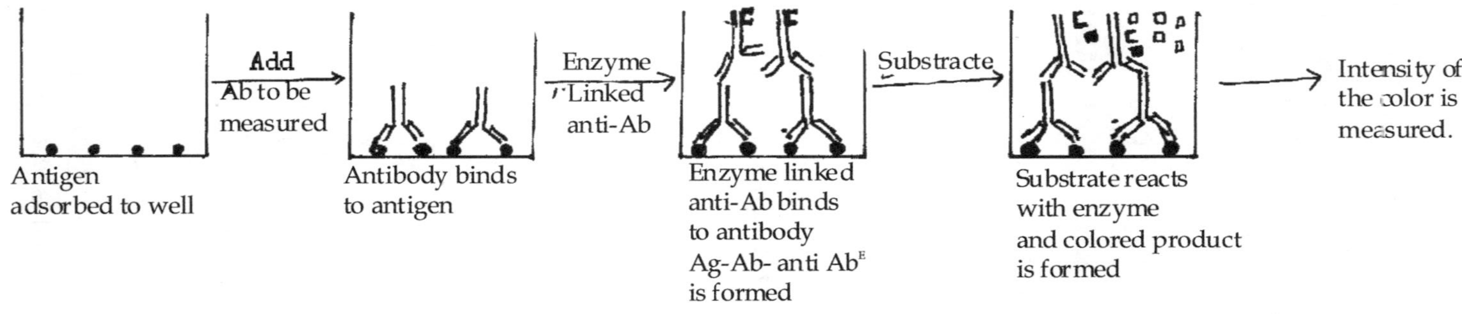

Fig. 20.12 : Steps involved in indirect ELISA

wells in the microtiter plate. The test serum is added. If specific antibodies are present they will bind with the antigens and antigen-antibody complex is formed. The wells are rinsed to remove any unbound antibody. Anti-antibodies (antibodies against the antibodies) linked to an enzyme are added. These enzyme linked anti-antibodies react with the antibodies and Ag-Ab-anti-Ab^E complex is formed **(Fig. 20.12).** A chromogen is then, added. The chromogen reacts with the enzyme and a colored product is produced. The intensity of the product is measured, as in the direct ELISA. Indirect ELISA is commonly used in detecting antibodies to HIV and rubella virus. The test is also used to detect the presence of certain drugs in the serum.

COMPLEMENT-FIXATION TEST

Complement fixation test was developed, in 1901, by Jules Bordet and Octave Gengou and is used to detect the presence of antibodies against a variety of pathogens in the body. *Wassermann test,* used for syphilis, is an example of complement fixation test.

Generally, during an antigen-antibody reaction, when antigen-antibody complex is formed, the complement is activated and it binds with the Fc portion of the antibody. Attachment of the complement facilitates the destruction of antigen-antibody complexes. This attachment of complement with the antigen-antibody complex is called *complement fixation* and can be used to detect the presence of antibodies or antigens.

Complement fixation test occurs in two stages. In the first stage, antigen and patient's serum are mixed and a standard amount of complement is mixed. If antigen specific antibodies are present in the serum it would bind with the antigen and antigen-antibody complex is formed. Complement binds (fixed) to this complex and is utilized. In the second stage, sheep RBC and antibodies to sheep RBC (anti-SRBC) are added. Sheep RBC and anti-SRBC form antigen-antibody complex. The test is observed for the presence and degree of red cell lysis. If complement is available (not utilized in the first stage) it would bind to the sheep RBC-anti-SRBC complex and cause hemolysis. In case, complement is fixed and utilized in the first stage (if antigen specific antibodies are present in the patient's serum), no hemolysis is observed **(Fig. 20.13).** Thus, the absence of hemolysis (no hemolysis) indicated the presence of specific antibodies in the patient.

POINTS TO REMEMBER

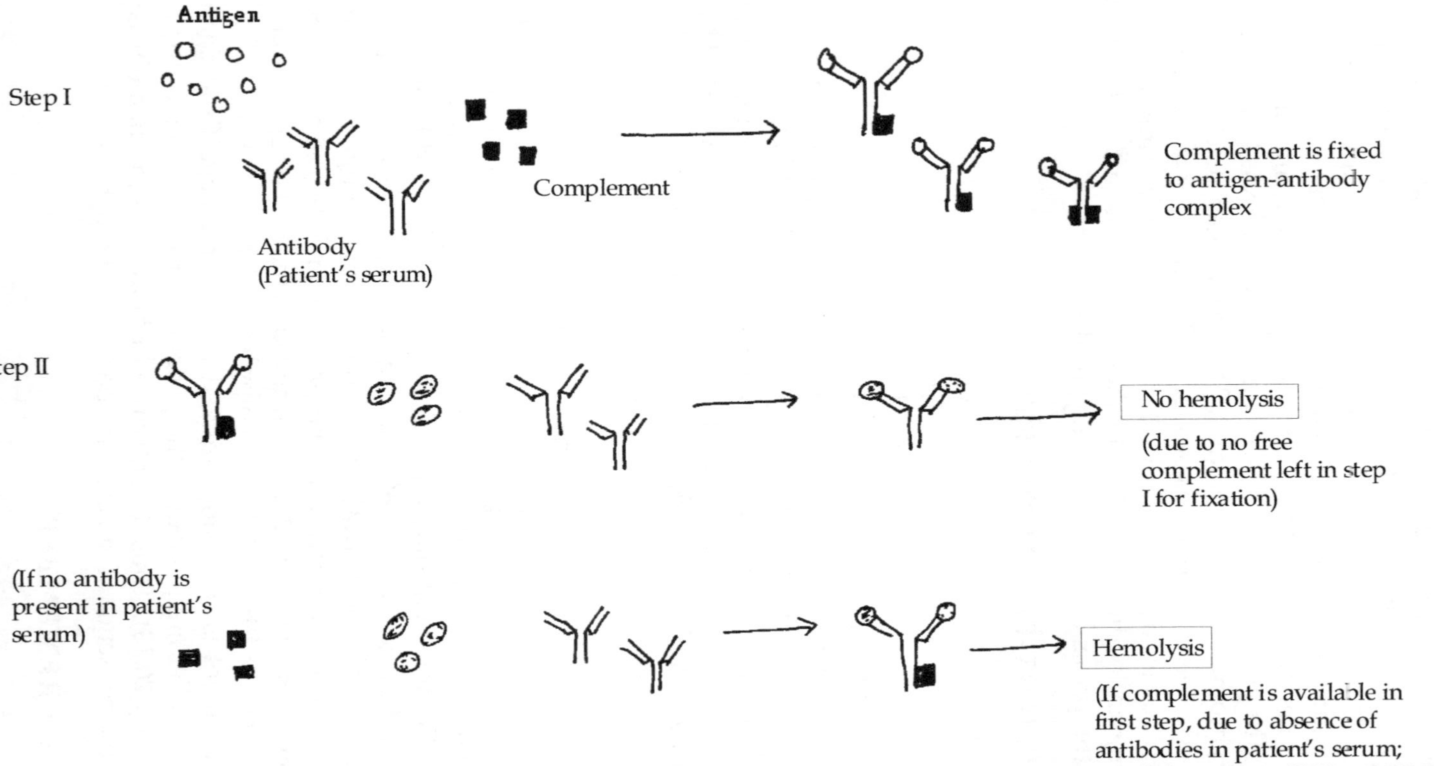

Fig. 20.13 : Complement fixation test

- There are many techniques used in immunology to estimate antigen or antibody qualitatively or quantitatively.
- Some of the commonly used techniques in immunology are: immunodiffusion, immunoelectrophoresis, immunofluorescence, hemagglutination, radioimmunoassay, ELISA and complement fixation test.
- Immunodiffusion is based on precipitation reaction.
- There are many variants of immunodiffusion such as double immunodiffusion and radial immunodiffusion.
- Immunoelectrophoresis is a combination of electrophoresis and double immunodiffusion of the antigen with the antiserum in the same gel.
- The two common variants of immunoelectrophoresis are: rocket immunoelectrophoresis and counter current immunoelectrophoresis.
- Immunofluorescence is an immunohistochemical technique, used for localization of antigens in the cells and tissues.
- In immunofluorescence, antibodies are conjugated with fluorescent dyes and conjugated antibodies are visualized under fluorescence microscope.
- Immunohemagglutination is used for determination of blood groups.
- Radioimmunoassay is one of the most reliable, precise and sensitive test, commonly used for quantification of hormones, drugs and steroids in a biological sample.
- ELISA is a commonly used method for detection of antigen-antibody reaction. Two types of ELISA are : direct and indirect methods.
- The attachment of complement with the antigen-antibody complex is called complement fixation. The test is used to detect the presence of antibodies or antigens.

REVIEW QUESTIONS

1. What do you understand by immunodiffusion? Describe different types of immunodiffusion techniques.
2. What is immunofluorescence? Explain its applications.

3. Describe the principle and various types of immunoelectrophoresis.
4. What is radioimmunoassay? Describe the principle and applications of RIA.
5. What is ELISA? Describe the principle and types of ELISA
6. Write notes on:
   i. Complement fixation test
   ii. Immunohemagglutination
7. Differentiate between:
   i. Agglutinogen and agglutinin
   ii. Direct and indirect ELISA

Glossary

Acquired Immunity: Immunity which is acquired after birth. It may be natural or artificial and active or passive.

Acute Phase Protein: A group of serum proteins which increase during infection.

Adjuvant: A substance which is used to increase the immune response to an antigen.

AFCs (antibody forming cells): See, *plasma cells.*

Affinity: Strength of binding between an antigen determinant (epitope) and antibody binding site.

Agglutination: Clumping of particles e.g. blood cells, in an antigen-antibody reaction.

AIDS: Refers to Acquired immune deficiency syndrome. It is a disease caused by the human immunodeficiency virus (HIV). The virus attacks on the CD4 cells of the immune system.

Allergen: An antigen responsible for producing allergic reactions by inducing IgE synthesis.

Allergy: Response of the body against the substance (allergen) which is normally harmless.

Alloantigens: Polymorphic antigen determinants which differ among members of the same species because of genetic variation.

Allogeneic: Genetically dissimilar within the same species.

Allograft: A tissue transplant between two genetically different individuals of the same species.

Allotype: A set of allotypic determinants which vary in different members of the same species.

Alpha-Feto-Protein (AFP): An antigen that remains present during fetal development and has immunosuppressive properties. High levels of AFP are found in liver, stomach and pancreas in cancer patients.

Alternative pathway: It is an antibody independent pathway of complement activation which is triggered by the binding of complement protein C3b to the surface of a pathogen. The alternative pathway also amplifies the classical pathway of complement activation.

Anaphylatoxin : A substance, capable of releasing histamine from mast cells and basophils.

Anaphylaxis: An acute, emergency immune reaction that occurs in some patients with allergies. It results in the release of toxins by the immune system that stops breathing, develop asthma and in some cases go into shock and die.

Anergy: A state of response failure towards antigens.

Antibody dependent cell mediated cytotoxicity (ADCC): A form of lymphocyte mediated cytotoxicity in which an effector cell kills an antibody coated target cell.

Antibody: A substance formed by the body against an antigen.

Antigen presentation: The display of an antigen as peptide fragments bound to MHC molecules on the surface of a cell.

Antigen processing: The degradation of antigens (proteins) into peptides. These peptides bind to MHC molecules for presentation to T cells.

Antigen: Any foreign body which can bind with an antibody.

Antigen-antibody complex: A molecular complex formed of antigen and antibody molecules binding together.

Antigen-binding site: Site found at the surface of the antibody molecule that makes physical contact with the antigen. Antigen-binding sites are made up of six hypervariable loops, three from the light-chain V region and three from the heavy-chain V region.

Antigenic determinant: The part of an antigen that contacts the antigen binding sites of an antibody or the T cell receptor, also known as *epitope.*

Antigenic drift: Characteristics of changing of an antigen through spontaneous point mutations and thus giving rise to new strains by pathogens.

Antigen-Presenting Cells (APCs): A heterogeneous group of immunocompetent cells that mediate cellular immune response by processing and presenting antigens to the T-cell receptor. These cells include macrophages, dendritic cells, and Langerhans cells.

Antiserum (plural: antisera): Fluid component of clotted blood from an individual that contains antibodies against the molecule used for immunization.

Antitoxin : An antibody specific for exotoxins produced by certain microorganisms.

Apoptosis: Cell death caused by activation of endogenous molecules, leading to the fragmentation of DNA . Also known as *programmed cell death*

Arthus reaction: A hypersensitivity reaction produced by local formation of antigen-antibody complexes that activate the complement cascade and cause thrombosis, hemorrhage, and acute inflammation.

Atopy: The clinical manifestation of type I hypersensitivity reactions, including hay fever, eczema, asthma and food allergy.

Autoantibody: An antibody, synthesized against an epitope of usual body component of an individual.

Autograft: Transplantation of tissue from one region to another region in the same individual.

Autoimmune disease: A clinical disorder caused by breakdown of self tolerance.

Autoimmunity: An abnormal immune response to autoantigens or self antigens.

Avidity: The overall functional binding strength of a multivalent antigen and multivalent antibody.

B cell co-receptor: It is a complex of three membrane bound transmembrane proteins: CR2, CD19 and CD8 associated with B cell receptor.

B cell receptor (BCR): B cell surface immunoglobulin molecule and two associated signal-transducing Ig-α/Ig-β molecules.

B cell: Lymphocyte derived from bone marrow and express membrane bound antibody.

Bare lymphocyte syndrome: An immunodeficiency disease in which MHC class II molecules are not expressed on cells as a result of one of several different regulatory gene defects.

Basophils : A type of white blood cells containing granules that stain with basic dyes. These cells have a function similar to mast cells.

Bence-Jones protein: Dimers of immunoglobulin light chains, found in the urine of patients with multiple myeloma.

Benign tumor: A tumor which is a nonmalignant form of neoplasm.

Bone marrow: It is a primary lymphoid organ and a soft tissue present within the cavities of bones.

Bradykinin: A nine amino acids long peptide, derived from serum α_2 globulin that produces an inflammatory response.

Bronchus -associated lymphoid tissue (BALT): Secondary lymphoid organs that are associated with the respiratory tract.

Bursa of Fabricius: A primary lymphoid organ in birds, where B cell maturation occurs.

Capsid: The protein coat around a virus.

Carcinoembryonic antigen (CEA): An oncofetal tumor associated antigen.

Carcinoma: A malignant tumor, which arises from cells of epithelium.

CD4: A glycoprotein molecule, present on T_H cells, that serves as a co-receptor. It binds to MHC class II molecules.

CD8: A dimeric protein molecule, present on Tc cells, that serves as co-receptor. It binds to MHC class I molecules.

Cell adhesion molecules (CAMs): A group of cell-surface molecules that mediates intracellular adhesion.

Cell mediated immunity (CMI): An immune response that does not involve antibodies. It is mediated by the activation of phagocytes, antigen-specific Tc cells and various cytokines in response to an antigen.

Cell-mediated cytotoxicity (CMC): Killing of a target cell by an effector lymphocyte.

Central tolerance: Tolerance of T or B cells, induced during their development.

Chemokines : Small chemo attractant proteins that stimulate the migration and activation of cells, especially phagocytic cells and lymphocytes

Chemotaxis: Movement of cell towards a substance, caused by a biochemical substance.

Class switching: The process by which an individual B cell stops secreting antibody of one isotype or class and starts producing antibody of a different isotype but with same antigenic specificity. Also known as *isotype switch.*

Clonal deletion: The loss of lymphocytes of a particular specificity due to contact with either self or foreign antigen.

Clonal selection: The fundamental basis of lymphocyte activation in which antigen selectively causes activation, division and differentiation only in those cells that express receptors.

Clone: A population of cells, all derived from a single progenitor cell.

Clusters of differentiation (CD) : Groups of monoclonal antibodies that identify the same cell-surface molecule. These are designated as CD followed by a number e.g. CD1, CD2, etc.

Complement: A group of 20 heat labile serum proteins which affect primarily antibody-mediated lysis of some cells. These proteins are involved in control of inflammation and activation of phagocytes.

Complementarity determining regions (CDRs): The complementarity-determining regions (CDR) are the parts of the variable regions of an antibody which participate in the binding of epitopes.

Conformational epitopes: These are the discontinuous epitopes on a protein antigen that are formed from several separate regions in the primary sequence of a protein brought together by protein folding.

Constant region (C region) : That part of an immunoglobulin or T-cell receptor which is relatively constant in amino acid sequence between different molecules.

Contact hypersensitivity: A delayed-type hypersensitivity (type IV) in which T cells respond to antigens that are introduced by contact with the skin. Poison ivy hypersensitivity is contact hypersensitivity.

Continuous epitopes: These are the antigenic determinants on proteins that are contiguous in the amino acid sequence.

Co-stimulation: Co-stimulation involves ligand-receptor interactions at the surfaces of a responder lymphocyte and an 'accessory' cell such as APCs for activation of T cells and helper T cells for activation of B cells.

C-reactive protein (CRP): A serum protein which is produced by liver cells as part of the acute phase response. C-reactive protein binds to phosphatidylcholine, a component of the surface of many bacteria and helps in the opsonization of the bacteria.

Cross reaction: The reaction of one antigen with antibodies developed against another antigen.

Cytokines: Low molecular weight proteins that mediate interactions between cells.

Cytotoxic T Cell: An effector T cell (CD8) that can recognize and kill virally infected host cells, bearing antigenic peptides, presented by MHC class I.

Degranulation: A process in which the contents of cytoplasmic granules of a cell are discharged out.

Delayed-type hypersensitivity(DTH): Type IV hypersensitivity response mediated by T lymphocytes.

Dendritic cells (DC): Antigen-presenting cells of the immune system, which play a critical role in the regulation of the adaptive immune response.

Diapedesis : The movement of blood cells, particularly leukocytes, from the blood across blood vessel walls into tissues.

DiGeorge's syndrome: A recessive genetic immunodeficiency disease in which there is a failure to develop thymic epithelium. The disease seems to be due to a defect in the development of neural crest cells.

Domain: Linear repeats of similar segments in H or L chains of antibody molecule. Each domain has an internal disulphide loop.

Double Negative (DN) Cells: A subset of developing T cells that do not express CD4 or CD8.

Double Positive (DP) Cells: A subset of developing T cells that express CD4 or CD8.

Effector cells: Lymphocytes that can mediate the removal of pathogens or antigens from the body without the need for further differentiation.

Endocytosis: Mechanism whereby substances are taken into a cell from the extracellular fluid by pinching off plasma membrane vesicles.

Endotoxins: Bacterial toxins that are released when bacterial cells are damaged or destroyed.

Enzyme linked immunosorbent assay (ELISA): It is a specific and sensitive assay for qualitative and quantitative estimation of serum antibody and antigen. The test uses antibodies and color change to identify a substance.

Eosinophils: A type of granulocyte that functions in antibody-dependent cell mediated cytotoxicity.

Epstein-Barr Virus (EBV): The causative agent of Burkitt's lymphoma and infectious mononucleosis. It can transform human B cell into stable cell lines.

Epitope: see, *antigen determinants.*

Erythroblastosis fetalis: A severe form of Rh hemolytic disease in which maternal anti-Rh antibody enters the fetus and produces a hemolytic anemia.

Erythropoietin: A glycoprotein produced by the kidney which stimulates red blood cell production.

Exocytosis: The process of the release of intracellular vesicle content to the exterior of the cell.

Extravasation : The movement of cells or fluid from/within blood vessels to the surrounding tissues.

Fab: Antigen binding fragment, containing a partial H and L chain, produced by enzymatic digestion of an IgG molecule with papain.

Fas ligand (Fasl): A membrane protein that is a member of the TNF family of proteins, expressed on activated T cells.

F_c: A fragment of antibody responsible for the binding of antibody and complement to antibody receptors of the cells.

Framework region (FR): A relatively conserved sequence of amino acid located on either side of the hyper variable regions in the variable domain of immunoglobulin heavy and light chains.

Freund's complete adjuvant: An oil containing killed mycobacteria and an emulsifier, which, enhances the immune response to that immunogen, when injected with an immunogen. If mycobacteria are not included, it is called incomplete Freund's adjuvant .

GALT (Gut associated lymphoid tissue): The accumulation of lymphoid tissue, associated within the gastrointestinal tract.

Germinal Centre: A region within lymph nodes and spleen where B cell activation, proliferation and differentiation occurs.

Goodpasture's syndrome: An autoimmune disease in which autoantibodies are produced against basement membrane or type IV collagen.

Graft Versus Host Disease (GVHD): A reaction that develops when a graft contains immunocompetent T cells that recognize and attack the host cells.

Granulocyte-macrophage colony-stimulating factor (GM-CSF): A type of cytokine which is involved in the growth and differentiation of myeloid and monocytic lineage cells.

Granulocytes: Leukocytes that contain granules in their cytoplasm. They are of three types-neutrophils, eosinophils and basophils.

Granuloma: A localized tumor like nodule, characterized by chronic inflammation with macrophages, epithelial cells, multinucleated giant cells and extensive fibrosis.

Graves' disease: An autoimmune disease in which antibodies are formed against the thyroid-stimulating hormone receptor. It is characterized by overproduction of thyroid hormone i.e. hyperthyroidism

H_2 complex: The MHC region in the 17 the chromosome of mouse.

Haplotype: Closely linked genes of a single chromosome.

Hapten: A small molecule containing a single epitope which is unable to cause immune response at its own.

Hemagglutination: Clumping of RBCs, caused by antibody.

Hematopoiesis: The process of formation of cellular elements of blood, including the red blood cells, leukocytes, and platelets.

Hemolytic disease of the newborn: see, *erythroblastosis fetalis.*

Herd immunity: The protection of unimmunized members of a community by the resistance of the majority of the population to a particular pathogen.

Hinge region: A flexible region, located between Fab and Fc segment of an antibody molecule that allows bending of the molecule. The region is susceptible to enzymatic cleavage.

Histamine: An amine, found in mast cell granules, which dilates capillaries and stimulates muscles of viscera.

Histocompatibility: Compatibility between the tissues of different individuals, so that one accepts a graft from the other without giving an immune reaction.

Human immunodeficiency virus (HIV): The causative agent of the acquired immune deficiency syndrome (AIDS). HIV is a retrovirus that attacks on macrophages and CD4 T cells.

Human Leukocyte Antigen (HLA): The human major histocompatibility complex that contains the genes coding for the polymorphic MHC class I and II class molecules and many other important genes.

Humoral immunity: The antibody-mediated specific immunity which can be transferred to unimmunized recipients by using immune serum containing specific antibody.

Humoral: Extracellular fluids, including serum and lymph.

Hybridoma: A hybrid cell, which forms from the fusion of an antibody secreting cell with a malignant cell.

Hyperacute graft rejection: It is an immediate reaction of an allogenic tissue graft which is caused by natural preformed antibodies that react against antigens on the graft.

Hypergammaglobulinemia : A disease in which there is elevation in the serum immunoglobulin levels. It may develop in any condition where there is continuous stimulation of the immune system, such as chronic infection, autoimmune disease or systemic lupus erythematosus.

Hypersensitivity: see, *allergy*

Hypervariable regions: Certain regions, within the variable region of Ig molecules, which exhibit hyper variability in amino acid composition from molecule to molecule.

Idiotope: A particular antigenic determinant on the variable region of an antibody.

Idiotype: The antigenic characteristics of the variable region of an antibody.

Immediate-type hypersensitivity: Hypersensitivity reaction which occurs within minutes after the interaction of antigen and IgE antibody.

Immune complex: A macromolecular complex of antigen-antibody reaction that also contains components of complement.

Immune response: The reaction of the body to an antigen.

Immunity: The ability to resist infection.

Immunization: see, *vaccination.*

Immunocompetent: Having a normal immune response i.e. the body is capable of mounting an appropriate immune response, when necessary.

Immunodeficiency: Decrease in immune response due to absence or defect of some component of the immune system.

Immunodiffusion: It is a technique used for the detection of antigen or antibody by formation of an antigen-antibody precipitate in agar gel.

Immunofluorescence: A technique used for detecting molecule using antibodies labeled with fluorescent dyes.

Immunogenic: The ability to produce B or T cell-mediated immune response.

Immunoglobulin: Glycoproteins of the gamma globulin fraction of serum.

Immunological tolerance: The loss of the capacity of the immune system to react on antigen.

Immunology: Branch of biology which deals with the study of all aspects of host defense against infection and of adverse consequences of immune responses.

Immunoreceptor typrosine-Based activation mutif (ITAM): Amino acid sequence in the intracellular portion of signal transuding cell surface molecules that interacts with intracellular kinesis.

Inflammation: A non-specific immune response to the tissue damage. The term used for the local accumulation of fluid, plasma proteins, and white blood cells. Inflammation is characterized by local swelling, tissue redness and increase in temperature and pain.

Innate immunity: The first line of defense mechanisms which consist of various components which non- specifically protect the body from foreign infection.

Integrins: A group of heterodimeric cell-adhesion molecules, present on leukocytes.

Interferons (IFN): Low molecular weight glycoproteins, produced by viral infected cells.

Interleukins: The regulatory proteins, secreted by T cells, that are involved in signaling between cells of the immune system.

Isograft : Tissue transplanted between two genetically identical individuals.

Isotype: An antibody class which is determined by the constant region sequence of the heavy chain.

J chain: A short peptide that joins two monomers in the polymeric immunoglobulins IgM and IgA.

J gene : A gene segment coding for the J or joining segment in immunoglobulin or T-cell receptor chains.

K cells: A group of lymphocytes that are able to destroy their target by antibody dependent cell-mediated cytotoxicity.

Kappa (κ) chains: One of two types of light chains in immunoglobulin molecules.

Karyotype: The chromosomal constitution of a cell.

Kinins: Protein causing vasodilatation, including smooth muscle contraction and increased vascular permeability.

Lamda (λ) Chain: One of two types of light chains in immunoglobulin molecules.

Langerhans' cells: The phagocytic dendritic cells, found in the epidermis. These cells can migrate from the epidermis to regional lymph nodes via the afferent lymphatics.

Leukemia : Uncontrolled proliferation of a malignant leukocyte.

Leukocytes: Refers to white blood cells. These include polymorphonuclear cells, lymphocytes and monocytes/macrophages.

Ligand: A molecule that binds to a receptor.

Light chain (L chain): The smaller of the two types of polypeptide chains that makes up all immunoglobulins. It consists of one V and one C domains. Light chain is bound to heavy chain by disulphide bonds.

Linear epitopes: see, *Continuous epitopes*

Lymph Node: An oval-shaped organ of the lymphatic system, distributed widely throughout the body It acts as a filter that traps foreign particles.

Lymph: A colorless Interstitial fluid derived from blood plasma and circulates in lymphatic vessels.

Lymphoblast: A proliferating lymphocyte.

Lymphoid organs: These are the organized tissues characterized by large numbers of lymphocytes interacting with a non-lymphoid stroma.

Lymphokines: Soluble, nonspecific substances released by lymphocytes.

Lymphomas: The tumors of lymphocytes that grow in lymphoid and other tissues but do not enter the blood in large numbers.

Lymphotoxin: A cytotoxic cytokine, secreted by lymphocytes.

Macrophage: A mononuclear phagocytic cell derived from the blood monocyte.

MALT (Mucosa-associated lymphoid tissues): Lymphoid tissues located along the mucous membranes, associated with gastrointestinal tract, bronchial and urinogenital tract.

Marginal Zone: A diffuse region of spleen, located between red pulp and white pulp.

Mast cell: Non-motile connective tissue cell, present along the capillaries throughout the body.

Membrane Attack Complex (MAC): The complex of complement components of the lytic pathway that becomes inserted into cell membrane.

Memory cells: Long lived lymphocytes, formed on exposure to an antigen for the first time. They react more actively then naive lymphocytes when re-stimulated with same antigen.

MHC (Major Histocompatibility Complex): A genetic region found in all mammals that are required for antigen presentation to T cells and for rapid graft rejection.

MHC restriction: Refers to the property of T lymphocytes to respond only when they are presented with the appropriate antigen in association with either self MHC class I or class II molecules.

Minor histocompatibility antigens (minor H antigens): The peptides of polymorphic cellular proteins that are bound to MHC molecules. These antigens can lead to graft rejection when recognized by T cells.

Mitogen : A substance that stimulates the proliferation of many different clones of lymphocytes.

Monoclonal antibodies: These are the antibodies that are produced by a single clone of B lymphocytes. They are usually produced by hybridoma technique produced from fusion of non secreting myeloma cells with immune spleen cells.

Monocytes : A type of white blood cell with a bean-shaped nucleus.

Motif : A pattern of amino acids in the sequence of a molecule which is critical for the binding of a ligand.

Multilineage CSF (M-CSF): A cytokine that supports the differentiation and growth of a number of myeloid cells lineages.

Multiple sclerosis: An autoimmune disease of the central nervous system in which an inflammatory response results in demyelination and loss of neurologic function.

Myasthenia gravis: An autoimmune disease in which autoantibodies, against the acetylcholine receptor on skeletal muscle cells, cause a block in neuromuscular junctions.

Myeloid cells: The lineages of bone marrow derived phagocytes.

Myeloma: A tumor of plasma cells.

Naive: Denoting B and T cells that have not encountered antigen.

Natural killer (NK) Cells: Granular lymphocytes that can kill target cells such as bacterial cells and tumor cells.

Necrosis: Death of cells or tissues due to chemical or physical injury.

Neoplasm: Any abnormal growth which is converted into benign or malignant tumor.

Neutralization: The capacity of an antibody to block or inhibit the effects of a virus.

Neutrophil: A short lived phagocytic, polymorphonuclear white blood cell.

Null cells: A small proportion of lymphocytes that do not possess surface markers. They are neither T nor B cells.

Oncofetal tumor antigens: Malignant cells expressing antigens which are also present on embryonic cells but not expressed in tissues except tumor cells.

Oncogenes: Genes causing cancer.

Opsonins: Substances such as antibody that facilitate phagocytosis.

Opsonization: Deposition of opsonins on antigen for the process of phagocytosis.

Paratope : The part of an antibody which recognizes an antigen, the antigen-binding site of an antibody.

Passive acquired Immunity: Immunity which is induced by introducing antibodies in the body. Antibodies are introduced in the body either naturally or artificially.

Pathogen: An organism that causes diseases.

Pattern-recognition receptors (PRRs): Receptors of the innate immune system that identify molecular patterns present on pathogen but absent in host.

Payer's Patches: Group of lymphoid nodules in the small intestine.

Periarteriolar lymphoid sheath (PALS): The part of the inner region of the white pulp of the spleen that contains mainly T cells.

Peripheral tolerance: Tolerance acquired by mature B or T lymphocytes in the peripheral tissues.

Phagocytosis: The process by which macrophages engulf other cells or solid particles.

Pinocytosis : Ingestion of liquid by vesicle formation in a cell.

Plasma : Fluid component of unclotted blood.

Plasma cell: The antibody producing B cell.

Positive Selection: A process that permits the survival of only those T cells whose receptors recognize self MHC.

Precipitin reaction: The formation of a visible precipitate when a soluble antigen reacts with an antibody.

Primary immune response: The immune response resulting from an individual's first contact with an antigen.

Primary Lymphoid Organs: Lymphoid organs in which lymphocytes mature. In mammals, bone marrow and thymus are the primary lymphoid organs.

Priming : The activation of naive lymphocytes by exposure to an antigen.

Privileged site: A location, within the body, where foreign grafts are not rejected. e.g cornea of eye.

Pro-B cell: Earliest stage of B cell differentiation.

Programmed cell death: see, *apoptosis*

Proto-oncogenes: Genes that encode a factor that regulates cell function in normal cells, but in certain circumstances they are converted into oncogenes.

Provirus : The DNA form of a retrovirus when it is integrated into the host cell genome.

Radioimmunoassay (RIA): An immunological technique for measuring antigen or antibody using radio labeled reagents.

Repertoire: The entire library of antigenic specificities generated by either B or T lymphocytes to respond to foreign antigen.

Rheumatoid arthritis: An autoimmune disease that results in a chronic, systemic inflammatory disorder that may affect many tissues and organs, especially the joints.

Rheumatoid factor: An autoantibody (usually IgM) that reacts with an individual's own IgG; present in rheumatoid arthritis.

Rhogam: The immunoglobulin given to Rh negative women, carrying Rh positive infants, to prevent her body from developing an immunity against the babies' red blood cells.

Sarcoma: Tumor of connective tissue.

SCID (Severe Combined Immunodeficiency): A condition in which adaptive immune response does not occur because of the absence of both B and T cells.

Secondary immune response: The immune response that follows a second exposure to an antigen.

Secretory component: Component of the poly-Ig receptor that attaches to dimeric IgA. It protects IgA from proteolytic cleavage as it is transported through an epithelial cell.

Selectins : A family of cell-surface adhesion molecules, found on leukocytes and endothelial cells.

Sensitization : Administration of an antigen to induce a primary immune response.

Serology : Study or diagnostic examination of blood serum, in particular with consideration to the response of the immune system, to pathogens.

Serum: The fluid component of clotted blood.

Somatic hypermutation: Change in the variable region sequence of an antibody, produced by a B cell following antigenic stimulation.

Suppressor cell: A lymphocyte that reduces the immune response of other cells to antigen.

Syngeneic graft : A graft between two genetically identical individuals . Same as *isograft.*

Systemic lupus erythematosus (SLE): An autoimmune disease which affects many organs of the body. The body produces high levels of antibodies against the components of cell nuclei, particularly DNA.

T cell: Thymus derived cell that expresses T cell receptor, CD3 and CD4. Sub groups of T cells are T_C cell and T_H cell.

T cytotoxic (T_C) cell: Class I MHC restricted T cells that kill target cells in an antigen specific manner.

T helper (T_H) cell: Class II MHC restricted cell which plays a central role in humoral and cell-mediated immunity.

TCR (T cell receptor): Antigen binding molecules which are present on the surface of T cell.

T-dependent antigen: An immunogen that requires T_H cells to interact with B cells in order to induce antibody synthesis.

Thymectomy: Surgical removal of the thymus.

Thymocytes: The lymphoid cells found in the thymus, mainly consist of developing T cells,

Thymus: A primary lymphoid organ, located in the thoracic cavity where T cell maturation occurs.

T-independent antigen: An immunogen that induces antibody synthesis in the absence of T cells.

Titer: A titer (or titre) is a way of expressing concentration. Antibody titer is a laboratory test that measures the level of antibodies in a blood sample.

TNF (Tumor Necrosis Factor): A lymphocyte derived cytokine which is cytotoxic to tumor cell.

Tolerance: Immunological state in which no immune response occurs.

Toxoids : Inactivated toxins that are no longer toxic but retain their immunogenicity. Toxoids can be used for immunization.

Transfection: The insertion of small pieces of DNA into cells.

Transgenic animal: An animal in which one or more new genes have been incorporated.

Transplantation antigens: The cell surface antigens responsible for graft rejection.

Transplantation: Grafting of a tissue or cells from one location to other or from one individual to another.

Tumor immunology: The study of host defenses against tumors.

Vaccination: The administration of antigenic material (a vaccine) to stimulate an individual's immune system to develop adaptive immunity to a pathogen. Also called *immunization.*

Vaccine: The therapeutic material used for vaccination.

Variable regions: The N-terminal portion of an Ig or TCR which contains antigen-binding site.

Viremia: The presence of virus in the blood of a host.

Xenogenic: Denoting individuals of different species.

Xenograft: Tissue or organs from an individual of one species transplanted into or grafted onto an organism of another species.

X-linked agammaglobulinemia: An immunodeficiency disease , found in males, manifesting as absence of mature B cells; due to defective *tyrosine kinase btk.*

Index

A

B

O

P

R

S

T

W

X

Zeitfracht Medien GmbH
Ferdinand-Jühlke-Straße 7
99095 Erfurt, Deutschland
produktsicherheit@kolibri360.de